Recent Advances in Semiconductor Device Research and Applications

Edited by

Girdhar Gopal
Department of Electronics & Communication
National Institute of Technology Patna
Patna, Bihar, India

Meena Panchore
Department of Electronics & Communication Engineering
National Institute of Technology Patna
Patna, Bihar, India

Arun Kishor Johar
Department of Electronics & Communication Engineering
Manipal University Jaipur, Jaipur
Rajasthan, India

&

Tarun Varma
Department of Electronics and Communication Engineering
Malaviya National Institute of Technology Jaipur
Jaipur, Rajasthan, India

Recent Advances in Semiconductor Device Research and Applications

Editors: Girdhar Gopal, Meena Panchore, Arun Kishor Johar & Tarun Varma

ISBN (Online): 979-8-89881-579-0

ISBN (Print): 979-8-89881-580-6

ISBN (Paperback): 979-8-89881-581-3

Published by Bentham Science Publishers Pte. Ltd. Singapore,

in collaboration with Eureka Conferences, USA.

First published in 2026.

Bentham Science Publishers Pte. Ltd.
No. 9 Raffles Place
Office No. 26-01
Singapore 048619
Singapore
Email: subscriptions@benthamscience.net

CONTENTS

Preface

The semiconductor industry stands at the forefront of technological progress, driving innovation and enabling the digital revolution that has transformed our world. From the microprocessors in our smartphones to the complex systems that manage our power grids and medical devices, semiconductors are the foundation for modern technology. As we push the boundaries of what is possible with the technology, the demand for higher performance, lower power consumption, and new functionalities continues to rise, prompting the need for advancements beyond traditional silicon-based semiconductor devices.

This book, **"Recent Advances in Semiconductor Device Research and Applications,"** arrives at a critical juncture in the evolution of semiconductor technology. It captures the essence of the latest innovations and research breakthroughs shaping this dynamic field's future. The emergence of wide bandgap semiconductors, two-dimensional materials, and novel transistor designs signifies a new era of possibilities, offering unprecedented speed, efficiency, and miniaturization. These advancements are enhancing existing applications and paving the way for new ones that were previously inconceivable. The purpose of this book is multifaceted. It aims to educate readers on the fundamental principles and recent developments in emerging semiconductor technologies, providing a solid foundation for understanding the current landscape. It also delves into the investigation of cutting-edge research findings and technological breakthroughs, offering insights into the challenges and opportunities that lie ahead. The book showcases the potential impact on the future of electronics and beyond by demonstrating the practical applications of these technologies across various industries. This work is a testament to the collaborative efforts of researchers, engineers, and industry professionals dedicated to advancing semiconductor technology. It is designed to inspire and inform, serving as a valuable resource for academic researchers, industry professionals, and graduate students. This book bridges the gap between research and application, fostering innovation and collaboration within the semiconductor community through a blend of theoretical insights, experimental results, and real-world examples. As you embark on this journey through the pages of **"Recent Advances in Semiconductor Device Research and Applications"**. I hope, as a reader, you will gain a deeper appreciation for the remarkable progress in this field and feel inspired to contribute to the ongoing advancements in semiconductor technology. The future of electronics is bright, and it is through the collective efforts of individuals like you that we will continue to push the boundaries of what is possible.

Girdhar Gopal
Department of Electronics & Communication
National Institute of Technology Patna
Patna, Bihar, India

Meena Panchore
Department of Electronics & Communication Engineering
National Institute of Technology Patna
Patna, Bihar, India

Arun Kishor Johar
Department of Electronics & Communication Engineering
Manipal University Jaipur, Jaipur
Rajasthan, India

&

Tarun Varma
Department of Electronics and Communication Engineering
Malaviya National Institute of Technology Jaipur
Jaipur, Rajasthan, India

List of Contributors

A. Theja	Department of Electronics and Communication Engineering, TKR College of Engineering and Technology, Hyderabad, Telangana, India
A. Vikas	Department of Electronics and Communication Engineering, TKR College of Engineering and Technology, Hyderabad, Telangana, India
Anil Kumar Pathakamuri	School of Electronics Engineering, VIT-AP University, Amaravati, Andhra Pradesh, India
Arun Kishor Johar	Department of Electronics and Communication Engineering, Manipal University Jaipur, Jaipur, Rajasthan, India
Arun Kumar Sharma	Department of Electronics and Communication Engineering, Malaviya National Institute of Technology Jaipur, Jaipur, Rajasthan, India
Avnish Bora	ECE Department, Jodhpur Institute of Engineering & Technology, Mogra, Rajasthan, India
B. Karthikeyan	Department of Electronics and Communication Engineering, Velammal College of Engineering and Technology, Viraganur, Tamil Nadu, India
Basudha Dewan	Department of Electronics & Communication Engineering, Manipal University Jaipur, Jaipur, Rajasthan, India
Brinda Bhowmick	Department of Electronics and Communication Engineering, National Institute of Technology Silchar, Silchar, Assam, India
Chandan Kumar Pandey	School of Electronics Engineering, VIT-AP University, Amaravati, Andhra Pradesh, India
Chithraja Rajan	Department of Computer Science and Engineering (AI-ML), Shri Ramdeobaba College of Engineering and Management, Nagpur, Maharashtra, India
Chitrakant Sahu	Department of Electronics and Communication Engineering, National Institute of Technology Raipur, Raipur, Chhattisgarh, India
Dasari Srikanya	Department of Electronics and Communication Engineering, Dayananda Sagar University, Harohalli, Karnataka, India
Devender Pal Singh	Department of Electronics and Communication Engineering, Punjab Engineering College, Chandigarh, India
Dharmendra Singh Yadav	National Institute of Technology Kurukshetra, Thanesar, Haryana, India
George Mili	Department of Electronics and Communication Engineering, National Institute of Technology Silchar, Silchar, Assam, India
Girdhar Gopal	Department of Electronics & Communication Engineering, National Institute of Technology Patna, Patna, Bihar, India
Hemant Jain	ECE Department, Jodhpur Institute of Engineering & Technology, Mogra, Rajasthan, India
Jai Kumar Bhatt	Department of Electronics and Communication Engineering, Malaviya National Institute of Technology Jaipur, Jaipur, Rajasthan, India
K. Kavitha	Department of Electronics and Communication Engineering, Velammal College of Engineering and Technology, Viraganur, Tamil Nadu, India

Man Mohan Shukla	Department of Electronics and Communication Engineering, Pranveer Singh of Technology Kanpur, Kanpur, Uttar Pradesh, India
Meena Panchore	Department of Electronics and Communication Engineering, National Institute of Technology Patna, Patna, Bihar, India
Menka Yadav	Department of Electronics and Communication Engineering, Malaviya National Institute of Technology Jaipur, Jaipur, Rajasthan, India
Mohit Kumar Srivastava	Department of Electronics and Communication Engineering, Pranveer Singh of Technology Kanpur, Kanpur, Uttar Pradesh, India
P. Suveetha Dhanaselvam	Department of Electronics and Communication Engineering, Velammal College of Engineering and Technology, Viraganur, Tamil Nadu, India
Prabhat Singh	Indian Institute of Information Technology Senapati, Manipur, India
Prajwal B. Pillewan	National Institute of Technology Kurukshetra, Thanesar, Haryana, India
Preeti Agarwal Mittal	Department of Electronics and Communication Engineering, Pranveer Singh of Technology Kanpur, Kanpur, Uttar Pradesh, India
Rajesh Saha	Department of Electronics and Communication Engineering, National Institute of Technology Silchar, Silchar, Assam, India
Rashi Chaudhary	Department of Electronics and Communication Engineering, Malaviya National Institute of Technology Jaipur, Jaipur, Rajasthan, India
Ritik Kumar	Department of Electronics and Communication Engineering, Malaviya National Institute of Technology Jaipur, Jaipur, Rajasthan, India
S. Nagarajan	Department of Electronics and Instrumentation Engineering, Easwari Engineering College, Chennai, Tamil Nadu, India
Shailendra Singh	Department of Electronics and Communication Engineering, Pranveer Singh of Technology Kanpur, Kanpur, Uttar Pradesh, India
Shreyas Tiwari	Department of Electronics and Communication Engineering, Pandit Deendayal Energy University, Gandhinagar, Gujarat, India
Tarun Varma	Department of Electronics and Communication Engineering, Malaviya National Institute of Technology Jaipur, Jaipur, Rajasthan, India
Utkarsh Pandey	Department of Electronics and Communication Engineering, Pranveer Singh of Technology Kanpur, Kanpur, Uttar Pradesh, India
Varnit Goswami	Department of Electronics and Communication Engineering, Malaviya National Institute of Technology Jaipur, Jaipur, Rajasthan, India
Vivek Kumar	Department of Electronics and Communication Engineering, Pranveer Singh of Technology Kanpur, Kanpur, Uttar Pradesh, India
Zohmingliana	Department of Electronics and Communication Engineering, National Institute of Technology Silchar, Silchar, Assam, India

CHAPTER 1

Introduction to Emerging Semiconductor Devices

A. Vikas[1,*], **A. Theja**[1] and **Meena Panchore**[2]

[1] *Department of Electronics and Communication Engineering, TKR College of Engineering and Technology, Hyderabad, Telangana, India*

[2] *Department of Electronics and Communication Engineering, National Institute of Technology Patna, Patna, Bihar, India*

Abstract: At the moment, the sector of electronics faces huge bottlenecks in power and performance; the most influential current semiconductor technologies are the FinFETs, TFETs, and devices made of 2-D materials. FinFETs provide better electrostatic control and alleviate short-channel effects, which enable the continuous scaling of ever-smaller technology nodes. TFETs use band-to-band tunneling to realize switching at very low voltages below the thermionic threshold and thus open the path to ultra-low power operation. Two-dimensional material devices, including graphene, transition metal dichalcogenides, and black phosphorus, are known for their extreme mechanical flexibility, high carrier mobility, and good scalability. These technological advances stand to drive the new generation of neuromorphic computing, quantum devices, and ultra-low power Internet-of-Things platforms. Consequently, future research should be oriented towards long-term integration of these enabling technologies to overcome the limitations of silicon - only electronics with sustainability and high performance.

Keywords: FinFETs, TFETs, 2-D materials, IoT, High-performance devices.

INTRODUCTION TO EMERGING SEMICONDUCTOR DEVICES

The ever-shrinking dimensions of semiconductors are exceeding the limitations of the traditional Complementary Metal Oxide Semiconductor (CMOS) technology due to inherent physical constraints. Calibrated substitutions such as FinFETs, Tunnel-FETs, spintronic devices, and transistors based on two-dimensional materials allow for enhanced electrostatic control and lower power consumption and therefore extend their usefulness in applications of electronics and optoelectronics. Although such technologies show a significant improvement in speed and efficiency, the deployment of these technologies on a large scale of integration remains a formidable challenge.

* **Corresponding author A. Vikas:** Department of Electronics and Communication Engineering, TKR College of Engineering and Technology, Hyderabad, Telangana, India; E-mail: vikasambekar@tkrcet.com

Girdhar Gopal, Meena Panchore, Arun Kishor Johar & Tarun Varma (Eds.)

Overview of Semiconductor Technology

Semiconductor materials are the backbone of the technology we use today in our modern world, enabling and engineering devices such as computational systems, telecommunications networks, medical diagnostics, and energy conversion. These semiconductors contain the fundamental parts - transistors, diodes, integrated circuits - the evolution of which has propelled technological development through the last several decades. The unremitting desire for improved performance, lower power consumption, and miniaturization, coupled with ongoing limitations in silicon-based architectures, has driven the development of novel semiconductor devices. These innovations leverage alternative materials, new design paradigms, and advanced fabrication methods to expand the functional landscape of integrated circuits.

Key Drivers for Innovation

A plethora of drivers are responsible for contemporary advances in semiconductor technology. The main difficulty comes from the limitations of scaling; maintaining Moore's Law is bound to face physical and cost limits. Silicon transistors with sub-10 nm dimensions face increased leakage and short-channel effects, necessitating new device architectures and materials to push performance boundaries [1]. Energy efficiency is of equal importance, as demand for sustainable, portable electronics increases. The need to decrease power consumption while maintaining performance has driven the development of FinFETs, gate-all-around transistors, and complex power management schemes [2].

Emerging applications like artificial intelligence, quantum computing, and the Internet of Things are accelerating the demand for more intelligent technological solutions than traditional designs of chipsets constructed with silicon or silicon on insulator (CMOS). AI accelerators, neuromorphic architectures, and secure edge devices demand architectures that depart from traditional paradigms. Quantum computing, with superconducting qubits and semiconductor spin qubits, is at the same time leading to the advancement of materials science and fabrication methods. Consequently, there is a severe need for materials-centric innovations to drive next-generation semiconductors, given the plethora of limitations faced by silicon-based devices [3]. Arrays of advanced materials, such as graphene, transition metal dichalcogenides, and perovskites, exhibit outstanding electrical, thermal, and mechanical performance and can offer superior efficiency and new functionalities for next-generation applications [4].

Semiconductors are used in a large number of applications ranging from automotive and medical devices to renewable energy systems. Optimal performance is essential as electric vehicles, autonomous technologies, and sophisticated diagnostic tools in medicine are likely to be adopted - all of which will depend on robust semiconductor solutions. Power electronics for renewable energy and grid management are also areas of continuing development [5]. New transistor designs, energy-lighting layout, special computation, and advanced materials will determine the future of electronics in this evolving industry, guiding us through rapidly changing times.

Types of Emerging Semiconductor Devices

FinFETs and TFETs are the new semiconductor devices that realize efficiency and scalability. The 3D fin structure in FinFETs increases gate control, reducing leakage and short-channel effects, thereby enabling high-speed, low-power operation of state-of-the-art processors and AI accelerators. TFETs apply quantum tunneling for carrier transport to realize steep subthreshold slopes below 60mV/decade, with ultra-low-power consumption optimally suitable for IoT devices and biosensors. Such deep-subthreshold-slope TFETs have much lower operating voltages than MOSFETs, thereby reducing power consumption by orders of magnitude. Heterojunction-based TFETs boost performance even further by offering higher energy efficiency. FinFET and TFET are the two most important devices to transcend the scaling limitations of semiconductors. These will open the door to future low-power and high-performance electronics.

2D material devices and spintronics are revolutionizing semiconductor technology. 2D materials like graphene and MoS2 are highly mobile, flexible, and bandgap tunable. These characteristics facilitate ultra-thin and high-speed transistors for flexible optoelectronics and electronics. They also eliminate short-channel effects, which are essential for next-generation devices. Spintronics utilizes electron spin rather than charge, allowing high-speed, low-power, and non-volatile memory. Technologies such as STT-MRAM provide better endurance and radiation hardness. Spintronic transistors are being investigated for low-power computing and AI. The combination of 2D materials and spintronics increases device performance and efficiency. These technologies are transforming the future of quantum computing and neuromorphic architectures. Fig. (**1**) shows a detailed overview of the types of emerging semiconductor devices.

In this chapter, the major emerging semiconductor devices such as FinFETs, TFETs, Spintronic devices, and 2-D material devices have been explored. This chapter also provides an outline of the working of these devices, along with their applications in the scientific world.

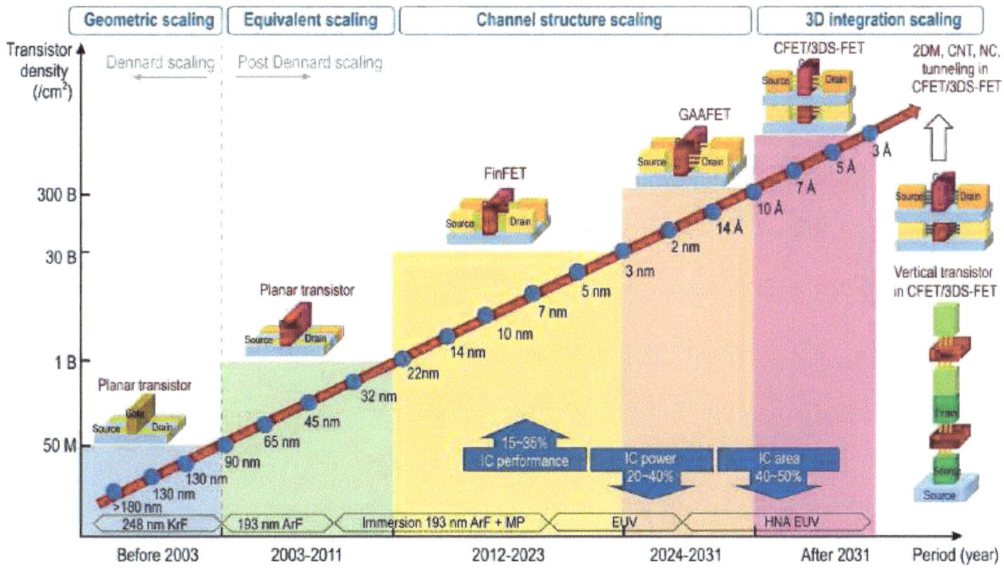

Fig (1). Evolution of modern nano-scale devices [5].

FinFETs (Fin Field-Effect Transistors)

FinFETs have revolutionized transistor design by introducing a three-dimensional fin-shaped channel. The architecture shown in Fig. (**2**) has better electrostatic control, which reduces leakage and increases efficiency. In a FinFET, the gate surrounds a thin slice of undoped silicon called a "fin" where current flows from one end of the fin to the other. The wraparound feature of the gate enhances gate control over the channel, reducing short-channel effects and, hence, leakage currents [6]. Thus, a FinFET has better gate control; the channel is controlled from its two sides. This reduces short-channel effects, leading to lower drain-t--source leakage current in the off state and a higher intrinsic gain. The current in FinFETs flows through three faces of the fin. Therefore, the equivalent channel width is $W = W_F + 2H_F$, where W_F is the width of the fin, and H_F is its height. Typically, $W_F = 6$ nm and $H_F = 50$ nm [7]. As H_F is not controlled by the circuit designer, W_F can be selected in such a way that $W_F + 2H_F$ gives the desired transistor width. However, the width of the fin impacts the series resistance of drain and source as well as the channel length modulation; hence, the width of the fin is kept fixed, which makes the transistor width discrete [8]. For instance, with $W = W_F + 2\ H_F = 120$ nm, wider transistors are implemented by adding the number of fins.

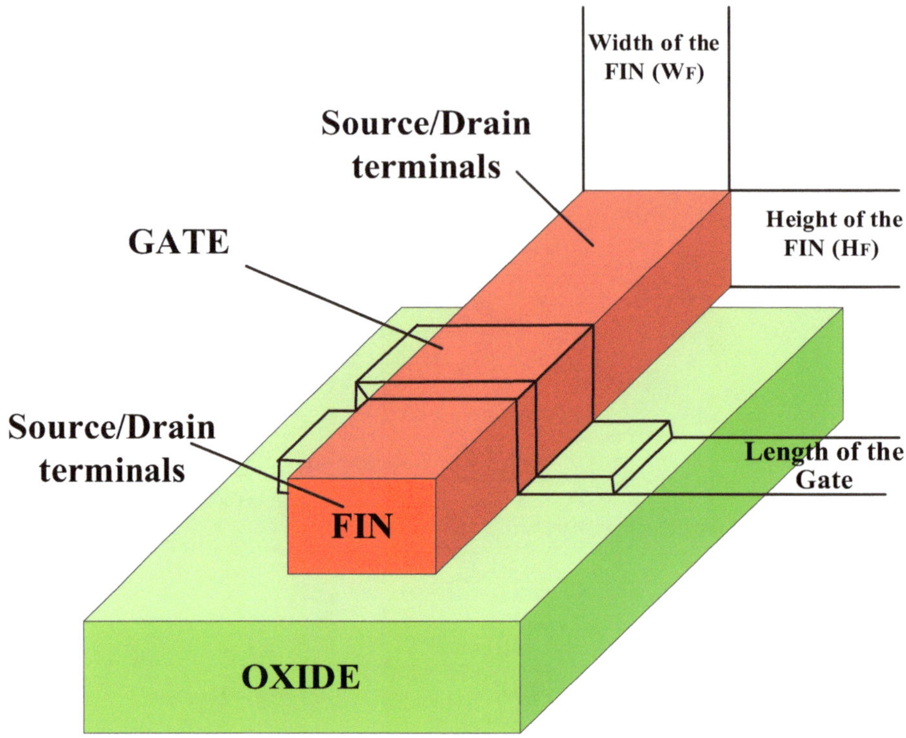

Fig. (2). Conventional FinFET structure [10].

The application of FinFETs extends broadly in high-performance computing, mobile devices, and advanced processors due to their efficient and superior performance. This architecture has proved to be a superior architecture with fast switching times, low power consumption, and high scalability. The three-dimensional physical construction is an efficient way to mitigate short-channel effects and is therefore preferred in today's semiconductor technologies [9, 10].

Design Innovations

FinFETs pave the way for technology design concepts that can boost performance, reduce power consumption, and improve scalability, surpassing conventional planar MOS field-effect transistors. The key innovation is their three-dimensional structure (fin-like), in which a ridge extends from the substrate. Via this geometry, the gate can surround the channel on three sides, and therefore increase the control of the electrostatic potential and reduce the occurrence of short channel effects (*e.g.* drain-induced barrier lowering and threshold voltage roll off) [10]. As a result, FinFETs are better suited to a lower-voltage supply and faster switching, due to better gate control that reduces sub-threshold leakage, resulting in reduced standby power for low-power applications [11]. Moreover, the three-dimensional form means that drive current density is higher, meaning that switching transients are quicker and overall operational throughput is increased. The reduction of parasitic resistance and capacitance further reduces the delay, making the whole circuit more efficient [12].

Transistors have seen scalability rise throughout their entire history as a design principle in their main use. While the industry has moved into the sub-10 nm regime, FinFETs have continued to offer an outstanding power-performance trade-off for multifins and allow fine-tuning of transistor characteristics [13]. Innovative fabrication techniques, notably the self-aligned double-patterning method (SADP), have facilitated process simplification and process variation prediction. In addition, both FinFETs and wrap.-around.-gate-mOSFETs have proven to be able to overcome short-channel effects, an important issue in scaling advanced transistors [14]. Therefore, FinFETs have formed the backbone of modern semiconductor devices and are widely used across all families of processors, including mobile processors, GPUs, FPGAs, AI accelerators, and supercomputing systems. This has resulted from a proper balance between power efficiency and performance that persisted and led FinFETs to be ranked as the front-runner transistor architecture in semiconductor manufacturing. Potential directions show that the FinFET will be a transition architecture toward more complicated transistor geometries, like GAAFET and nanosheet transistors, which will advance miniaturization limits with efficiency in semiconductor products.

Tunnel Field-Effect Transistors (TFETs)

Alternative techniques were found to enhance device efficiency, which makes the MOSFET replaceable. The conduction of the TFET, a form of PIN diode, is regulated by an electric field [15]. TFETs are unlike normal MOSFETs in that they employ band-to-band tunneling (BTBT) to transport current rather than thermionic diffusion. TFETs perform better in small-channel networks, minimize leakage current, and enable a subthreshold swing (SS) of less than 60 mv/dec due to the fact that they employ the BTBT conduction mechanism [16]. The aforementioned advantages of TFET demonstrate its utility as an ultra-low power configuration in the foreseeable future. However, unresolved challenges still hinder the performance and use of TFETs. Differentiating them from conventional MOSFETs is their reduced drive current (ION) and high ambipolarity (conductivity under both positive and negative gate voltages) [17 - 19]. Over the last few years, many techniques have been explored to boost the ON-state currents of the devices. They are employing low-bandgap materials, high-κ dielectrics, heterostructures, and nanowire tunnel FETs [20 - 22]. In order to achieve effective

BTBT, TFETs must have sharp junctions. Because of thermal annealing and ion-implantation, doped sources/drains require a high thermal budget. It is very difficult to achieve an abrupt junction profile for physically doped structures because charge carriers diffuse from the drain (or source) region to the channel region. Increasing random dopant fluctuations (RDF) are also observed as a result of continuous size reductions of doped TFETs with heavy doping profiles [23 - 25].

Mechanism of Operation

In quantum mechanical terms, TFETs are gated p-i-n devices, which operate on the principle of band-to-band tunneling. TFETs use a reverse-bias gate control mechanism to operate as solid-state semiconductor devices. A simple schematic of a TFET is illustrated in Fig. (**3**). The corresponding energy band diagram of a TFET is shown in Fig. (**4**). Under reverse bias, electrons tunnel from the source valence band into the channel conduction band due to the Zener tunneling mechanism [26, 27]. Due to heavy doping of the n+ drain and p+ source, the channel region is completely depleted in the OFF state. Application of the gate voltage reduces the depletion regions, increases the electric field, and hence makes the device conducting. The applied gate voltage drives the flow of minority carriers from the source to the channel.

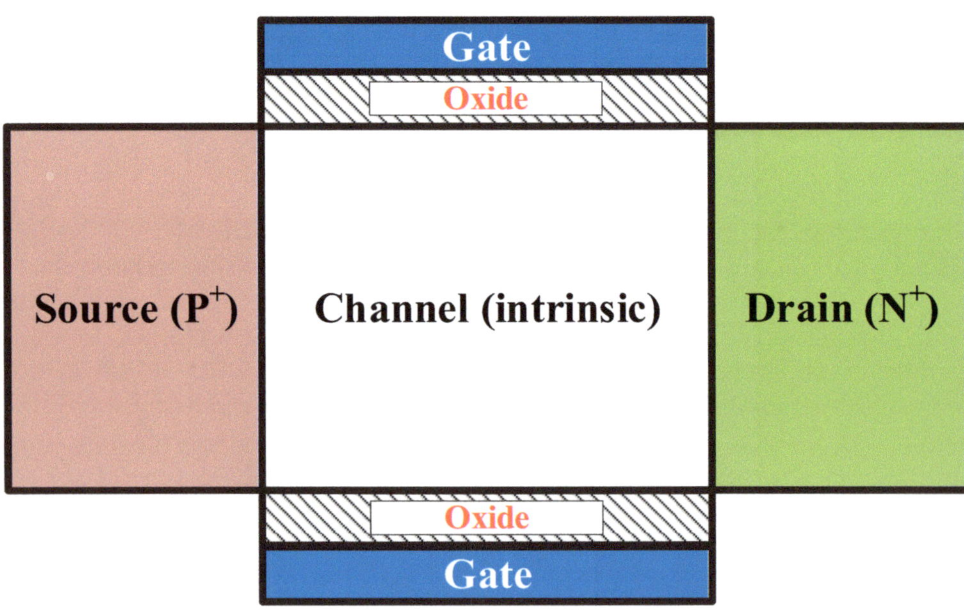

Fig (3). Schematic diagram of a conventional Double Gate TFET [20].

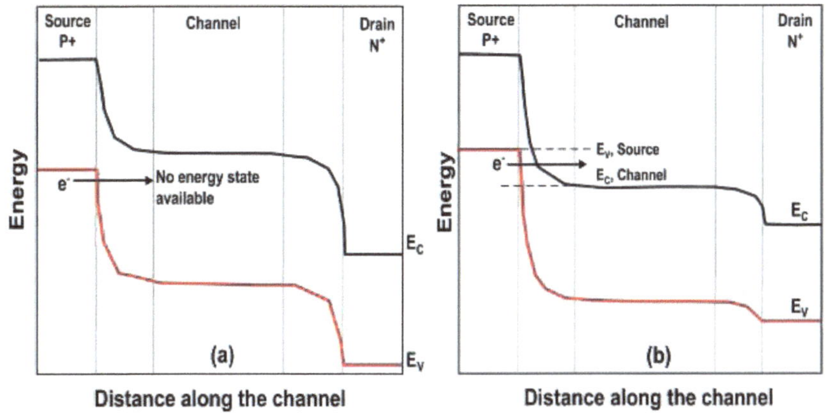

Fig. (4). Energy band diagram of a TFET in (a) OFF state, and (b) ON state [27].

A significant gate voltage is necessary in MOSFETs to overcome the barrier due to thermionic emission impeding injected charges. The process of transferring charge carriers through tunneling from source to channel medium of the device is known as the band-to-band tunneling (BTBT) process [28 - 31]. The energy band gap due to tunneling suppresses the Boltzmann tail of the electron distribution in the p-type source region. Hence, thermal dependence decreases, allowing steeper SS less than about 60 mV/decade. Therefore, the current due to tunneling may be given by Eq. (1).

$$I_{BTBT} \cong A\varepsilon^2 \exp[\frac{B}{\varepsilon}] \quad \textbf{(1)}$$

A and B are the material-dependent parameters here. TFETs are p-i-n structures; i.e., they are reverse-biased in the OFF state and thus have lower leakage current than MOSFETs. Thus, TFETs can suppress SCEs, possess large I_{ON}/I_{OFF} ratios, and are geometry-insensitive scaling.

Spintronic Devices

Spintronics, a short form of spin transport electronics, is an exciting field that extends traditional electronics to take advantage of not only the charge but also the intrinsic spin and the magnetic properties of electrons. The unique approach promises faster, more energy-efficient data storage and processing due to the fact that it influences next-generation technology. Unlike conventional electronics, which rely on moving electrical charge, spintronics introduces a different way of controlling and manipulating information; besides the advantages of low power consumption, higher speed, and higher data retention [32], the current flow direction is determined by a magnetic field, as shown in Fig. (**5**). The discovery of giant magnetoresistance and tunneling magnetoresistance has already transformed hard drives and magnetic random-access memory for increased storage density and faster access [33]. While spintronics deals with controlling spin in metallic systems, its insulating counterpart is known as multiferroics, which explores the very same interactions in materials exhibiting both magnetic and electric properties. All these advances have opened up new possibilities in low-power electronics, neuromorphic computing, and quantum technologies [34].

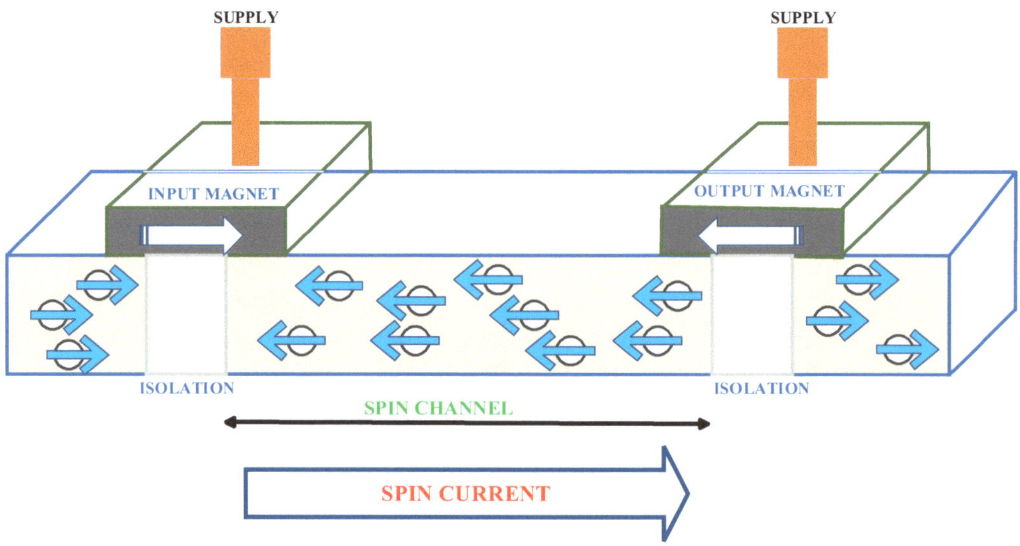

Fig. (5). Flow of current in a conventional Spintronic device setup [35].

The big win with spintronics is really in non-volatile memory, which keeps data even when the power is off. Its low power consumption and ability to be powered independently make it ideal for energy-efficient computing, which is a critical challenge in electronics today [35]: how to reduce power consumption without reducing performance. Spintronics is gradually expanding its horizons as researchers search for new materials, such as ferromagnetic semiconductors, topological insulators, and two-dimensional compounds. Such material innovations have the potential to develop high-performance ultrafast logic circuits and memory architectures for enhanced computational intelligence and efficiency [36]. As conventional semiconductor technology is reaching the physical limits of its ability to scale, spintronics is emerging as a powerful and novel candidate for future technology architectures. The incorporation of spintronic devices into existing frameworks with modern computer (CMS) chips offers new options for spin-based transistor and quantum computing pipelines, potentially even advancing brain-inspired processing units [37]. Moreover, it is spintronics that combines charge and spin synergistically to increase electronic functionality and update mechanisms of information processing and storage.

Memory technologies such as MRAM, the neuromorphic computing platform, and sensor systems have begun to stand out from the rest, thanks to their non-volatility, exceptional durability, and extraordinary energy efficiency. Emerging concepts, such as racetrack memory and spin-transfer torque (STT) MRAM, push the limits of storage density, access speed, and energy efficiency even higher [37, 38]. Racetrack memory uses magnetic domain walls to achieve high-speed data storage, while STT-MRAM uses state change to be effectively switched via spin-current manipulation.

2-D Material-Based Devices

Recent in-depth studies of 2D materials have shed light on their extraordinary electrical, thermal, and mechanical properties, thereby boosting their development. They are among the most promising candidates for future electronic and optoelectronic technologies. Unlike large bulk materials, these 2D materials are only a few atoms thick; they have high carrier mobility, tunable bandgaps, and are quite mechanically flexible [39]. Applications under consideration include FETs, photodetectors, flexible electronics, and energy storage devices. Graphene is so far the most studied 2D material and consists of one layer of carbon atoms arranged in a hexagonal lattice. It exhibits exceptional properties: ultrahigh electrical conductivity (~10^6 S/cm), high thermal conductivity (~5000 W/mK), and very high mechanical strength (~1 TPa) [40]. However, the absence of a bandgap in graphene prohibits the use of this material in digital electronics, which leads researchers to the study of other 2-D materials: TMDs, BP, and h-BN [41].

TMDs such as MoS_2 and WSe_2 have tunable bandgaps in the range 1.0–2.0 eV and show promising semiconducting characteristics for low-power electronics and optoelectronics [42]. Unlike graphene, TMDs change from indirect to direct bandgap when reduced to a single layer, which significantly enhances their optical absorption and luminescence. Another highly promising 2-D material is black phosphorus, which has a thickness-dependent bandgap of ~0.3 eV for bulk and ~2.0 eV for a monolayer, a high carrier mobility of ~1000 cm^2/Vs, and prominent in-plane anisotropy, which makes it beneficial for high-performance transistors and photonic devices [43]. There are problems with their integration into functional electronic devices due to issues with large-scale synthesis, interfacing, and ambient stability. Recently, CVD and ALD have demonstrated huge potential toward scalable, high-quality fabrication of these materials [44]. The detailed classification of 2-D materials is shown in Fig. (**6**). The existing research focuses on improving material stability and optimizing device performance while considering a wide range of 2-D material heterostructures for various multifunctional applications [45].

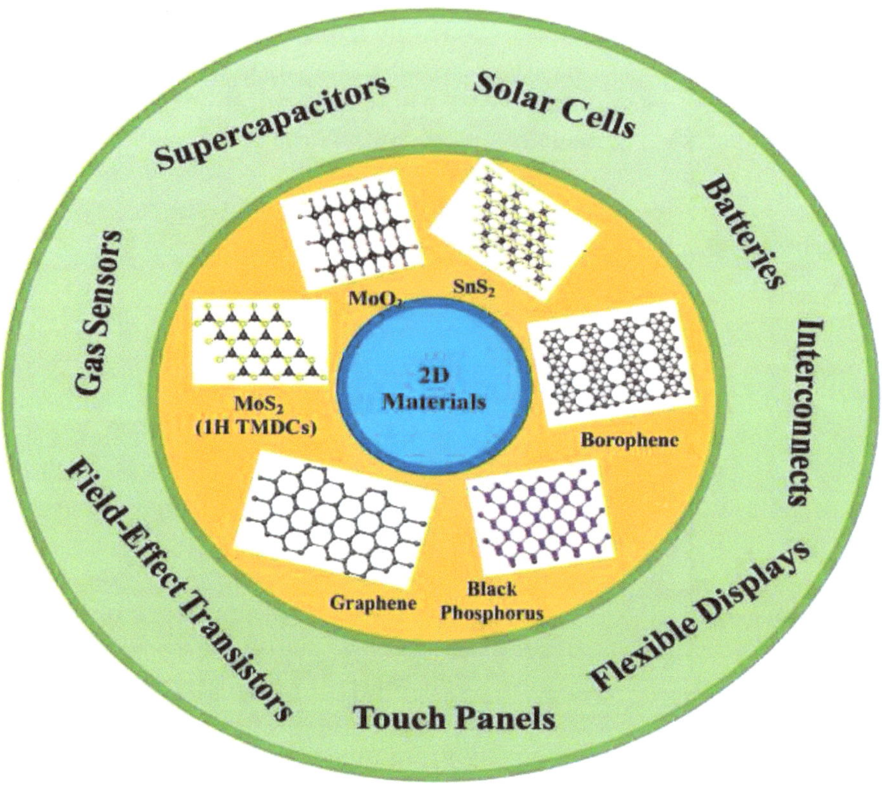

Fig (6). 2-D materials in electronic applications [45].

Applications in Electronic Devices

The integration of 2D materials into electronic devices has transformed high-performance and energy-efficient applications. Graphene boasts exceptionally high electron mobility and is used in ultrafast transistors, flexible displays, and transparent electrodes. TMDs offer tunable bandgaps; hence, they are ideal for FETs, photodetectors, and memory devices. Black phosphorus also offers high carrier mobility, enabling high-performance transistors. Novel device architectures with tailored electronic and optical properties are enabled by the stacked 2D heterostructures. These progresses open up new opportunities for next-generation nanoelectronics, optoelectronics, and flexible computing technologies [44, 45].

Optoelectronic Advancements

Novel optical properties of 2D materials have opened up new opportunities in the field of optoelectronics. TMDs have a strong interaction with light, being suitable for photonic applications such as LEDs, photovoltaics, and photodetectors operating in the visible and infrared wavelengths [46]. Graphene offers ultrafast carrier dynamics, along with a broadband optical response, thus enhancing speed and efficiency in optoelectronic devices for high-speed optical modulators and ultrafast photodetectors.

Beyond optoelectronics, 2D materials have excellent performance in sensing applications because of their high surface-to-volume ratio and strong environmental sensitivity. Such features make possible the development of ultrasensitive gas sensors, biosensors, and wearable health monitors. For example, graphene-based sensors detect biomolecules with very high specificity, thereby making it possible to improve medical diagnostics and environmental monitoring. Recently, graphene-TMD heterostructures have been used to enhance the sensitivity of SPR biosensors for DNA hybridization detection [47]. More recently, nanophotonic innovation has used 2D materials to develop magnetic field sensors with ultra-high sensitivity and to detect glucose in biological samples. Their future contributions to optoelectronic and sensing technologies will definitely be revolutionary as research is continuously being carried out.

Challenges and Future Outlook

Scaling, result reproduction, and compatibility of these 2D materials with the necessary technology that can plug them into actual devices still pose challenges. The limitations that we have highlighted require a number of formidable undertakings, with innovative manufacturing techniques and sound interdisciplinary collaboration. Further investigation into novel two-dimensional materials, cutting-edge device architectures and scalable fabrication methods is promising to achieve big technological breakthroughs in the near future. Nonetheless, despite the potential greatness of two-dimensional materials, multiple obstacles remain in translating these materials to commercial viability, in particular challenges related to scaling, long-term stability under ambient conditions, and seamless integration into existing manufacturing infrastructures.

Advantages of Emerging Semiconductor Devices

The rapid progress in semiconductor technology has enabled the development of semiconductor devices that significantly outperform conventional devices (CMOS structures) in performance, energy efficiency, and scalability. Such devices are the foundation of next-generation applications - such as artificial intelligence, high-performance computing, 6G infrastructure, and biomedical engineering. Principally, the performance of these novel semiconductor technologies is superior. Tri-gate field-effect transistor (TFET), negative capacitance field-effect transistor (NC-FET), and ferroelectric field-effect transistor (FeFET) are examples of this advancement, based on a revolutionary operating principle that enables very low sub-threshold swing, much better drive current, and reduced short-channel effect. For these reasons, the aforementioned attributes guarantee quick switching and very low power dissipation in both digital and analog domains that require high velocity [48].

Primarily, an overwhelming benefit is energy efficiency. Leakage currents and power density limit conventional scaling of the field-effect transistor, whereas new devices, such as TFET and spintronic transistors, make use of other switching mechanisms: band-to-band tunneling and spin-based logic, with sub-60 mV/decade sub-threshold swing, allowing for significant reduction of power consumption. This minimization of energy draw is indispensable for the Internet of Things, edge computing, and, generally, for battery-powered systems, since energy efficiency is the governing factor for stability and operational lifetime [46]. Nevertheless, scalability is the primary driving force. Beyond the realities that have curtailed Dennard scaling and Moore's Law, novel device architectures aim to further decrease, on the one hand, and expanding functionality, on the other hand, through three-dimensional integration, monolithic stacking, and heterogeneous integration. Transition metal dichalcogenides, black phosphorus, and III-V semiconductors represent other material platforms with higher carrier mobilities, allowing the realization of ultra-scaled transistors with better electrostatic control.

Challenges in Developing Emerging Semiconductor Devices

Several practical barriers that have been brought in transitioning a laboratory prototype to a mass-market implementation of future semiconductor technologies that must be addressed to enable the field to realize its massive potential. First and foremost is the incorporation of new materials -two-dimensional semiconductors, III-V compounds, and ferroelectric layers into the current CMOS process architecture. This requires a lot of research and development power and detailed validation that these novel materials could perfectly blend with silicon fabrication tools and processes. The second, and possibly the most powerful challenge, is in the manufacturing economics. Technologically impressive processes like the atomic-layer deposition, extreme-ultraviolet lithography or the molecular-beam epitaxy are increasing capital as well as operational costs. The attained cost ceiling discourages large-scale expansion that is crucial to commercial feasibility. Reliability cannot be left aside, along with the cost factor. Unconventional device topologies and new materials must demonstrate high endurance under long-term operations with low defect rates and respond to realistic operating loads without any decline. In the absence of these guarantees, the system's reliability would be worse off, as it would break consumer confidence and regulatory approval. Another limitation is the absence of uniform design regulations and process principles for these new devices. The absence of universal specifications causes large-scale integration to be a major challenge to the industry players and makes it harder to collaborate hindering corporate investment. It is education, industry, and government agencies that are necessary to overcome such obstacles. Organized partnerships that promote public-private partnerships, strategic capital, and funded research projects can hasten the creation of new, high-performance, energy-saving semiconductor technology.

Future Directions

Heterogeneous integration, like coincidental integration of two-dimensional semiconductors and III-V epitaxial stuff, serves as a forerunner of semiconductor technologies to new epochs [48]. Quantum devices are expected to make significant advances in computing, secure communication, and ultra-sensitive sensing. Neuromorphic computing makes use of memristive elements and ferroelectric field effect transistors to enable low-power artificial intelligence acceleration. Flexible electronics will be set to produce innovations in wearable systems, biosensors, foldable displays, and smart textiles. Advanced manufacturing paradigms will include atomic-layer deposition and additive manufacturing methodologies, including additive layer 3D printing, to achieve improved manufacturing scalability and fabrication precision [49]. Collectively, these developments are expected to establish prevailing trajectories for next-generation electronics which will impact artificial intelligence, healthcare, and secure communication.

CONCLUSION

The development of new semiconductor devices promises a major paradigm shift in electronics by demonstrating ways to overcome the limitations of the traditional chip-making process, complementary metal oxide semiconductor (CMOS). FinFETs, TFETs, and transistors based on two-dimensional materials offer opportunities to boost performance and reduce power dissipation and continue down the path of miniaturisation envisioned in many future technology roadmaps that include artificial intelligence, neuromorphic computing, quantum systems, and the Internet of Things. Despite ongoing challenges of integration, technological constraints, and non-standardized processes, these challenges are steadily overcome through collaborative research and technological development.

The strategy for the immediate future includes materials diversification combined with refined device design, the creation of innovations in fabrication, the development of fast, compact, efficient electronics that transcend conventional silicon, and the creation of interoperable, intelligent systems.

AI DISCLOSURE

During the preparation of work, author(s) used AI tools & Grammarly tool to enhance the writing style and readability of the content. The authors clarify that AI was not involved in generating the substantive content and that they take full responsibility for the publication.

REFERENCES

[1] G.E. Moore, "Cramming more components onto integrated circuits", *Electronics (Basel),* vol. 38, no. 8, pp. 114-117, 1965.
[http://dx.doi.org/10.1109/JPROC.1998.658762]

[2] Chenming Hu, J. Bokor, Tsu-Jae King, E. Anderson, C. Kuo, K. Asano, H. Takeuchi, J. Kedzierski, Wen-Chin Lee, and D. Hisamoto, "FinFET-a self-aligned double-gate MOSFET scalable to 20 nm", *IEEE Trans. Electron Dev.,* vol. 47, no. 12, pp. 2320-2325, 2000.
[http://dx.doi.org/10.1109/16.887014]

[3] S. Borkar, and A.A. Chien, "The future of microprocessors", *Commun. ACM,* vol. 54, no. 5, pp. 67-77, 2011.
[http://dx.doi.org/10.1145/1941487.1941507]

[4] H. Iwai, "Roadmap for 22nm and beyond (Invited Paper)", *Microelectron. Eng.,* vol. 86, no. 7-9, pp. 1520-1528, 2009.
[http://dx.doi.org/10.1016/j.mee.2009.03.129]

[5] K.S. Novoselov, A.K. Geim, S.V. Morozov, D. Jiang, Y. Zhang, S.V. Dubonos, I.V. Grigorieva, and A.A. Firsov, "Electric field effect in atomically thin carbon films", *Science,* vol. 306, no. 5696, pp. 666-669, 2004.
[http://dx.doi.org/10.1126/science.1102896] [PMID: 15499015]

[6] B. Yu, "FinFET scaling to 10 nm gate length", *Proc. IEEE Int. Electron Devices Meeting (IEDM),* pp. 251-254, 2002.
[http://dx.doi.org/10.1109/IEDM.2002.1175843]

[7] J-P. Colinge, *FinFETs and Other Multi-Gate Transistors.* Springer, 2008.
[http://dx.doi.org/10.1007/978-0-387-71752-4]

[8] T-J. King Liu, "FinFET: History, fundamentals, and future", *IEEE Solid-State Circuits Mag.,* vol. 9, no. 3, pp. 25-30, 2017.
[http://dx.doi.org/10.1109/MSSC.2017.2739748]

[9] A. Theja, A. Vikas, M. Panchore, and K. Cecil, "FinFET process technology for RF and millimeter wave applications", In: *RF Circuits for 5G Applications.,* S. Singh, Ed., Wiley, 2023.
[http://dx.doi.org/10.1002/9781119792949.ch11]

[10] M. Rostami, and K. Mohanram, "Dual-Vth independent-gate FinFETs for low power logic circuits", *IEEE Trans. Comput. Aided Des. Integrated Circ. Syst.,* vol. 30, no. 3, pp. 337-349, 2011.
[http://dx.doi.org/10.1109/TCAD.2010.2097310]

[11] E. Leobandung, J. Gu, L. Guo, and S.Y. Chou, "Wire-channel and wrap-around-gate MOSFETs with a significant reduction of short-channel effects", *IEEE Electron Device Lett.,* vol. 18, no. 12, pp. 571-573, 1997.

[http://dx.doi.org/10.1109/55.643383]

[12] T. Nirschl, P-F. Wang, W. Hansch, and D. Schmitt-Landsiedel, "The tunneling field effect transistor (TFET): Temperature dependence, simulation model, and its application", *Proc. IEEE Int. Symp. Circuits and Systems (ISCAS),* vol. 3, p. III–713, 2004. [http://dx.doi.org/10.1109/ISCAS.2004.1328894]

[13] R. Gandhi, Z. Chen, N. Singh, K. Banerjee, and S. Lee, "CMOS-compatible vertical-silicon-nanowire gate-all-around p-type tunneling FETs with ≤50 mV/decade subthreshold swing", *IEEE Electron Device Lett.,* vol. 32, no. 11, pp. 1504-1506, 2011. [http://dx.doi.org/10.1109/LED.2011.2165331]

[14] C. Convertino, C.B. Zota, H. Schmid, D. Caimi, L. Czornomaz, A.M. Ionescu, and K.E. Moselund, "A hybrid III–V tunnel FET and MOSFET technology platform integrated on silicon", *Nat. Electron.,* vol. 4, no. 2, pp. 162-170, 2021. [http://dx.doi.org/10.1038/s41928-020-00531-3]

[15] A. Shaker, M. El Sabbagh, and M.M. El-Banna, "Influence of drain doping engineering on the ambipolar conduction and high-frequency performance of TFETs", *IEEE Trans. Electron Dev.,* vol. 64, no. 9, pp. 3541-3547, 2017. [http://dx.doi.org/10.1109/TED.2017.2724560]

[16] S. Hussain, N. Mustakim, S. Singha, and J.K. Saha, "A comprehensive study on tunneling field effect transistor using non-local band-to-band tunneling model", *J. Phys. Conf. Ser.,* vol. 1432, no. 1, p. 012028, 2020. [http://dx.doi.org/10.1088/1742-6596/1432/1/012028]

[17] Mingshan Liu, Yan Liu, Hongjuan Wang, Qingfang Zhang, Chunfu Zhang, Shengdong Hu, Yue Hao, and Genquan Han, "Design of GeSn-based heterojunction-enhanced n-channel tunneling FET with improved subthreshold swing and on-state current", *IEEE Trans. Electron Dev.,* vol. 62, no. 4, pp. 1262-1268, 2015. [http://dx.doi.org/10.1109/TED.2015.2403571]

[18] Q. Smets, A.S. Verhulst, S. El Kazzi, D. Gundlach, C.A. Richter, A. Mocuta, N. Collaert, A.V-Y. Thean, and M.M. Heyns, "Calibration of the effective tunneling bandgap in GaAsSb/InGaAs for improved TFET performance prediction", *IEEE Trans. Electron Dev.,* vol. 63, no. 11, pp. 4248-4254, 2016. [http://dx.doi.org/10.1109/TED.2016.2604860]

[19] K. Alam, "Orientation engineering for improved performance of a Ge–Si heterojunction nanowire TFET", *IEEE Trans. Electron Dev.,* vol. 64, no. 12, pp. 4850-4855, 2017. [http://dx.doi.org/10.1109/TED.2017.2766681]

[20] S. Kanungo, "2D materials-based nanoscale tunneling field effect transistors: Current developments and future prospects,", *npj 2D Materials and Applications,* vol. 6, 2022.

[21] A. Maurya, K. Koley, B.C. Mech, J. Kumar, and P. Kumar, "Investigation of source region random doping fluctuation effects on analog and RF performance in all-Si DG-TFET", *IEEE Trans. Electron Dev.,* vol. 69, no. 9, pp. 5330-5336, 2022. [http://dx.doi.org/10.1109/TED.2022.3193992]

[22] U.E. Avci, D.H. Morris, and I.A. Young, "Tunnel field-effect transistors: Prospects and challenges", *IEEE J. Electron Devices Soc.,* vol. 3, no. 3, pp. 88-95, 2015. [http://dx.doi.org/10.1109/JEDS.2015.2390591]

[23] K.H. Kao, A.S. Verhulst, W.G. Vandenberghe, B. Soree, G. Groeseneken, and K. De Meyer, "Direct and indirect band-to-band tunneling in germanium-based TFETs", *IEEE Trans. Electron Dev.,* vol. 59, no. 2, pp. 292-301, 2012. [http://dx.doi.org/10.1109/TED.2011.2175228]

[24] W. Cao, "Designing band-to-band tunneling field-effect transistors with 2D semiconductors for next-generation low-power VLSI", *Proc. IEEE Int. Electron Devices Meeting (IEDM),* p. 3, 2015.

[http://dx.doi.org/10.1109/IEDM.2015.7409682]

[25] R. Asra, M. Shrivastava, K.V.R.M. Murali, R.K. Pandey, H. Gossner, and V.R. Rao, "A tunnel FET for VDD scaling below 0.6 V with a CMOS-comparable performance", *IEEE Trans. Electron Dev.,* vol. 58, no. 7, pp. 1855-1863, 2011. [http://dx.doi.org/10.1109/TED.2011.2140322]

[26] V. Ambekar, and M. Panchore, "Realization of high-speed logic functions using heterojunction vertical TFET", *Appl. Phys., A Mater. Sci. Process.,* vol. 129, no. 3, p. 166, 2023. [http://dx.doi.org/10.1007/s00339-023-06419-1]

[27] Y.R. Jhan, Y.C. Wu, and M.F. Hung, "Performance enhancement of nanowire tunnel field-effect transistor with asymmetry-gate based on different screening length", *IEEE Electron Device Lett.,* vol. 34, no. 12, pp. 1482-1484, 2013. [http://dx.doi.org/10.1109/LED.2013.2285156]

[28] H. Qu, S. Zhang, and H. Zeng, "Two-dimensional MSI2N4 heterostructure p-type transistors with sub-thermionic transport performances", *IEEE Electron Device Lett.,* vol. 44, no. 9, pp. 1492-1495, 2023. [http://dx.doi.org/10.1109/LED.2023.3294593]

[29] S.A. Wolf, D.D. Awschalom, R.A. Buhrman, J.M. Daughton, S. von Molnár, M.L. Roukes, A.Y. Chtchelkanova, and D.M. Treger, "Spintronics: A spin-based electronics vision for the future", *Science,* vol. 294, no. 5546, pp. 1488-1495, 2001. [http://dx.doi.org/10.1126/science.1065389] [PMID: 11711666]

[30] M.N. Baibich, J.M. Broto, A. Fert, F.N. Van Dau, F. Petroff, P. Etienne, G. Creuzet, A. Friederich, and J. Chazelas, "Giant magnetoresistance of (001)Fe/(001)Cr magnetic superlattices", *Phys. Rev. Lett.,* vol. 61, no. 21, pp. 2472-2475, 1988. [http://dx.doi.org/10.1103/PhysRevLett.61.2472] [PMID: 10039127]

[31] N.A. Spaldin, and M. Fiebig, "Materials science. The renaissance of magnetoelectric multiferroics", *Science,* vol. 309, no. 5733, pp. 391-392, 2005. [http://dx.doi.org/10.1126/science.1113357] [PMID: 16020720]

[32] I. Žutić, J. Fabian, and S. Das Sarma, "Spintronics: Fundamentals and applications", *Rev. Mod. Phys.,* vol. 76, no. 2, pp. 323-410, 2004. [http://dx.doi.org/10.1103/RevModPhys.76.323]

[33] Y. Xu, D. Awschalom, and J. Nitta, *Handbook of Spintronics.* Springer, 2016. [http://dx.doi.org/10.1007/978-94-007-6892-5]

[34] S. Bandyopadhyay, and M. Cahay, *Introduction to Spintronics.* CRC Press, 2019.

[35] S.S.P. Parkin, M. Hayashi, and L. Thomas, "Magnetic domain-wall racetrack memory", *Science,* vol. 320, no. 5873, pp. 190-194, 2008. [http://dx.doi.org/10.1126/science.1145799] [PMID: 18403702]

[36] K.S. Novoselov, "Two-dimensional materials: Beyond graphene", *Science,* vol. 353, no. 6298, pp. 461-465, 2016. [http://dx.doi.org/10.1126/science.aac9439] [PMID: 27471306]

[37] A.K. Geim, and K.S. Novoselov, "The rise of graphene", *Nat. Mater.,* vol. 6, no. 3, pp. 183-191, 2007. [http://dx.doi.org/10.1038/nmat1849] [PMID: 17330084]

[38] M. Chhowalla, H.S. Shin, G. Eda, L.J. Li, K.P. Loh, and H. Zhang, "The chemistry of two-dimensional layered transition metal dichalcogenide nanosheets", *Nat. Chem.,* vol. 5, no. 4, pp. 263-275, 2013. [http://dx.doi.org/10.1038/nchem.1589] [PMID: 23511414]

[39] H. Wang, "Transition-metal dichalcogenides for energy storage and electronics", *Chem. Soc. Rev.,* vol. 47, no. 16, pp. 6101-6127, 2018. [http://dx.doi.org/10.1039/C8CS00314A] [PMID: 30022215]

[40] L. Li, Y. Yu, G.J. Ye, Q. Ge, X. Ou, H. Wu, D. Feng, X.H. Chen, and Y. Zhang, "Black phosphorus field-effect transistors", *Nat. Nanotechnol.,* vol. 9, no. 5, pp. 372-377, 2014. [http://dx.doi.org/10.1038/nnano.2014.35] [PMID: 24584274]

[41] Z. Cai, B. Liu, X. Zou, and H.M. Cheng, "Chemical vapor deposition growth and applications of two-dimensional materials and their heterostructures", *Chem. Rev.,* vol. 118, no. 13, pp. 6091-6133, 2018. [http://dx.doi.org/10.1021/acs.chemrev.7b00536] [PMID: 29384374]

[42] T. Dutta, *Electronic properties of 2D materials and their junctions.* Nano Materials Science, 2023. [http://dx.doi.org/10.1016/j.nanoms.2023.05.003]

[43] D. Akinwande, N. Petrone, and J. Hone, "Two-dimensional flexible nanoelectronics", *Nat. Commun.,* vol. 5, no. 1, p. 5678, 2014. [http://dx.doi.org/10.1038/ncomms6678] [PMID: 25517105]

[44] A. Yadav, M. Mishra, S.K. Tripathy, A. Kumar, O.P. Singh, and P. Sharan, "Improved surface plasmon effect in Ag-based SPR biosensor with graphene and WS2: An approach towards low-cost urine-glucose detection", *Plasmonics,* vol. 18, no. 6, pp. 2273-2283, 2023. [http://dx.doi.org/10.1007/s11468-023-01945-3]

[45] M.H. Kim, A. Aziz, and S.K. Lim, "Design and optimization of tunnel FET-based circuits for ultra-low power applications", *IEEE Trans. Electron Dev.,* vol. 65, no. 8, pp. 3404-3411, 2018. [http://dx.doi.org/10.1109/TED.2018.2846479]

[46] S. Salahuddin, and S. Datta, "Use of negative capacitance to provide voltage amplification for low power nanoscale devices", *Nano Lett.,* vol. 8, no. 2, pp. 405-410, 2008. [http://dx.doi.org/10.1021/nl071804g] [PMID: 18052402]

[47] A.C. Ferrari, "Science and technology roadmap for graphene, related two-dimensional crystals, and hybrid systems", *Nat. Nanotechnol.,* vol. 15, pp. 870-884, 2020. [http://dx.doi.org/10.1038/s41565-020-0685-9]

[48] A.K. Geim, and I.V. Grigorieva, "Van der Waals heterostructures", *Nature,* vol. 499, no. 7459, pp. 419-425, 2013. [http://dx.doi.org/10.1038/nature12385] [PMID: 23887427]

[49] C.A. Ross, "Patterned nanostructures in thin films", *Nat. Nanotechnol.,* vol. 4, no. 4, pp. 253-254, 2009. [http://dx.doi.org/10.1038/nnano.2009.67]

CHAPTER 2

Basic Concepts of TFETs

A. Theja[1,*], **A. Vikas**[1], **Meena Panchore**[2] and **Chithraja Rajan**[3]

[1] *Department of Electronics and Communication Engineering, TKR College of Engineering and Technology, Hyderabad, Telangana, India*

[2] *Department of Electronics and Communication Engineering, National Institute of Technology Patna, Patna, Bihar, India*

[3] *Department of Computer Science and Engineering (AI-ML), Shri Ramdeobaba College of Engineering and Management, Nagpur, Maharashtra, India*

Abstract: This chapter explores the fundamental concepts of tunnel field-effect transistors (TFETs) as an alternative to MOSFETs for energy consumption. TFETs rely on Band-to-Band Tunnelling (BTBT) to achieve a subthreshold swing below 60 mV/decade, thereby enabling ultralow-power operation. We investigate the operating mechanisms of TFETs, the energy band lineup, and various device architectures, such as heterojunction, double-gate, and doping-less TFETs. The role of materials (such as silicon, germanium, III-V compounds, and two-dimensional materials) in enhancing tunnelling performance is discussed. The major performance figures, such as I–V characteristics and leakage currents, are compared to those reported for MOSFETs. The potential applications of TFETs for low-power digital circuits, analog/RF circuits, biomedical electronics, and hardware security are highlighted. Finally, future prospects of TFET applications in future electronics are also discussed at the end of this chapter.

Keywords: Band-to-Band Tunneling (BTBT), III-V Semiconductors, Tunnel Field-Effect Transistor (TFET), Valence Band (VB), Conduction Band (CB), Subthreshold Swing (SS).

INTRODUCTION

Tunnel Field-Effect Transistors (TFETs) are now considered a novel contender to challenge conventional Metal-Oxide-Semiconductor Field-Effect Transistors (MOSFETs) in ultra-low-power applications. Unlike MOSFETs based on thermionic emission, TFETs utilise band-to-band tunnelling (BTBT), in which charge carriers quantum mechanically tunnel through a potential barrier [1, 2]. A typical TFET structure, depicted in Fig. (**1**), is an n-channel TFET and a p-channel TFET, both distinguished by doping configurations. While an n-channel

* **Corresponding author A. Theja:** Department of Electronics and Communication Engineering, TKR College of Engineering and Technology, Hyderabad, Telangana, India; E-mail: thejaarkala@tkrcet.com

Girdhar Gopal, Meena Panchore, Arun Kishor Johar & Tarun Varma (Eds.)

TFET is made up of a heavily p-doped (p+) source and n-doped drain for electron tunneling from the valence band (VB) of the source to the conduction band (CB) of the channel [3], a p-channel TFET is made up of an n+ source and p-doped drain for hole tunneling. This basic operational distinction enables TFETs to have steeper subthreshold swing (SS) and operate at much lower supply voltages, leading to immense power consumption savings. TFETs are of specific interest for next-generation energy-efficient electronics such as biomedical implants, Internet of Things (IoT) devices, and neuromorphic computing systems. TFETs have also been of interest for security applications such as Physical Unclonable Functions (PUFs) because of their intrinsic variability in tunneling behaviour.

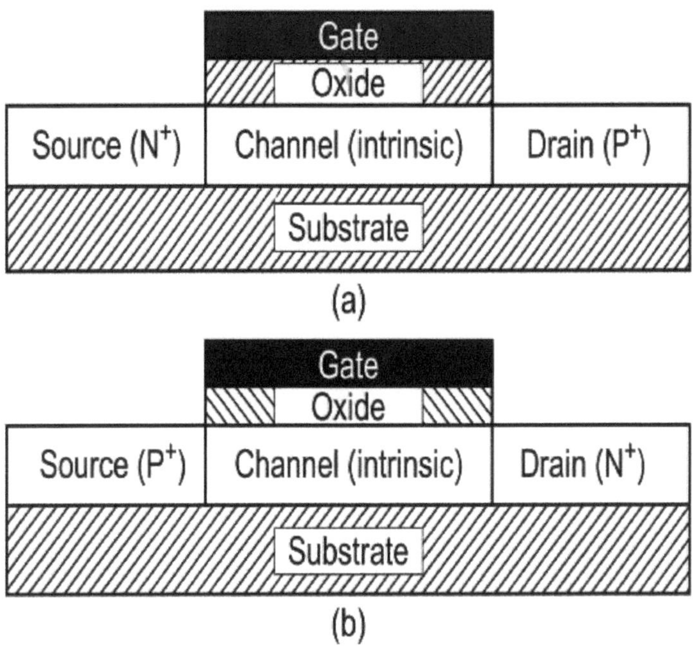

Fig. (1). Cross-sectional diagrams of (a) p-channel and (b) n-channel Tunnel FETs.

COMPARISON WITH MOSFETS

MOSFETs have been utilized as a fundamental building block in semiconductor technology for many years; however, their limitations in power scaling have resulted in increased leakage current and power loss. The key difference between MOSFET and TFET is in the switching mechanism. Table **1** illustrates the primary difference between MOSFET and TFET.

Table 1. Performance comparison of MOSFETs and TFETs.

Feature	MOSFET	TFET
Carrier Injection	Thermionic emission	Band-to-band tunneling (BTBT)
Subthreshold Swing (SS)	Limited to 60 mV/decade	Much lower than 60 mV/decade
Operating Voltage (V_{DD})	Requires higher voltage	Can operate at a lower voltage
Power Consumption	Higher due to leakage current	Lower due to suppressed leakage
ON-Current (I_{ON})	High	Lower than MOSFETs (a challenge)
Leakage Current (I_{OFF})	Higher leakage at small dimensions	Lower leakage, leading to better energy efficiency

One of the most significant benefits of TFETs is that they can achieve a SS below 60 mV/decade, which is the thermal intrinsic limit for traditional MOSFETs when operated thermally. This enables TFETs to be operated at lower voltage levels (V_{DD}), leading to a considerable reduction in power dissipation. The low ON-current (I_{ON}) and physical complexity must be overcome for large-scale usage [2].

IMPORTANCE IN LOW-POWER APPLICATIONS

With growing demand for energy-efficient electronic devices, TFETs are being seen as powerful contenders for a range of low-power applications. In IoT and wearable electronics, TFETs offer ultra-low-power consumption, with battery life being greatly enhanced in edge-computing and IoT applications. Low-voltage operation of TFETs is also well-suited for biomedical electronics, where high power efficiency is required for implantable devices in order to achieve long lifetimes. TFETs are also well-suited for neuromorphic computing, where their steep switching behavior is exploited in brain-inspired systems [4]. In hardware security, the intrinsic variability of TFETs can be leveraged in Physical Unclonable Functions (PUFs) for secure authentication and cryptographic applications. TFETs are also promising for ultra-low-power analog and RF circuits, *e.g.*, low-noise amplifiers (LNAs), where power dissipation needs to be minimized.

The low-voltage operation and minimal leakage currents of TFETs have them in the lead for the post-MOSFET era, addressing the power crunch of semiconductor technology. The intrinsic advantage of TFETs is their unique carrier transport mode, which is fundamentally different from the thermionic emission mode of MOSFETs. To understand the superior energy efficiency of TFETs, one needs to explore their working mechanism, i.e., the BTBT mechanism governing their operation. The physics of TFETs, the contribution of quantum mechanical tunneling towards achieving steep switching and low-power operation, is the subject of the following section.

Working Principle of TFETs

TFETs rely on a different carrier-transport mechanism than traditional MOSFETs. Unlike MOSFETs, which rely on thermionic emission—charge carriers tunneling across a potential barrier via thermal excitation—TFETs rely on band-to-band tunneling (BTBT), a quantum-mechanical effect that allows carrier transport even under low-voltage bias. Tunneling is the mechanism by which TFETs can produce a subthreshold swing (SS) below the 60 mV/decade limit of MOSFETs at 300K, rendering them of special interest for ultra-low-power applications [5]. TFET operation can be explained exhaustively by an analysis of their band-t--band tunneling mechanism, energy band alignment in ON and OFF states, and tunneling probability factors. The subsequent sections will give a detailed discussion of these concepts.

BTBT MECHANISM

The underlying operation mechanism of TFETs relies on BTBT, which occurs when a high electric field allows charge carriers to tunnel from the VB of the source to the CB of the channel. This quantum-mechanical phenomenon is determined by the electric field at the source–channel junction [6].

In contrast to traditional MOSFETs, where thermal energy is required by the carriers to tunnel through the energy barrier in the source-channel junction, tunnel field-effect transistors (TFETs) make use of band-to-band tunneling (BTBT) to achieve current conduction at much lower voltages, hence making them extremely energy-efficient [7]. Fig. (**2a**) shows the energy band diagram of a MOSFET, where thermionic emission enables current conduction and switching at higher voltages. In contrast, Fig. (**2b**) shows the energy band diagram of a TFET, where the tunneling mechanism enables subthreshold swing improvement and operation at lower voltages.

The BTBT mechanism in TFETs begins with gate-voltage modulation, in which a suitable gate voltage narrows the potential barrier width at the source-channel junction, thereby favoring tunneling. As the applied electric field at the source-channel junction increases, the width of the tunneling barrier decreases, increasing the quantum tunneling probability [8]. Charge carriers tunnel directly from the VB of the source to the CB of the channel without depending on thermionic emission. After tunneling, the applied drain voltage also facilitates carrier transport through the channel to the drain terminal, allowing current to flow. The efficiency of tunneling can be enhanced by using high-k dielectrics, heterojunctions, and strain-engineering methods to increase the electric field and improve ON-state characteristics.

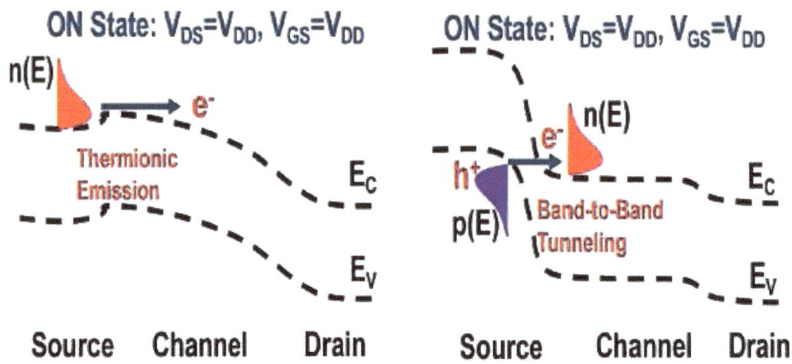

Fig. (2). Energy band profiles of (a) MOSFET and (b) TFET [9].

ENERGY BAND ALIGNMENT IN OFF AND ON STATES

The functionality of a TFET can be effectively demonstrated through its output and transfer characteristics, which graphically represent the device's drain current behavior. The energy band diagram of a TFET provides insight into its performance under different biasing conditions. In the basic operation of a TFET, a voltage is provided to the drain terminal while the source is maintained at ground voltage. Traditional Tunnel Field-Effect Transistors (TFETs) can be operated for both positive and negative gate voltages. In an n-type TFET, a positive gate-to-source voltage (V_{GS}) is essential for its operation, whereas ambipolar operation is obtained due to a negative V_{GS}. In a p-type TFET, the V_{GS} polarity is reversed.

When, in the OFF-state, a positive voltage (V_{DS}) is supplied to the drain while V_{GS} is zero, the tunneling field-effect transistor (TFET) is in its non-conductive state. In this state, the energy barrier at the interface between the source and channel remains wide, thus inhibiting carrier tunneling and limiting electron transfer from the source to the channel, as depicted in Fig. (**3a**). As the TFET enters the ON-state, the creation of a suitable positive V_{GS} in addition to a positive V_{DS} leads to the generation of carriers in the channel, thus forming a tunneling junction at the source-channel interface. With the increase in V_{GS}, the channel's conduction band (CB) aligns with the source's valence band (VB), as shown in Fig. (**3b**). The alignment enables band-to-band tunneling (BTBT), allowing electrons to tunnel through the potential barrier and for current to flow in the device. Tunneling enables charge transport at low voltages, resulting in a considerably enhanced subthreshold swing (SS) compared to conventional field-effect transistors (FETs). TFETs also have low OFF-state current (I_{OFF}), which significantly inhibits leakage current (Ileakage) and static power dissipation, making them very energy-efficient [10].

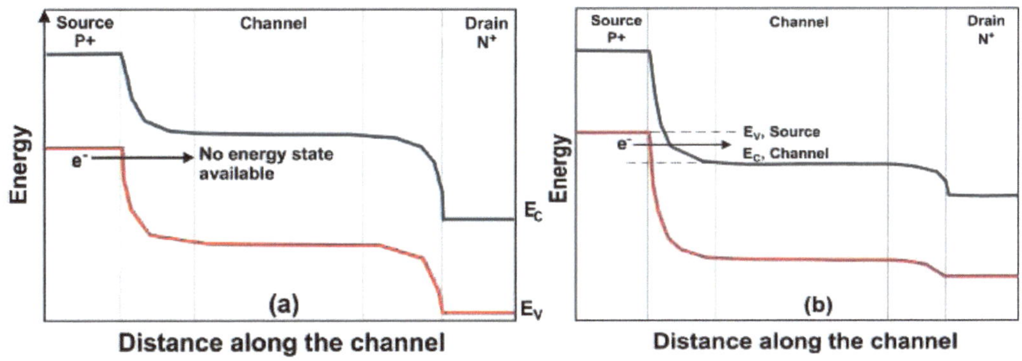

Fig. (3). Energy band profile of TFET when it is (a) in turned OFF (i.e., at $V_{GS} = 0$ V) and (b) in deep ON state [10].

Fig. (**4a**) illustrates the energy band profile of the TFET under the application of a negative gate voltage ($V_{GS}<0$), a condition referred to as the ambipolar state of the TFET. Initially, whenever the V_{GS} is zero, electron tunneling does not occur at either the source-channel or channel-drain junctions. Although as the V_{GS} becomes increasingly negative, the energy bands at the channel-drain junction shift upwards. With an additional increase in negative gate voltage, the VB of the channel aligns with the CB of the drain, thereby increasing the device's drain current. Fig. (**4b**) illustrates this ambipolar behaviour, showing that electrons continue to tunnel from left to right. This indicates that TFETs exhibit ambipolar conduction, in which the device current increases as the negative bias is increased.

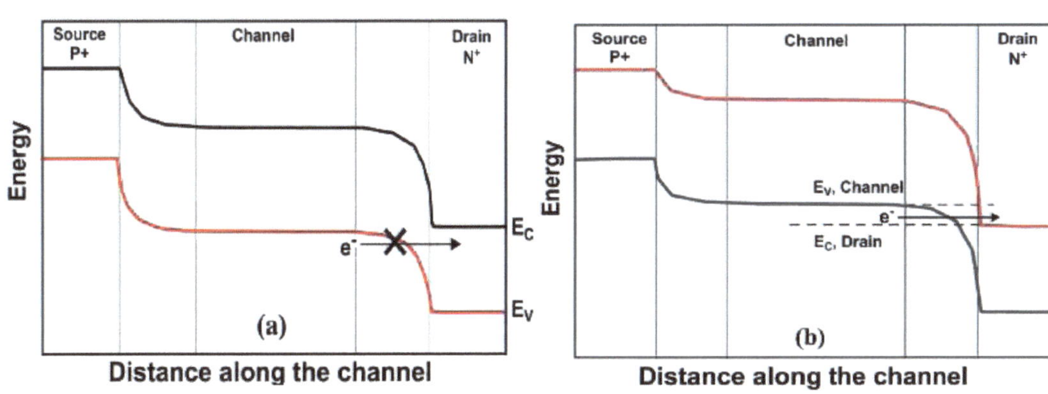

Fig. (4). Energy band profile of TFET for (a) negative gate voltage (b) more negative gate voltage.

The transfer characteristic ($I_D - V_{GS}$) of a TFET is shown in Fig. (**5**), depicting its operation across different regions. It is evident that in the OFF state, the device exhibits minimal current flow, signifying that the CB of the channel does not coincide with the VB of the source. While the gate bias increases in magnitude, the tunneling barrier width reduces, enhancing the drain current. With an additional rise in gate voltage, the rate of current increase slows due to diminishing

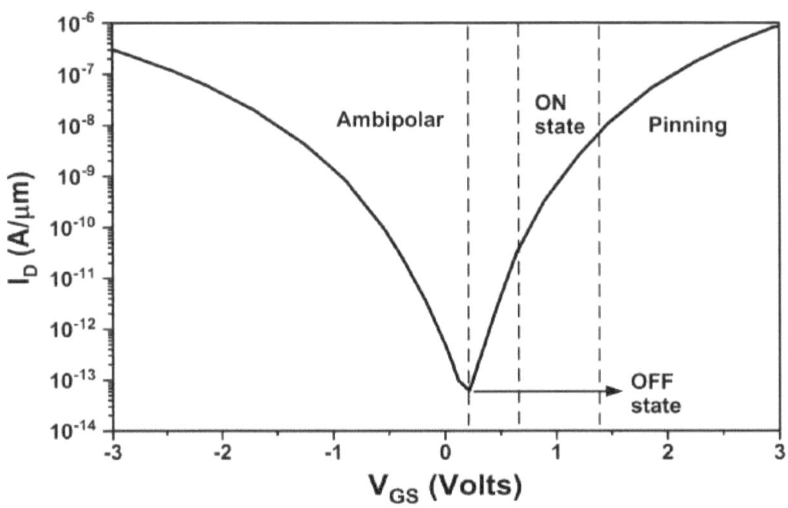

Fig. (5). The transfer characteristic ($I_D - V_{GS}$) of an n-channel TFET indicating various operating regions [10].

availability of additional initial states in the source VB, whenever the gate voltage gets closer to the drain voltage, current growth reduces further due to band pinning. An identical phenomenon occurs under negative gate bias, where the current in this region, known as ambipolar current, results from tunneling at the channel-drain junction. This behavior, characterized by an increase in drain current under negative gate bias, is depicted in Fig. (5).

TUNNELING PROBABILITY AND ITS INFLUENCING FACTORS

Tunneling probability in a Tunnel Field-Effect Transistor (TFET) is a significant parameter governing the efficacy of charge carrier transport through the source-channel junction. Unlike traditional Metal-Oxide-Semiconductor Field-Effect Transistors (MOSFETs), where carriers require thermal energy to tunnel through an energy barrier, TFETs rely on quantum mechanical band-to-band tunneling (BTBT) [11]. The successful tunneling probability is governed by a host of factors, including the electric field at the source-channel junction, energy band alignment, material, doping concentration, gate dielectric properties, and temperature. These parameters must be optimized to maximize ON-state current, reduce leakage current, and obtain a low subthreshold swing (SS) [12].

The electric field at the source-channel junction is mostly responsible for enhancing the probability of tunneling. An increased electric field narrows the tunneling barrier, thereby increasing the probability of charge carriers tunneling from the source's valence band to the channel's conduction band. The intensity of this electric field is proportional to the gate voltage and drain voltage used. Increasing the gate voltage enhances electrostatic control, increases energy band bending, and improves efficient tunneling. The use of a high-k dielectric material increases gate capacitance and the electric field intensity, thereby improving tunneling efficiency.

Energy band alignment is another significant parameter that determines tunneling probability. For effective tunneling, the conduction band (CB) of the channel should be at the same energy level as the valence band (VB) of the source [13]. This is based on the gate-source voltage (VGS), which induces band bending to achieve good tunneling conditions, and on the doping concentrations in the source and channel, which determine the level of band bending. The use of heterojunction materials enhances band alignment, thereby minimizing the tunneling barrier width and improving carrier transport efficiency. With these energy bands aligned, tunneling is more effective, increasing the ON-state current.

Material characteristics play an important role in determining the likelihood of tunneling in TFETs. The selection of the semiconductor material is most important, as it determines the bandgap, which in turn affects tunneling. Narrow-bandgap materials such as Indium Arsenide (InAs) and Germanium (Ge) support tunneling due to their reduced energy barriers between the VB and CB. This process reduces the operating voltage and enhances the subthreshold performance. Furthermore, the utilization of heterojunction TFETs, in which source and channel areas are composed of distinct materials, favors the efficiency of tunneling by creating a sharp band transition. Additionally, bandgap engineering through choosing materials wisely can enhance the ON-state performance of TFETs a long way [14].

The doping concentration in the source region has a profound impact on the number of charge carriers available for tunneling. Increasing the doping concentration in the source region increases the carrier density, which in turn sharpens band bending across the source-channel junction, thereby reducing the tunneling barrier width and increasing the tunneling probability. This increased tunneling probability leads to increased ON-current and overall device performance. Overdoping, however, causes detrimental effects such as increased junction leakage and higher parasitic capacitances, which adversely affect overall device efficiency. Proper optimization of the doping concentration is therefore necessary in order to achieve an optimal symbiotic balance between high tunneling efficiency and low leakage current.

The gate dielectric material and its corresponding thickness are important parameters governing the electrostatic control of the TFET. High-k dielectrics such as Aluminum Oxide (Al_2O_3) and Hafnium Oxide (HfO_2) enhance the gate capacitance, which further enhances the electric field at the source-channel junction. This enhanced electric field results in more abrupt bending of the energy bands, thus enhancing tunneling efficiency. A thinner gate dielectric thickness further enhances this effect by further enhancing the control of the energy bands by the gate. A too-thin dielectric, however, can result in counterproductive gate leakage, which can adversely affect the device's reliability. Therefore, engineering of the gate dielectric has to be carefully done to enhance TFET performance [14].

Temperature is also a critical factor in the tunneling probability of TFETs. Unlike MOSFETs, where carrier transport is thermally controlled, TFETs are quantum mechanically dominated and are therefore relatively insensitive to temperature. However, at very low temperatures, the contribution of phonon-assisted tunneling can be suppressed, which would have a very small effect on carrier transport. However, at high temperatures, one can expect a small increase in leakage current. As TFETs are not thermionic emission dependent, they possess better subthreshold behavior even at low temperatures and are therefore best applicable in low-power and cryogenic applications [15]. By tuning these control parameters, TFETs exhibit significantly lower leakage currents and better switching characteristics, further making them compatible with low-power applications. The interdependency of tunneling efficiency and device structure fundamentally governs TFET behavior, requiring creative structural adaptations to improve electrostatic control and scalability. This transition from knowledge of tunneling processes towards optimising device structures provides the context for the subsequent discussion of TFET design variations.

Device Structures and Variants

The structural architecture of TFETs plays a crucial role in determining important parameters such as scalability, electrostatic control, and tunneling efficiency. To increase ON-state current, decrease subthreshold swing, and enhance device performance, various TFET architectures have been introduced. Planar versus vertical TFETs, heterojunction TFETs (HTFETs), double-gate and gate-all-around (GAA) TFETs, and doping-less TFETs are the most significant structural differences in TFETs. Each of these architectures is appropriate for various low-power and high-performance applications and has distinct advantages and disadvantages.

PLANAR VS. VERTICAL TFETS

Planar and vertical TFETs are two basic device architectures that both have their unique strengths. Planar TFETs, as shown in Fig. (**6a**), are made up of a lateral structure where the source, channel, and drain are aligned in a straight line on a silicon substrate, similar to the structure of traditional MOSFETs. The ease of fabrication using CMOS-compatible processes makes planar TFETs a perfect candidate for direct integration with conventional circuits. However, planar TFETs are plagued by reduced electrostatic control of the channel and longer tunneling path length, parameters that can impact their efficiency and ON-state driving capability.

Conversely, vertical TFETs [16], as shown in Fig. (**6b**), are defined by a perpendicular arrangement of the source, channel, and drain regions. This setup reduces the tunneling distance considerably and enhances electrostatic control, hence leading to improved switching performance and drive current. Because of their large tunneling area, which allows for greater current conduction, vertical TFETs are especially suited for low-power and high-performance applications. Additionally, because of their small size, they can be formed in three-dimensional (3D) structures, thus raising the state-of-the-art semiconductor technologies' area efficiency. The formation of vertical TFETs requires advanced techniques, i.e., accurate layer deposition and etching processes, to ensure good device performance. Fig. (**6**) illustrates the comparison of the structural features between vertical and planar TFETs; for example, the advantage of the vertical structure in enhancing electrostatic control and tunneling efficiency, both being the most significant features of next-generation energy-efficient semiconductor devices, are highlighted.

HETEROJUNCTION TFETS (HTFETS)

Heterojunction TFETs [17] (HTFETs) enhance tunneling efficiency as they employ different materials for source and channel regions with an abrupt energy band transition. As depicted in Fig. (**7**), the employment of heterojunctions, i.e., III-V materials (InAs, GaSb, Ge, *etc.*), leads to improved band alignment, which significantly reduces the tunnelling barrier width and significantly increases the ON-state current [18]. These materials have a smaller effective mass and a smaller bandgap, enabling efficient BTBT and reducing leakage current. HTFETs enjoy numerous benefits over homojunction TFETs, *e.g.*, higher drive current, higher switching speed, and better scalability. With the employment of material engineering, HTFETs overcome the constraint of low ON-current of traditional TFETs, rendering them highly beneficial for ultra-low-power operations in next-generation semiconductor technologies. Nevertheless, employment of III-V materials in silicon-based CMOS technology is held back by fabrication challenges, *e.g.*, lattice mismatch, interface defects, and complicated process integration.

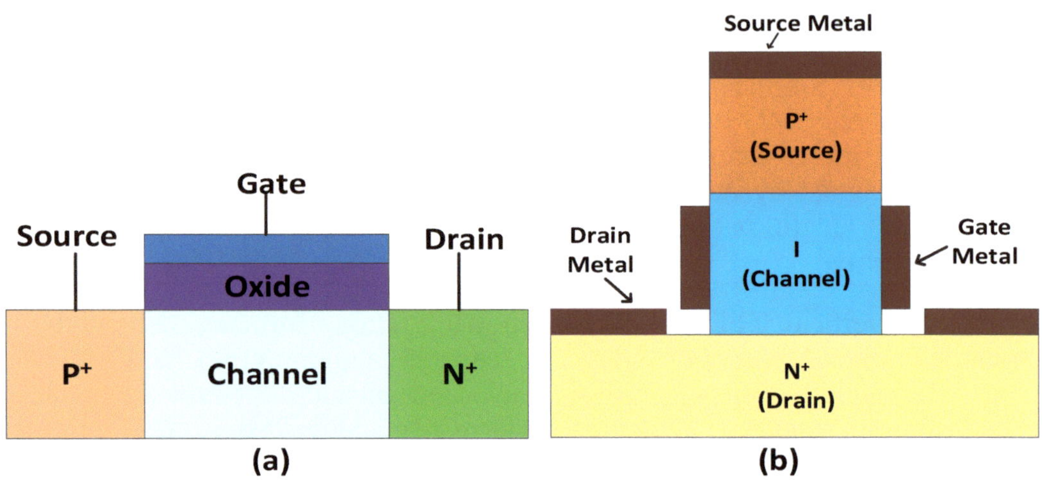

Fig. (6). Device structure of (a) Lateral TFET and (b) Vertical TFET.

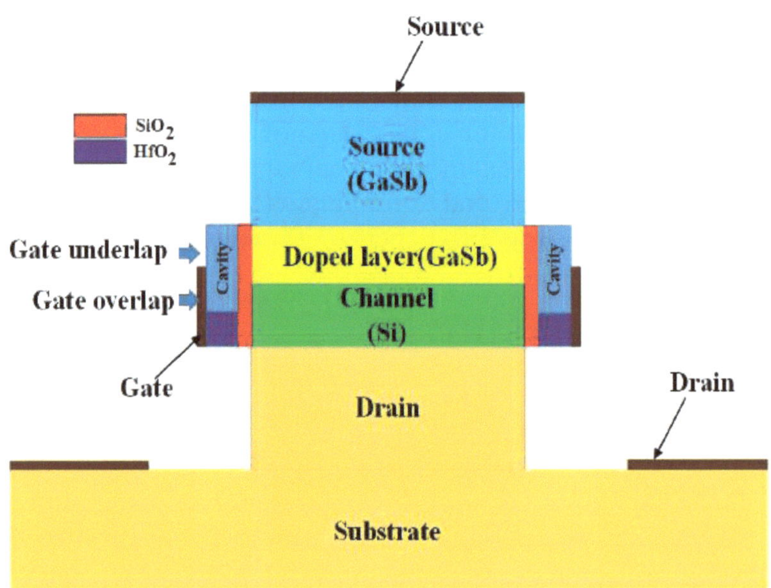

Fig. (7). Device structure of Heterojunction Vertical TFET [17].

DOUBLE-GATE & GATE-ALL-AROUND TFETS

For better electrostatic control and short-channel effect minimization, novel multi-gate architectures such as gate-all-around (GAA) TFETs and double-gate (DG) TFETs have been proposed. DG-TFET and GAA TFET structures have been illustrated in Fig. (**8a** and **8b**), respectively. They provide better gate control over the channel, which minimizes the leakage current and improves subthreshold behaviour.

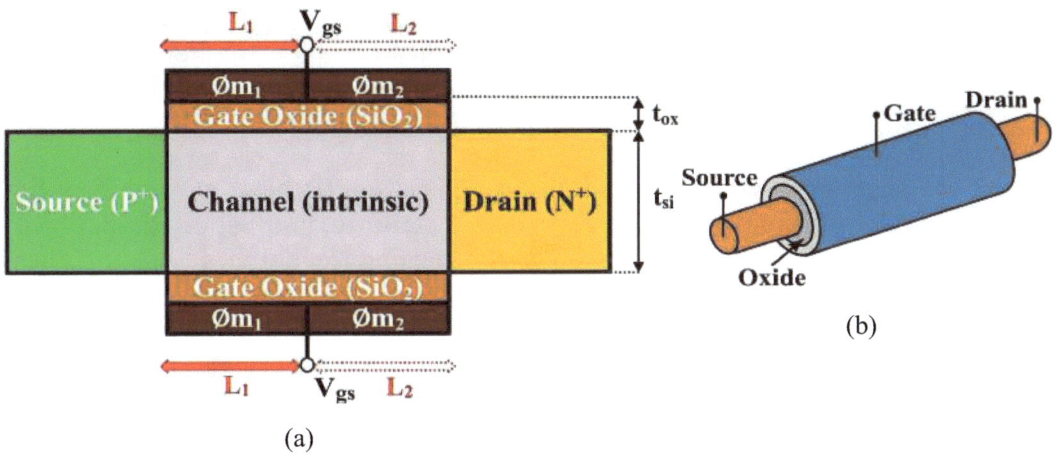

Fig. (8). Device structure of (a) Double-Gate, (b) Gate-All-Around TFET.

DG TFETs [19] are defined by the confinement of two gate electrodes that encircle the channel, as this greatly increases the electric field at the junction of the source-channel. This increased electric field enhances better tunneling, as a result of which ON-currents are higher and switching performance is improved. DG-TFETs improve subthreshold swing (SS) and energy efficiency by effectively modulating the tunneling barrier width. Gate-all-around (GAA) TFETs [20] improve on this concept by completely surrounding the channel with the gate electrode, thus ensuring maximum electrostatic control. This configuration minimizes short-channel effects and maximizes the efficiency of tunneling, making GAA-TFETs particularly well-suited to ultra-scaled transistor technology. As devices are scaled down to sub-10 nm, GAA-TFETs provide unparalleled gate control, improving performance and energy efficiency. However, the complexity of fabrication involved in multi-gate TFETs requires sophisticated processing techniques, thus making their fabrication more difficult and expensive.

DOPING-LESS TFETS

Doping-free TFETs offer a novel solution to the problems of conventional doping processes, such as dopant level variations and process variations. Unlike conventionally used TFETs, which are built from heavily doped source and drain regions, doping-free TFETs make use of intrinsic or lightly doped materials; in such devices, charge carriers are created electrostatically instead of being added by chemical doping. By eliminating the complexity of doping, the devices are more uniform, less variable, and more reliable after fabrication. The absence of impurity scattering allows for enhanced carrier mobility, which helps to enhance consistency in performance. Doping-free TFETs are implemented using dielectric-engineered junctions, where charge carriers are created by work-function engineering instead of by impurity diffusion.

As can be seen from (Fig. **9**), doping-less TFETs [20] have several benefits, including lower leakage current, higher device stability, and better compatibility with new nanomaterials and advanced processing techniques. All of these characteristics make them an extremely promising material for use in future transistor technologies, particularly because the accurate control of doping concentrations becomes increasingly difficult at nanometer scales.

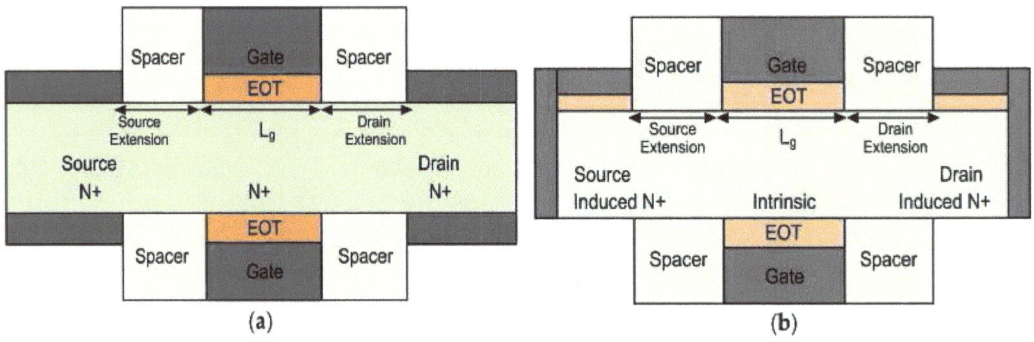

Fig. (9). Cross-sectional representations of (a) Conventional JLET (b) Doping less JLFET.

Electrical Characteristics of TFETs

The performance and use of TFETs in ultra-low-power devices are determined by their electrical characteristics. Because TFETs operate on the basis of BTBT, they differ from conventional MOSFETs in terms of their power consumption, subthreshold swing, and current-voltage characteristics. TFETs are a viable option for upcoming energy-efficient electronic devices because of all these factors.

CURRENT-VOLTAGE (I-V) CHARACTERISTICS

The tunneling-based carrier transport mechanism of TFETs distinguishes them from MOSFETs on the basis of their I-V characteristics. In the OFF state, the tunneling barrier between the source-channel junction is considerable, thus restricting carrier flow and leading to an ultra-low leakage current (I_{OFF}). When, however, the V_{GS} is increased, the CB of the channel aligns with the VB of the source, thus triggering BTBT, as indicated in Fig. (**10**). The effect leads to a steep rise in the drain current (I_D), enabling switching transition at high speed. Unlike MOSFETs, which rely on thermionic emission, TFETs enjoy exponential current rise due to tunneling, leading to greatly enhanced switching behavior. Despite this, however, a serious challenge confronting TFETs is ambipolar conduction, whereby negative gate bias reduces the tunneling width at the drain-channel junction, thus leading to undesirable leakage currents. Sophisticated device engineering, such as heterojunction schemes and work-function engineering, is employed to offset this problem while maintaining high ON-state performance [21].

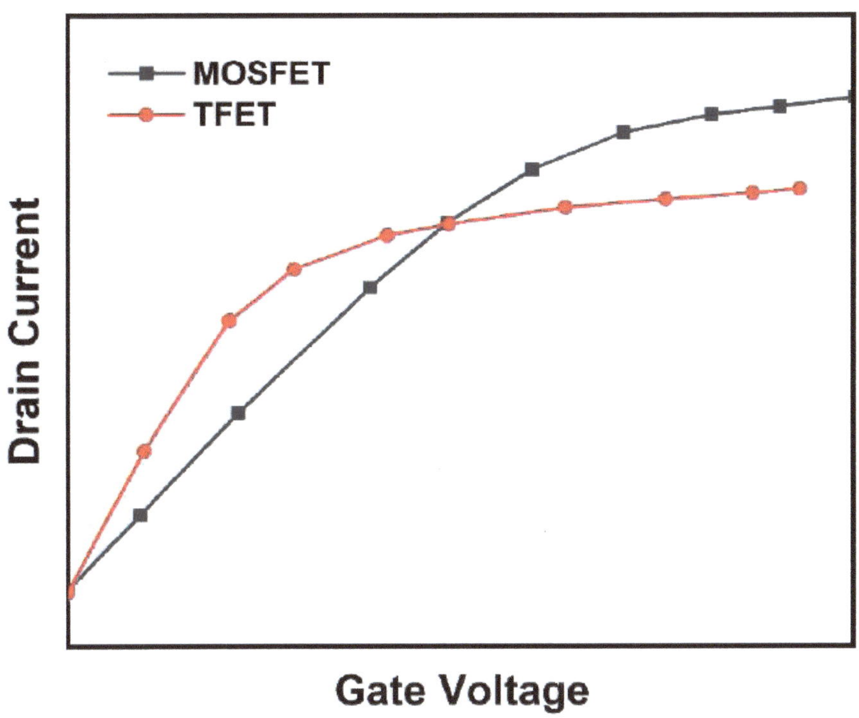

Fig. (10). I-V characteristics of MOSFET and TFET.

SUBTHRESHOLD SLOPE (SS) AND ITS ADVANTAGES

The ability of TFETs to achieve an SS below the theoretical limit of 60 mV/dec at room temperature is one of their most important advantages over MOSFETs. MOSFETs can only operate at ultra-low voltages due to the thermionic transport of carriers, which adheres to the Boltzmann distribution and limits subthreshold swing. TFETs circumvent this limitation by using quantum mechanical tunneling, which makes it a possibility to achieve subthreshold swing values lower than 30 mV/dec in optimized designs, as shown in Fig. (**10**).

A larger subthreshold swing (SS) in TFETs has some benefits. Firstly, it supports operation at lower supply voltages (V_{DD}), leading to huge power savings and making TFETs an excellent choice for battery-powered devices and Internet of Things (IoT) devices. Secondly, the rapid OFF-to-ON state switching minimizes standby leakage currents, hence improving the energy efficiency of low-power circuits. Thirdly, the improved subthreshold swing reduces dynamic and static power consumption, hence making TFETs an interesting choice for future computing systems. Moreover, TFETs exhibit greater scalability relative to MOSFETs since they provide strong electrostatic control at sub-5-nanometer nodes, where short-channel effects (SCEs) become a bottleneck for traditional transistors. However, variability in subthreshold swing due to material imperfections, interface traps, and process-induced defects can deteriorate TFET performance. To overcome these limitations, researchers are exploring alternative material systems like III-V semiconductors and two-dimensional (2D) materials, and using strain engineering techniques to allow uniform and reproducible sub-60 mV/decade operation.

LEAKAGE CURRENT AND POWER CONSUMPTION

Another major benefit of TFETs is their ultra-low OFF-state leakage current, which greatly improves their energy efficiency over MOSFETs. TFETs' OFF-state current (I_{OFF}) is governed mainly by the tunneling barrier width, which is high when there is no proper gate bias, leading to very low leakage. This parameter is of great use in applications that demand ultra-low standby power consumption. The leakage suppression in TFETs is governed by numerous variables such as tunneling barrier engineering, doping profile optimization, and gate work-function selection. Coupling heterostructure TFETs with III-V or 2D materials, the tunneling barrier width can be minimized, improving ON-current (ION) while maintaining I_{OFF} at low values. In a similar manner, doping profile optimization in the drain and source regions suppresses parasitic leakage channels while providing high drive current. Gate work-function engineering is also essential in preventing ambipolar conduction, which otherwise raises the drainage leakage currents. Aside from their low static power dissipation, TFETs also exhibit lower dynamic power consumption due to their lower operating voltage requirement. Equation (1) represents the total power consumption (P) of a transistor [22], which is given by:

$$P = CV_{DD}^2 f + I_{OFF}V_{DD} \tag{1}$$

where V_{DD} is the supply voltage, C is the capacitance, f is the switching frequency, and I_{OFF} is the leakage current. Since TFETs are capable of operating well at reduced V_{DD} values without compromising on high-speed switching, they offer significant power savings, hence being well-suited for power-restricted applications like IoT devices, biomedical electronics, and wearable systems.

Although these benefits accompany some TFET structures, ambipolar conduction is one of their disadvantages, and BTBT at the drain-channel junction results in increased leakage currents for a negative gate bias. The phenomenon is particularly significant in symmetric TFET structures, wherein source and drain junctions have the same tunneling properties. To negate this, methods like asymmetric drain doping, gate-underlap structures, and heterojunction engineering are employed to suppress ambipolar leakage and enhance device performance.

MATERIALS FOR TFETS

Material choice is a key factor that defines the efficiency of TFETs because it directly affects the BTBT process, ON-current (ION), and scalability. As TFETs are dependent on quantum mechanical tunneling for carrier transport, material selection with proper electronic properties can dramatically improve device performance. Material parameters controlling TFET operation are bandgap energy, carrier effective mass, and heterojunction band alignment. Different material systems like Silicon (Si), Germanium (Ge), III-V compound semiconductors, and two-dimensional (2D) materials have been investigated by researchers to minimize the BTBT process. Strain engineering has also been proposed as a promising method to modulate material properties and improve tunneling characteristics [23].

Silicon (Si) and Germanium (Ge) TFETs

Si is the most widely used semiconductor material because of its established process of manufacturing and compatibility with CMOS technology. However, its relatively higher bandgap of 1.12 eV and higher effective mass limit the probability of tunneling, resulting in lower ON-current (I_{ON}) and lower switching efficiency. In spite of these limitations, Si-based TFETs are still a promising candidate for low-power logic circuits because of their scalability and integration with existing semiconductor technologies. Then again, Germanium (Ge) exhibits improved tunneling efficiency due to its reduced bandgap of 0.66 eV and lower effective mass. This leads to increased BTBT rates, which work to increase ON-current (I_{ON}) and subthreshold values significantly. Ge TFETs, however, have larger leakage currents, which can prove to be problematic in affecting standby power consumption. Additionally, oxide interface quality and process integration issues pose challenges to the application of Ge in TFET technology across the board. Ge-based heterojunctions and high-k dielectrics are being eyed by research to bypass these challenges and enhance device performance [19].

III-V Compound Semiconductor TFETs

III-V compound semiconductors such as gallium antimonide (GaSb) [17, 18], indium arsenide (InAs), and indium gallium arsenide (InGaAs) have attracted a lot of attention because of their direct bandgap and reduced effective mass, which both increase BTBT efficiency. These materials have better carrier mobility than silicon (Si) and germanium (Ge) TFETs, enabling higher drive currents and quicker switching. One of the most significant benefits of III-V materials is their capacity to form heterojunctions with broken or staggered band alignments, which further increases tunneling efficiency. For example, the band alignment optimization of InAs/GaSb heterojunction TFETs [17] allows for steep subthreshold swing and superior ON-current. However, the mass production of III-V TFETs is hindered by issues of lattice mismatch, high interface defect densities, and process complexities in terms of fabrication. In addition, the integration of III-V materials with existing Si-based CMOS technology remains an issue, which requires sophisticated processing techniques like Molecular Beam Epitaxy (MBE) and Metal-Organic Chemical Vapor Deposition (MOCVD). Despite all these issues, III-V TFETs have tremendous potential for high-performance, ultra-low-power applications.

TWO-DIMENSIONAL (2D) MATERIAL TFETS

TFET technology has advanced with the introduction of two-dimensional (2D) materials such as Molybdenum Disulfide (MoS2), Tungsten Diselenide (WSe2) [9], and Black Phosphorus (BP). The atomically thin nature of these 2D materials minimizes short-channel effects and encourages electrostatic control. Additionally, tunable bandgaps and high carrier mobility make them very attractive for TFET devices. One of the major strengths of 2D materials is their capability to form van der Waals heterostructures, thus enabling engineered band alignments, which enhance BTBT efficiency. Unlike bulk materials, 2D TFETs are extremely scalable and possess improved subthreshold characteristics, which make them suitable for application in flexible electronics, low-power wearable electronics, and emerging computing technologies. However, issues like challenges in material synthesis, high contact resistance, and defect sensitivity need to be resolved before 2D TFETs can be industrially produced on a large scale.

STRAIN ENGINEERING FOR PERFORMANCE ENHANCEMENT

Strain engineering has been shown to be a valuable means of modifying the electronic properties of tunneling field-effect transistor (TFET) materials and maximum tunneling efficiency optimization [24]. Tensile strain in silicon (Si) and germanium (Ge) decreases the effective charge carrier mass, resulting in higher band-to-band tunneling (BTBT) rates and enhanced ON-current (ION). Strain engineering in III-V and two-dimensional (2D) materials [9] has the potential to greatly improve carrier transport and band alignment, further boosting TFET performance. Strained TFETs exhibit better switching characteristics, reduced threshold voltages, and elevated drive currents, making them a promising candidate for future semiconductor devices. Current research is targeted towards strain-induced bandgap modulation and lattice engineering in order to unlock the full potential of TFETs for ultra-low-power devices. Consequently, current research on new materials, such as III-V compounds, two-dimensional materials, and strain-engineered structures, is crucial to break through the limitations of traditional TFETs. Such developments are bound to make it possible to create high-performance, low-power electronic devices, hence transforming TFETs into a viable candidate for next-generation semiconductor technologies.

Challenges and Limitations

Although material and engineering advancements have improved the performance of TFETs considerably, numerous underlying challenges and obstacles prevent their large-scale deployment. The intrinsic dependence on band-to-band tunneling is accompanied by performance trade-offs, leakage, and process issues that must be addressed for TFETs to become a suitable alternative to MOSFETs in commercial applications.

LOW ON-CURRENT (ION) AND PERFORMANCE TRADE-OFFS

One of the most significant drawbacks of TFETs is their intrinsically small ON-current (I_{ON}) compared to traditional MOSFETs. Since TFETs function based on quantum mechanical tunneling and not thermionic emission, the current driving capability is limited by the tunneling probability. This lower ION impacts both switching speed and overall circuit performance, rendering TFETs less attractive for high-speed digital applications without optimization. To circumvent this limitation, researchers have explored a range of methods, including the use of low-bandgap materials like Germanium and III-V compounds, strain engineering, and new gate engineering to enhance tunneling efficiency.

AMBIPOLAR CONDUCTION AND LEAKAGE ISSUES

One of the biggest issues with TFETs is ambipolar conduction, which occurs when parasitic BTBT takes place at the drain-channel junction under reverse gate bias. This leads to unnecessary leakage currents, thus reducing power efficiency and affecting device reliability. The ambipolar effect is worse in symmetric TFET structures, where both electron and hole tunneling will be responsible for the leakage. Several approaches have been employed to limit ambipolar conduction, such as asymmetric source-drain doping, heterojunction structures, and gate work-function engineering.

FABRICATION PROBLEMS AND CMOS COMPATIBILITY

The incorporation of TFETs into conventional CMOS technology is plagued by serious fabrication issues. While MOSFETs have well-developed process flows, TFETs require tight doping control, defect-free heterointerfaces, and ultra-sharp junctions to realize optimal tunneling efficiency. Moreover, multi-material TFETs, especially III-V compound or two-dimensional material-based ones, are plagued by material deposition, etching, and contact formation issues. Large-scale deployment is a function of CMOS manufacturing process compatibility, and attempts are being made to develop new fabrication schemes to bridge the gap.

RELIABILITY ISSUES (BTI, NBTI, VARIABILITY)

Reliability is a major issue for TFETs due to their vulnerability to a variety of degradation mechanisms such as bias temperature instability (BTI), negative bias temperature instability (NBTI), and process-induced variability [20]. BTI and NBTI have the capability to induce threshold voltage shifts over long periods of operation that can lead to performance degradation and reduced circuit stability. In addition, variability in tunneling junctions due to fabrication variability impacts reproducibility and scaling of the device. Towards better TFET reliability, researchers are working on high-quality dielectric materials, optimal process control, and hard device structures that can counteract variability effects.

In spite of these challenges, ongoing progress in material, device, and fabrication technology is enabling the achievement of high-performance TFETs that can deliver the requirements of future low-power electronics. Overcoming them with innovative solutions will be essential to unleash the full potential of TFETs in future semiconductor technology.

Applications of TFETs

In spite of TFET challenges, their novel benefits like an ultra-low subthreshold swing, reduced leakage current, and energy-efficient operation make them extremely well-suited to an entirely massive number of new applications. With semiconductor technology shifting towards ultra-low-power and energy-efficient computing, TFETs are being researched in a very broad range of applications, ranging from digital and analog circuits to biomedical electronics and hardware security.

LOW-POWER DIGITAL CIRCUITS

Its optimal use is in ultra-low power digital circuits, where power efficiency and leakage minimization are primary issues. Since TFETs can be switched on at very low supply voltages without sacrificing reasonable switching speeds, they are optimally used in low-power microprocessors, logic circuits, and memory chips. TFETs are more power-efficient than traditional MOSFETs, with higher leakage currents and lower performance at low voltages. TFET integration onto ultra-lo--power computer platforms, i.e., edge computing platforms and Internet of Things (IoT) devices, will likely drive power-sensitive application innovation.

ANALOG & RF APPLICATIONS

Besides digital circuits, TFETs have also shown encouraging performance in radio frequency (RF) and analog circuits as a result of their steep subthreshold swing and improved electrostatic control. Low-power RF circuits such as voltage-controlled oscillators (VCOs), low-noise amplifiers (LNAs), and millimeter-wave transceivers can be improved using the low power consumption and improved scalability of TFETs. However, high-performance analog circuit design with TFETs is difficult because of their reduced ON-current and non-linear characteristics. Research focus, therefore, is on the optimization of TFET structure to enhance their gain, bandwidth, and noise performance in order to make them applicable in next-generation wireless communication and sensor networks.

BIOMEDICAL ELECTRONICS

The demand for ultra-low-power and energy-efficient devices has spurred the study of TFETs for their application in biomedical electronics. Implantable devices such as pacemakers, neural interfaces, and biosensors require long battery life and low heat dissipation. Low leakage current and the ability to operate at sub-0.5V for TFETs are ideal for such devices. TFETs' biocompatibility with bio-compatible materials and flexible substrates would also render them an ideal candidate for wearable health monitoring systems. Efforts are underway to integrate TFET-based circuits on lab-on-chip platforms for real-time health diagnosis and personalized medicine.

PHYSICAL UNCLONABLE FUNCTIONS (PUFS) AND HARDWARE SECURITY

With the increasing need for secure hardware in contemporary electronics, TFETs have become increasingly important in hardware security, specifically Physical Unclonable Function (PUF) design. PUFs harness process variation inherent in semiconductor devices to create unclonable, unique signatures for secure authentication and encryption purposes. Process variation and ultra-low power sensitivity make TFET-based PUFs offer improved security for secure authentication, anti-counterfeiting, and cryptographic key generation hardware. TFETs' dependability in energy-constrained applications also places them in secure computing systems, particularly in the IoT and embedded security spaces.

As the semiconductor business moves towards energy-efficient and secure computing models, TFETs are seen to be the technology that will shape the future of electronics. Their use in low-power digital circuits, radio frequency (RF) devices, biomedical devices, and hardware security devices is a sign of their status as a next-generation semiconductor design technology that is revolutionary.

CONCLUSION

TFETs serve as a compelling alternative to traditional MOSFETs, characterized by an ultra-low subthreshold swing, reduced leakage current, and minimal power consumption. Their capacity to achieve a sub-60 mV/decade subthreshold swing through BTBT renders them highly appropriate for low-power applications. The selection of novel materials like III-V compounds and two-dimensional (2D) materials has been instrumental in attaining ON-current optimization, tunneling efficiency, and scalability. TFETs are, however, confronted with low drive current, ambipolar conduction, and production challenges, requiring further investigation and optimization. In spite of this, TFETs are being studied appreciatively for their prospective role in ultra-low-power digital circuits, radio frequency (RF) technology, biomedical electronics, and hardware security. Their unique properties make them a vital part of future energy-efficient and secure semiconductor technologies.

AI DISCLOSURE

During the preparation of the work, the author(s) used AI tools & Grammarly tool to enhance the writing style and readability of the content. The authors clarify that AI was not involved in generating the substantive content and that they take full responsibility for the publication.

REFERENCES

[1] A.M. Ionescu, and H. Riel, "Tunnel field-effect transistors as energy-efficient electronic switches", *Nature,* vol. 479, no. 7373, pp. 329-337, 2011. [http://dx.doi.org/10.1038/nature10679] [PMID: 22094693]

[2] A.C. Seabaugh, and Q. Zhang, "Low-voltage tunnel transistors for beyond CMOS logic", *Proc. IEEE,* vol. 98, no. 12, pp. 2095-2110, 2010. [http://dx.doi.org/10.1109/JPROC.2010.2070470]

[3] U.E. Avci, and D. Morris, "Tunnel FETs: Prospects and challenges", *IEEE J. Explor. Solid-State Comput. Devices Circuits,* vol. 1, pp. 23-37, 2015.

[4] Y. Zhang, and N. Balasubramanian, "Energy-efficient applications of tunnel field-effect transistors", *IEEE Trans. Electron Dev.,* vol. 63, no. 4, pp. 1520-1527, 2016.

[5] Q. Zhang, "A Surface-Potential-Based Analytical Model for Double-Gate Tunnel-FETs", *IEEE Trans. Electron Dev.,* vol. 58, no. 7, pp. 1776-1782, 2011.

[6] S.K. Saha, "Analysis of Band-to-Band Tunneling and Its Application in Tunneling FETs", *IEEE Trans. Electron Dev.,* vol. 60, no. 1, pp. 280-286, 2013.

[7] S. Agarwal, "Energy-Efficient Tunnel FET Design for Ultra-Low-Voltage Applications", *IEEE J. Solid-State Circuits,* vol. 51, no. 3, pp. 701-710, 2016.

[8] M. Luisier, and G. Klimeck, "Simulation of nanowire tunneling transistors: From the Wentzel–Kramers–Brillouin approximation to full-band phonon-assisted tunneling", *J. Appl. Phys.,* vol. 107, no. 8, p. 084507, 2010. [http://dx.doi.org/10.1063/1.3386521]

[9] S. Kanungo, G. Ahmad, and P. Sahatiya, "2D materials-based nanoscale tunneling field effect transistors: current developments and future prospects", *npj 2D Mater Appl.,* vol. 6, no. 83, 2022. [http://dx.doi.org/10.1038/s41699-022-00352-2]

[10] J.K. Mamidala, R. Vishnoi, P. Pandey. *Tunnel Field-effect Transistors (TFET): Modelling and Simulation*, John Wiley & Sons, Inc. p. 208.,

[11] S. Mookerjea, and S. Datta, "Comparative Study of Si, Ge and InAs based Steep SubThreshold Slope Tunnel Transistors for 0.25V Supply Voltage Logic Applications", *2008 Device Research Conference,* pp. 47-48, 2008. Santa Barbara, CA, USA. [http://dx.doi.org/10.1109/DRC.2008.4800730]

[12] J. Knoch, and J. Appenzeller, "Tunneling phenomena in carbon nanotube field-effect transistors", *phys. stat. sol.,* vol. 205, pp. 679-694, 2008. [http://dx.doi.org/10.1002/pssa.200723528]

[13] H. Lu, and A. Seabaugh, "Tunnel Field-Effect Transistors: State-of-the-Art", *IEEE J. Electron Devices Soc.,* vol. 2, no. 4, pp. 44-49, 2014. [http://dx.doi.org/10.1109/JEDS.2014.2326622]

[14] G. Zhou, S. Lee, and K. Banerjee, "Metal contacts on two-dimensional semiconductors", *Nat. Nanotechnol.,* vol. 13, pp. 194-198, 2018.

[15] M. Ghosh, S. Saurabh, and M. Saxena, "Temperature Dependent Performance of TFETs: A Simulation

Study", *IEEE Trans. Electron Dev.,* vol. 61, no. 7, pp. 2512-2518, 2014.

[16] W. Hansch, C. Fink, J. Schulze, and I. Eisele, "A vertical mos-gated esaki tunneling transistor in silicon," Thin Solid Films, vol. 369, no. 1–2, p. 387–389", *Online (Bergh.),* 2000. [http://dx.doi.org/10.1016/S0040-6090(00)00896-8]

[17] A. Theja, and M. Panchore, "Performance Investigation of GaSb/Si Heterojunction Based Gate Underlap and Overlap Vertical TFET Biosensor", *IEEE Trans. Nanobiosci.,* vol. 22, no. 2, pp. 284-291, 2023. [http://dx.doi.org/10.1109/TNB.2022.3183934] [PMID: 35709121]

[18] A. Theja, and M. Panchore, "Performance assessment of charge plasma based hetero□structure VTFET biosensor for linearity enhancement", *Int. J. Numer. Model.,* vol. 37, no. 1, p. e3142, 2024. [http://dx.doi.org/10.1002/jnm.3142]

[19] V. Ambekar, and M. Panchore, "Realization of high-speed logic functions using heterojunction vertical TFET", *Appl. Phys., A Mater. Sci. Process.,* vol. 129, no. 3, p. 166, 2023. [http://dx.doi.org/10.1007/s00339-023-06419-1]

[20] M. Panchore, J. Singh, and S.P. Mohanty, "Impact of Channel Hot Carrier Effect in Junction-and Doping-Free Devices and Circuits", *IEEE Trans. Electron Dev.,* vol. 63, no. 12, pp. 5068-5071, 2016. [http://dx.doi.org/10.1109/TED.2016.2619621]

[21] Shubham Tayal, Dr Rahi, Jay Srivastava, and Sandip Bhattacharya, *Recent Trends in Compact Modeling of Negative Capacitance Field-Effect Transistors.*, 2021. [http://dx.doi.org/10.1201/9781003200987-9]

[22] J. Rabaey, A. Chandrakasan, and B. Nikolic, *Digital Integrated Circuits: A Design Perspective.* 2nd ed. Prentice Hall, 2003.

[23] S. Chahar, P. Sharma, K.K. Mishra, and R. Sharma, "A Review Article on the usage of Different Materials and Different Device architectures in the Design of Tunnel Field Effect Transistor", *2022 IEEE International Conference of Electron Devices Society Kolkata Chapter (EDKCON),* pp. 592-597, 2022. Kolkata, India. [http://dx.doi.org/10.1109/EDKCON56221.2022.10032894]

[24] J. Huang, Y. Wang, P. Long, Y. Tan, M. Povolotskyi, and G. Klimeck, *High-Performance Complementary III-V Tunnel FETs with Strain Engineering. arXiv:1605.00955*, 2016.

CHAPTER 3

Tunnel Field Effect Transistor Photosensor Review: Key Characteristics

Jai Kumar Bhatt[1,*], **Varnit Goswami**[1] and **Tarun Varma**[1]

[1] *Department of Electronics and Communication Engineering, Malaviya National Institute of Technology Jaipur, Jaipur, Rajasthan, India*

Abstract: Recently, optical sensors that consume low power and provide higher sensitivity are in high demand for various applications, including photodetectors, target tracking, *etc.*The Field-Effect Transistor (FET) as a photosensor has become increasingly important in human life in this fast-growing modern world. This flexibility with integrated circuits, the ability to downsize geometries, and the ability to provide enhanced sensitivity. This paper reviews an effective photosensor based on TFET, which provides lower power for applications that utilize the BTBT mechanism. Under exposure to incoming radiation, the proposed devices exhibit a higher illumination current, a low threshold voltage (V_{th}), a sharp sub-threshold swing (SS), and a high I_{ON}/I_{OFF} ratio. These characteristics make them highly suitable for achieving optimal efficiency while consuming low power. This paper provides an in-depth examination of TFET-based photosensors, covering everything from device assessment to their application in photo sensing. It includes both qualitative and quantitative analyses of parameters, such as sensitivity, and explores various factors that influence sensitivity by comparing alternative mechanisms and their transfer characteristics.

Keywords: Photosensor, TFET, BTBT, Responsivity, Sensitivity, Subthreshold swing.

INTRODUCTION

Industrial sensors have become the key to enhancing productivity and safety as the manufacturing industry becomes increasingly automated [1, 2]. Industrial sensors are the new factory floor's surveillence tool, available in a variety of sizes, shapes, and technologies. Inductive, capacitive, photoelectric, magnetic, and ultrasonic technologies are the most frequent [3, 4]. Because each technology has its advantages and disadvantages, the application's requirements will determine which technology should be chosen. Here, photosensors and their characteristics using TFET are discussed.

* **Corresponding author Jai Kumar Bhatt:** Department of Electronics and Communication Engineering, Malaviya National Institute of Technology Jaipur, Jaipur, Rajasthan, India; E-mail: 2018rec9109@mnit.ac.in

Girdhar Gopal, Meena Panchore, Arun Kishor Johar & Tarun Varma (Eds.)

Photodetectors, also known as photosensors, are sensors that detect light or electromagnetic radiation. Photodetectors are classified based on their detecting mechanism (photovoltaic) or other performance parameters (spectrum response) [5 - 8]. In photodetectors that rely on semiconductors to transform light photons into electrical energy, a p–n junction is identified. Within the depletion region, absorbed photons produce electron-hole pairs (EHP) [9, 10]. Photodiodes and phototransistors serve as examples of such photodetectors [11].

The photoelectric sensor determines whether an object is present or absent, or measures the distance between two points. It also has a photoelectric receiver and uses visible red or infrared light from a transmitter [12, 13]. Because they deliver quick and reliable results without requiring physical contact with the object, photoelectric sensors have become helpful in automation [14]. The main categories of photoelectric sensors include through-beam, retro-reflective, and diffuse sensors, as depicted in Figs. (**1**, **2**, and **3**), respectively [15]. Photoelectric sensors are widely utilized in various industries, including food and beverage, automotive, pharmaceuticals, and automated doors and gates [16, 17].

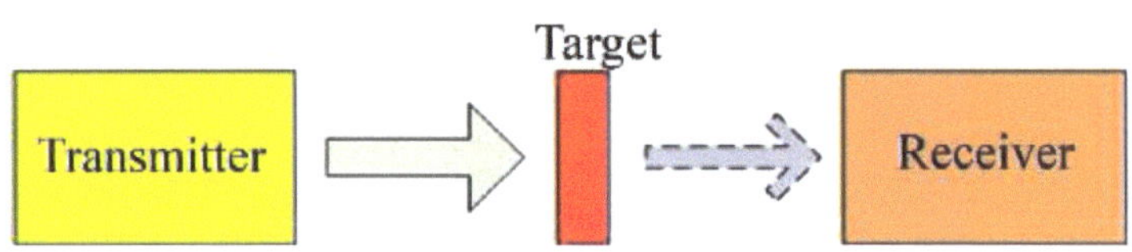

Fig. (1). Through-beam photoelectric sensor.

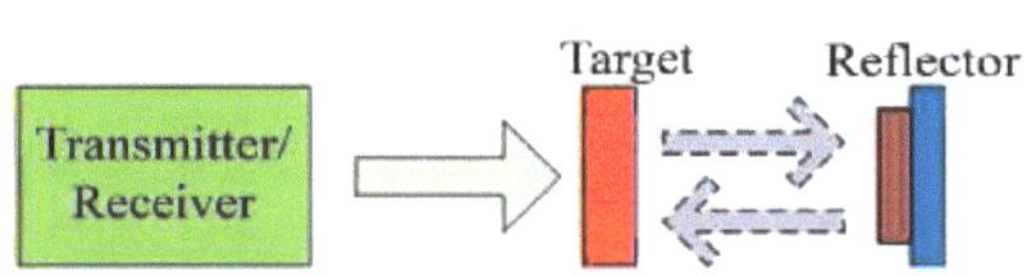

Fig. (2). Retro-reflective photoelectric sensor.

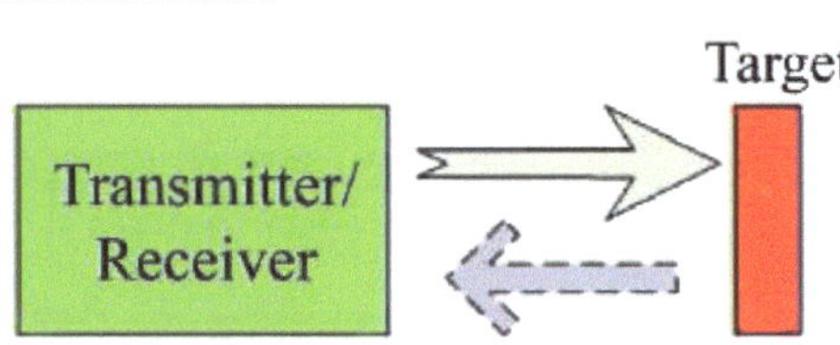

Fig. (3). Diffuse photoelectric sensor respectively.

The tunnel field-effect transistor (TFET) is a type of transistor used in academic research [18]. TFETs as photosensors have recently attracted much attention because of their compatibility with integrated circuits, the possibility of reducing dimensions, and the ability to give improved sensitivity [19 - 24]. Higher performance has been achieved in recent years due to better FET structure and channel material [25, 26]. The development of a photosensor using TFET is the result of earlier research.

A highly sensitive and low-power photosensor for NIR light detection in the wavelength spectrum of 0.7m to 1m employing twin MOS-capacitor (MOSCAP) Vertical TFETs was proposed in a study [27]. This redesigned TFET-based hybrid photosensor has the potential to be an innovative, highly sensitive photosensor and a suitable choice for sensing tightly spaced spectral lines. S. Joshi *et al.* [19] investigated an extremely sensitive and low-power photosensor for visible light detection in the 450-750nm wavelength band based on constructing a split-gate TFET. Through the DMG approach in a TFET, this hybrid design provides a next era incredibly sensitive photo-detector. In photovoltaics and optoelectronics, this WF engineering-based photo-sensing offers the prospect of merging ultrathin TMD with some other oxide-semiconductor systems. A study [19] outlines a novel approach for creating a waveguide-integrated photodetector. This method incorporates the characteristics of an atomically thin MoS_2 TFET along with the infrared detection capabilities of Germanium. MoS_2 TFETs' less power operation, low OFF-state current, and better subthreshold characteristics effectively improve the detector functioning for on-chip techniques [20]. There are some recent achievements regarding TFET as a photosensor. Therefore, the primary goal is determined by studying prior related work, and the present challenges for an effective p-Photosensor using TFET are: high sensitivity and minimal power operation, improved subthreshold swing, and high spectral sensitivity and responsivity.

TUNNEL FIELD EFFECT TRANSISTOR (TFET) THEORY

The Tunnel Field-Effect Transistor (TFET) is highly anticipated as the next-generation advancement in nanotechnology, primarily due to its low OFF current and subthreshold swing. This has led to the exploration of TFETs as potential substitutes for MOSFETs [29]. The comparison of subthreshold swing (SS) between TFETs and MOSFETs hinges on the fact that MOSFETs operate based on thermionic emission, whereas TFETs utilize the band-to-band tunneling (BTBT) mechanism [30].

Band-to-Band Tunneling (BTBT) takes place when the conduction band of an intrinsic region overlaps with the valence band of a P-type region under a sufficiently large gate bias. This overlap allows electrons from the valence band of the p-type region to tunnel into the conduction band of the intrinsic area, facilitating current flow throughout the transistor [31, 32]. However, when the gate bias reduces, the bands misalign, and the current cannot flow. The Tunnel

Field-Effect Transistor (TFET) functions as a gated p-i-n diode, as schematically shown in Fig. (**4**). The energy band diagrams at various gate voltages are presented in Fig. (**5**) [33]. In certain situations, BTBT current can also flow between the source and the drain. A gate is said to be "ON" when the tunnelling barrier decreases in response to a positive voltage being applied to it. However, when a negative potential is applied to a gate, BTBT current can flow, indicating an ambipolar (AMB) state. A point where BTBT arises differs between an ON and an AMB state. BTBT happens between a source and a channel when the ON state. When a channel and a drain are in the AMB condition, unintended BTBT occurs between them. Since the TFET exhibits ambipolar behavior, it remains ON even when the gate voltage is set to a negative value, preventing it from conducting [34].

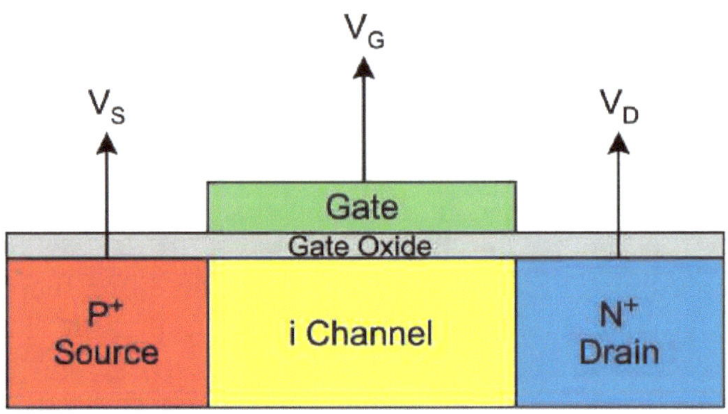

Fig. (4). TFET Structure schematic view [33].

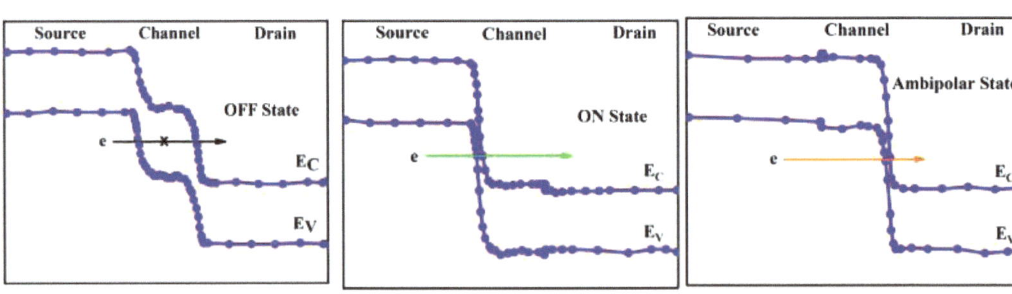

Fig. (5). TFET Energy Band Diagram.

Ambipolarity is a characteristic of a device that allows it to function under both positive and negative gate voltages. The current generated when a negative gate voltage is applied, is termed the ambipolar current (IAMB). This property is generally considered undesirable because it can compromise the performance of inverter-based logic circuits. In the case of tunnel field-effect transistors (TFETs), ambipolarity is linked to the device's internal structure. There is a limited body of research addressing the suppression of ambipolar current, as depicted in Fig. (**6**) [34 - 36].

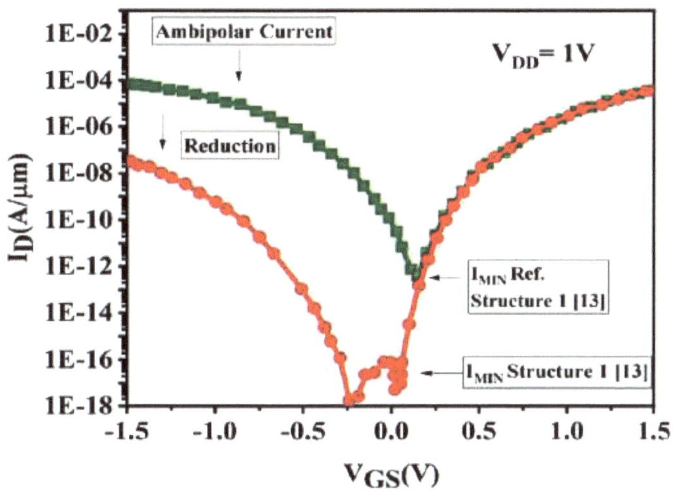

Fig. (6). The I_{AMB} and I_{MIN} currents are reduced with an aligned gate-drain, accentuating the transfer characteristics.

The figure of merit that controls a transistor's behavior in the subthreshold region is called subthreshold swing (SS). The thermal voltage (kT/q) limits the MOSFET performance, resulting in a theoretical minimum of SS = 60mV/decade. When modeling the weak inversion region, the subthreshold swing plays a vital role, particularly in applications such as high-gain analog circuits, imaging systems, and low-voltage technologies [37, 38]. $S=(d(\log10 I_{ds})/dV_{gs})^{-1}$ is the general formula for subthreshold slope (swing). Additionally, the subthreshold swing is defined as the inverse of the subthreshold slope. At room temperature (300K), a subthreshold swing (SS) of 52.8 mV/decade was observed, suggesting typical transistor functionality [39]. Fig. (**7**) illustrates the transfer characteristics of the fabricated 70 nm -long n-channel TFET, demonstrating exponential increases in drain current as the gate and drain biases are raised.

TUNNEL FIELD EFFECT TRANSISTOR (TFET) BASED PHOTOSENSOR

There is a superior choice: TFETs have higher sensitivity than MOSFET-based photosensors.

Here, four parameters of photosensors are considered within the visible wavelength range. Out of which first is the (i) Spectral Sensitivity (S_n), which is used to assess its utility. It is computed by changing the illumination wavelength from the second instance to the first instance and determining the associated drain current. It may also be written as [40]:

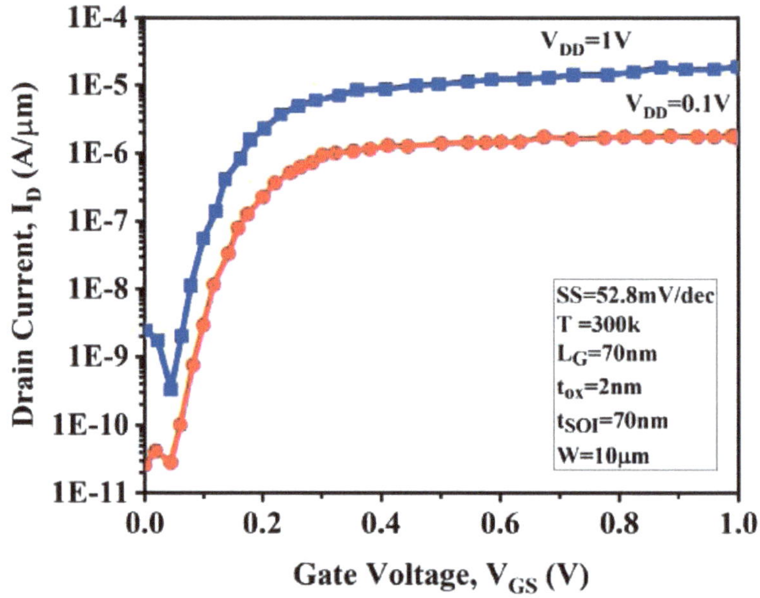

Fig. (7). Drain Current vs Gate Voltage curve [39].

$$S_n = \frac{I_D|_{\lambda_2} - I_D|_{\lambda_1}}{I_D|_{\lambda_1}}$$

Here $I_D|_{\lambda 2}$ is the Drain Current at light wavelength λ_2, and $I_D|_{\lambda 1}$ is the Drain Current at light wavelength λ_1.

Sensitivity plays a crucial role in assessing the effectiveness of photosensors, particularly in their ability to differentiate between closely grouped spectral wavelengths (100 nm) at lower light intensities (1 W/cm^2). Another significant factor is the Signal-to-Noise Ratio (SNR), which is determined by the ratio of photocurrent to dark current. This ratio is vital for evaluating how noise impacts the performance of photosensors.

It may be mathematically represented as [41]

$$SNR = 20\log\left(\frac{I_D|_{light}}{I_D|_{dark}}\right)$$

where, $I_D|_{light}$ and $I_D|_{dark}$ are the drain current under illumination and dark conditions, respectively.

(iii) The third most important factor is the Quantum Efficiency (η), which is described as the proportion of electron-hole pairs (EHPs) generated to the number of photons absorbed at a specific light wavelength.

It may be mathematically represented as [23]:

$$\eta = \frac{No.of EHPs\ generated}{No\ of\ Incident\ Photons}$$

(iv) Fourth factor is the Responsivity (R_n), which tells the essential photosensor characteristic. The photosensor's responsivity is a fundamental statistic that specifies its most essential qualities (R_n) [42]. The responsiveness of the photosensor is directly related to its quantum efficiency (η) [31] and can be described by the following expression:

$$\mathrm{R}_n = \frac{\text{Generated Photocurrent } \mathrm{I}_P}{\text{Incident Optical Power } P_O} = \frac{q\eta\lambda}{hc}$$

Here, the electronic charge is represented by q ($=1.6\times 10^{-19}$ C), Planck's constant is represented by h ($=6.62\times 10^{-34}$ J.s), and the velocity of light is represented by c ($=3\times10^8$ m/sec).

Some previous research on TFET-based photosensors and their characteristics is discussed here.

NIR Photosensor using Optically Gated DMOS Vertical TFET

A dual-MOSCAP vertical TFET (D-MOS VTFET) employs two MOS capacitors to vertically separate the channel and drain regions, resulting in an inverted configuration compared to the traditional TFET design. The intrinsic epi-layer ($1\text{x}10^{15}$ cm^{-3}) is found between a gate stack and a source. The MOSCAP region is made up of these three elements. A raised canal separates the source (which is p-type: 1x10^{20} cm^{-3}) and drain (which is n-type: 1x10^{18} cm^{-3}) parts. Photoconductivity is achieved using optical gates constructed from n-type silicon, which facilitate photon absorption. Fig. (**8**) presents a cross-sectional view of a proposed Tunnel Field-Effect Transistor (TFET) designed for photo-sensing applications. Optical generation occurs inside the device when it is stimulated with a light intensity of I = 0.5W/cm2 and a wavelength of 0.7m, as shown in Fig. (**9**).

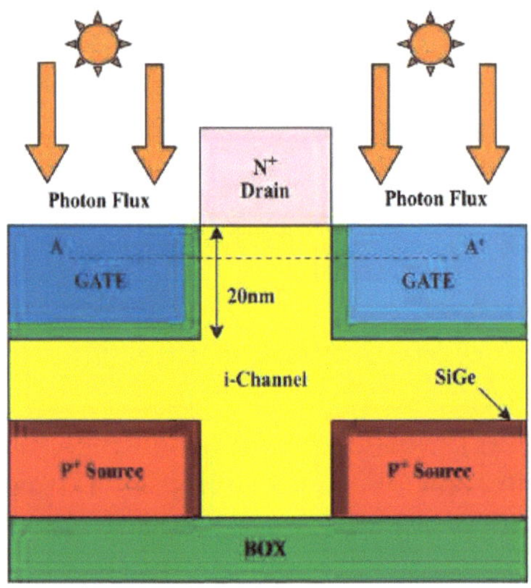

Fig. (8). Optically gated Dual MOSCAP [21].

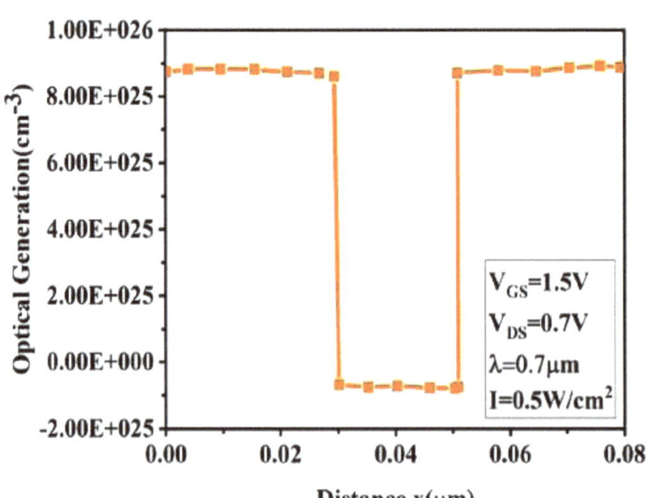

Fig. (9). Optical generation versus distance.

The novel D-MOS VTFET-based photosensor's practicality is calculated on the basis of Spectral Sensitivity (S_n), which is found by considering a drain current under light and dark situations. For six distinct values of V_{GS}, a S_n obtained through equation (1) is presented in Fig. (**10**).

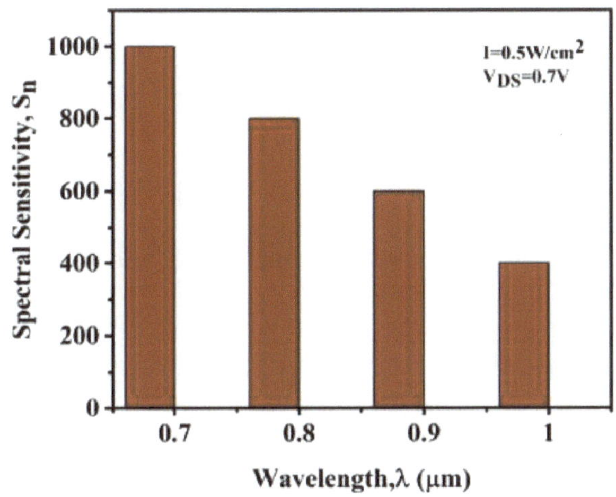

Fig. (10). S_n for different λ [21].

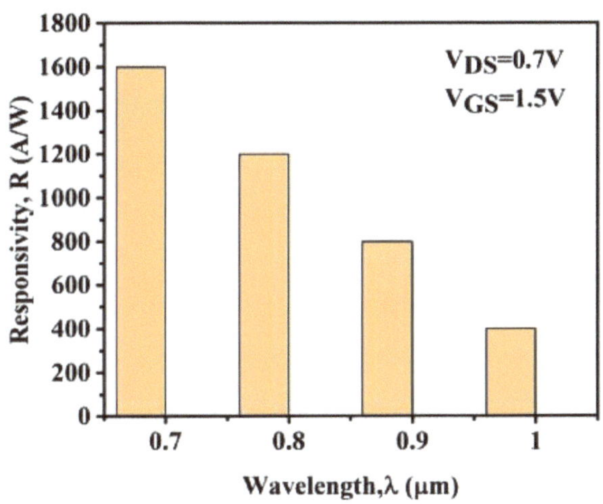

Fig. (11). Respective Responsivity for λ [21].

$$S_n = (I_{Photo} - I_{Dark})/I_{Dark} \quad \textbf{(1)}$$

The S_n has a greater value in the subthreshold regime, with larger responses at V_{GS}=0.5V for every wavelength range studied. As the wavelength of incoming light lowers, the sensitivity increases. Fig. (**11**) shows the photosensor's responsivity(R) when the incident light intensity is I_0=0.5W/cm^2. Equation (2) is used to calculate it.

$$R = I_{Photo}/P_{opt} \quad \textbf{(2)}$$

P_{opt} stands for incident optical power. In all cases, a responsivity is determined at V_{GS}=1.5V & in the order of 10^3. The peak value of R is 1.6x10^3 A/W and is obtained for λ= 0.7μm.

Split Gate TMD TFET-based Photosensor Using Photogating Effect

Photogating is a technique that uses a light-yielded gate field or voltage to regulate the conductance of a channel. Fig. (**12**) depicts the photosensor under examination. The proposed detector employs a DMG-TFET configuration, featuring a tunneling gate that acts as the photogate, alongside a secondary gate affected by metal. In line with the principles of photoconductivity, the likelihood of tunneling rises when the barrier height is reduced.

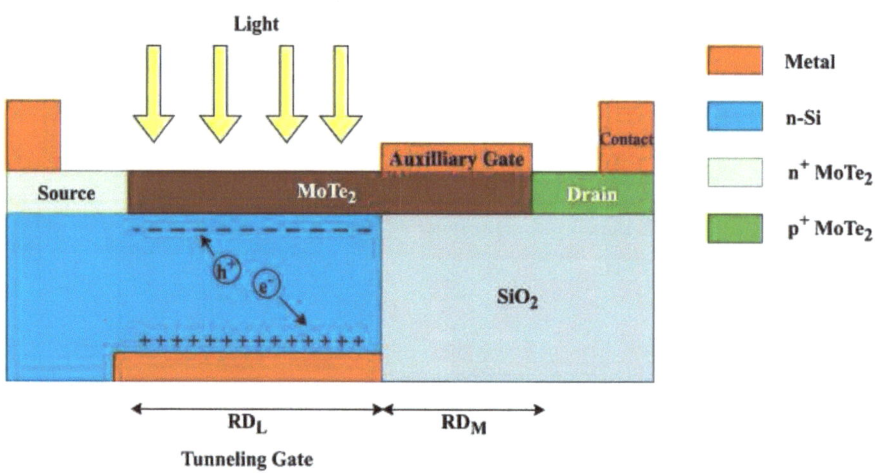

Fig. (12). Diagrammatic form of split gate TMD TFET based photodetector [19].

To comprehend the increase in work function (WF) induced by light in an oxide-silicon junction, it is useful to examine its energy band diagram and its effects on $MoTe_2$. The energy level shift, denoted by (3), is a result of the charge generated through photoexcitation.

$$\Delta E_{op} = qV_{op} = q^2 n_{op}/C_o \quad \textbf{(3)}$$

Under the influence of the metal gate, the illuminated slit of width Lt covering the tunneling region (RDL) and the additional region (RDM) of length Lm is initially set to 50nm. The surplus EHP created in a Si area is employed to photogate channel under normal illumination. Because of the importance of photodetection in the visible region, silicon has been chosen as the sensing element. The absorption coefficient falls from 10^5 cm^{-1} to 850 cm^{-1} as the wavelength increases from 400 nm to 800 nm. An analytical computation of a WF shift as a function of light intensity and wavelength is depicted in Fig. (**13**) [2].

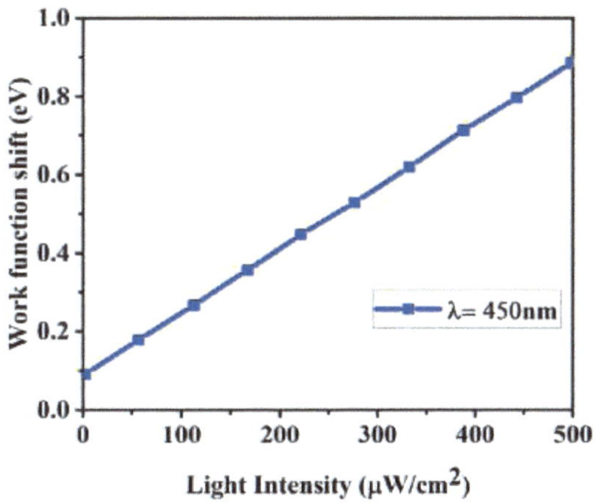

Fig. (13). Variations of the work function shift with light intensity.

The device's responsivity ($R = I_{ph}/P$, where I_{ph} is the photocurrent, and P is the incident light power) is shown in Fig. (**14a**) as a function of light intensity.

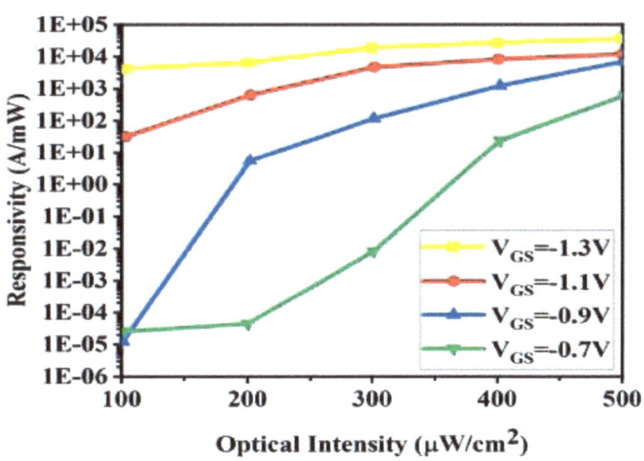

Fig. (14a). Responsivity of the device with Optical Intensity for different V_{GS}.

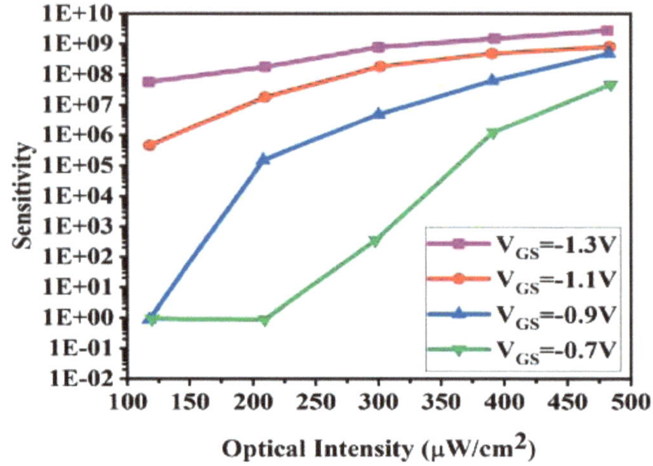

Fig. (14b). Relation between Sensitivity & respect to optical power density(λ=450nm).

The device shows excellent responsivity in the range of 103 AmW-1 to 105 AmW-1 at Vgs = -0.9 V, confirming the device's high sensitivity. For bias voltages under 0.7 V and when light intensity is below 200 W/cm^2, photosensitivity, defined as the ratio ($|I_{photo}-I_{dark}|/I_{dark}$), rises with an increase in V_{gs}. Fig. (**14b**) illustrates the relationship between the dark current and the current under illumination.

A Waveguide-Integrated Photodetector Employing a Transition Metal Dichalcogenide TFET with Germanium for Near-Infrared Sensing

According to the ITRS, TFET is identified as a promising substitute for MOSFET in low-power applications, offering improved sub-threshold characteristics and reduced. In the device illustrated in Fig. (**15**), a thin dielectric spacer and a MoS_2 layer serve to separate the Ge absorption layer from the Si rib waveguide. To provide strong responsivity with a minimal footprint, the optical mode solver is utilized to improve the provision of modes with an integrated Si and Ge layer. The optical restriction on guided-mode propagation in a SOI rib waveguide is suitable for an infrared-absorbing Ge layer. MoS_2 is deposited on top of the silicon waveguide in the TFET structure, and the MoS_2 layer can be grown using high-quality chemical vapor deposition.

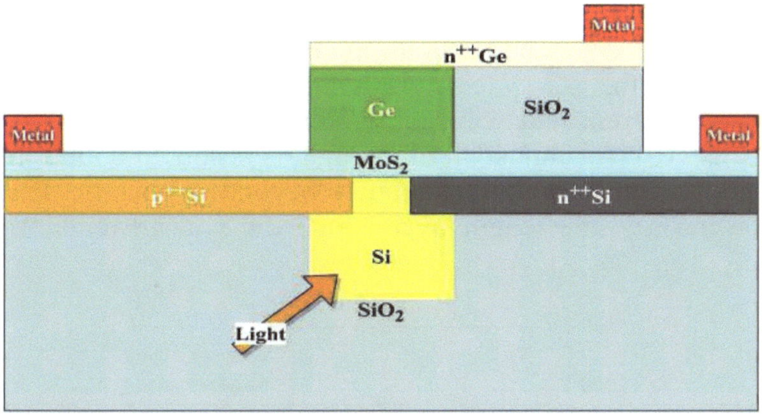

Fig. (15). Schematic view of a waveguide-MoS_2 TFET for different V_{GS}.

A charged plasma doped p^{++} and n^{++} regions of a TFET. The charge carrier passes through the underlying highly doped Si slab into the CP-doped MoS2 layer. Over four decades, the average SS = $(d \log_{10} I_{DS}/dV_{GS})^{-1}$, derived, i.e., 20 mV/decade. Fig. (**16**) shows the ID S-VD S properties of TFETs for various gate voltages [43].

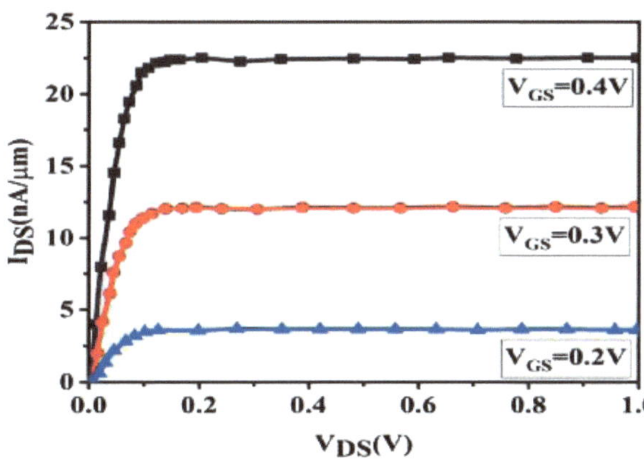

Fig. (16). I_{DS} vs V_{DS} characteristic of an integrated photodetector [43].

To replicate TFET behavior, a dynamic, non-local BTBT model was developed. Because of the high effective masses and wide bandgap of MoS_2, it has an exceptionally low I_{OFF} of the order of 10^{-13} to $10^{-15}\mu Am^{-1}$, which is a critical characteristic in the photodetector model, at which the strong dark current is the key limiting constraint. In the subthreshold region, the current exhibits greater variation than in the linear and saturation regions of operation.

As illustrated in Fig. (**17**), the drop in responsivity over 3 W cm-2 is attributable to a slight rise in current rather than an increase in incident power. At 3Wcm^{-2} incident light power, the gadget has quite a high responsivity of ~2700 AW^{-1}. The findings between current and waveguide lengths for various input powers are presented in Fig. (**18**). As almost all of the light is absorbed within this section of the device, the current reaches saturation beyond a specific length. As the incoming power rises, the length over which the current remains steady also extends.

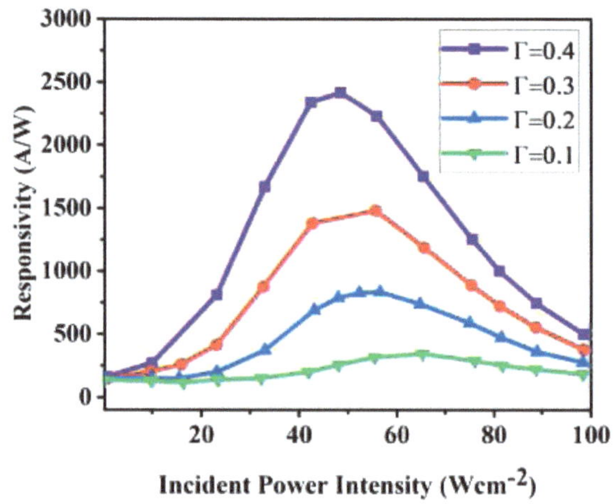

Fig. (17). Responsivity v/s Optical power.

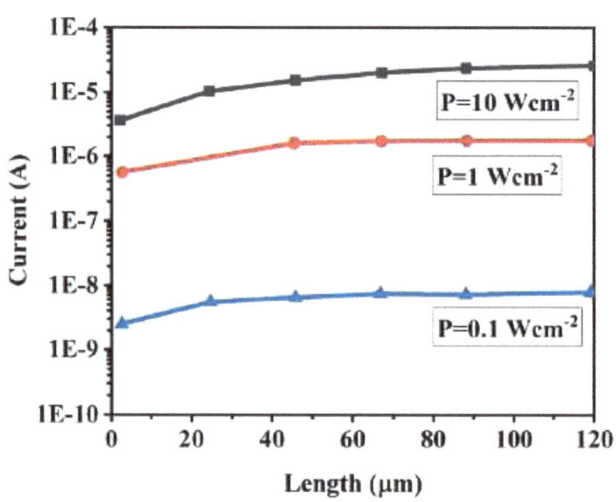

Fig. (18). Current v/s Waveguide Length.

A phototransistor operating in the infrared spectrum utilizing a JL-TFET (Junction-less Tunnel Field-Effect Transistor) architecture.

The electrical and optical performance of the sensor in question is analyzed through numerical methods, highlighting its optoelectronic characteristics. This study examines the influence of germanium (Ge) gate doping levels and high-k gate dielectrics on modifying optical Figures-of-Merit (FoMs), including responsivity, I_{ON}/I_{OFF} ratio, and optical switching speed. Unlike traditional FET-based sensors, the prototype model identifies promising avenues to enhance the FoMs of phototransistors. In the second part of our inquiry, we evaluate the proposed phototransistor's transition qualities and compare them to the standard model. The gadget is placed in the optical inverter circuit. Observed outcomes show that a proposed model has better optoelectronic performance than traditional designs, with respect to speed and gain. A new phototransistor architecture is created by combining the tunnel effect and the JL structure. Fig. (**19**) depicts the proposed sensor, built on a Ge-gate TFET device from a JL perspective. To prevent the high thermal expenditure required for the formation of abrupt p-n junctions, a SOI architecture is proposed, with a homogeneous channel n+-doping rather than a p-i-n structure. Furthermore, as illustrated in Fig. (**19**), the gated source aspect creates a tunneling effect at the source side [9].

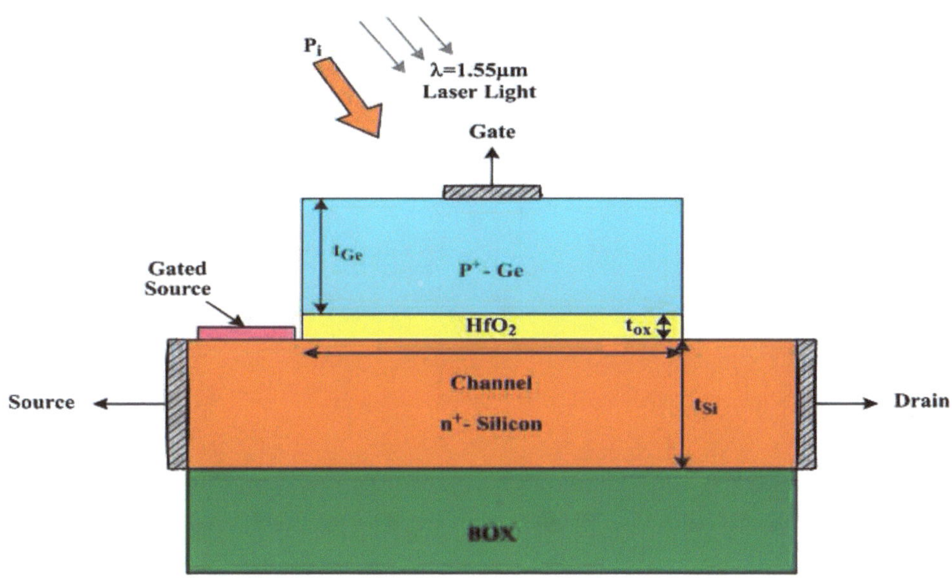

Fig. (19). Schematic structure of JL-TFET design [24].

Fig. (**20a**) investigates how varying the doping level of a Ge-gate affects the figures of merit (FoMs) of a device, such as responsivity, within the context of a projected Ge-based phototransistor model intended for optical communication applications.

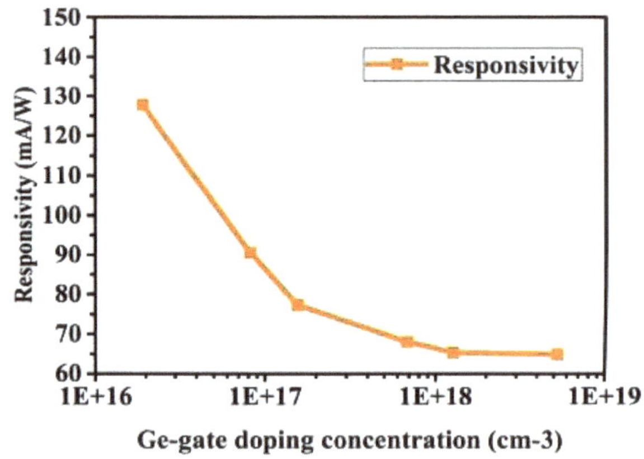

Fig. (20a). Responsivity with respect to the Ge-gate doping concentration.

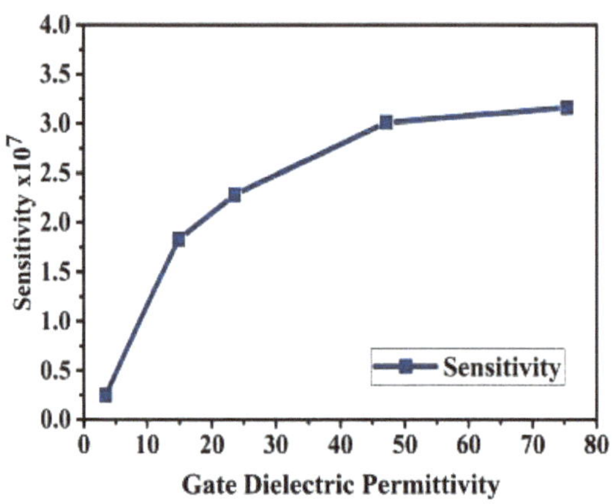

Fig. (20b). Sensitivity with respect to the high-k dielectric permittivity.

Consequently, Fig. (**20b**) illustrates how sensitivity varies with the permittivity of the gate dielectric, assuming an equivalent oxide thickness of 0.5 nm. The findings indicate that the proposed structure outperforms the inverter circuit's switching characteristics, resulting in enhanced optical gain. This study offers a novel perspective on designing high-performance, cost-effective, and ultra-low-power optical communication systems.

Optically Gated TFET for NIR Sensing

The optically gated TFET is developed to detect clustered spectral wavelengths (~100nm) in a NIR spectrum (1–0.7 m) at a minimum illumination intensity (1W/cm^2). Photogenerated voltage, gate charge, and semiconductor charge models were created to demonstrate device functioning. Under incoming radiation, the suggested device's high illumination current, low Vth, steep SS, and high current ratio make it well-suited for low-power operation with optimal efficiency. By exploiting the benefits of TFETs, the device reached a high sensitivity of 26.52. The performance of uniform and the comparison of Gaussian interface traps under varying concentrations are examined with respect to I_{ON}, I_{OFF}, and subthreshold swing (SS). Fig. (**21**) illustrates the structure of a photo-sensing transistor that operates on the basis of a band-to-band tunneling (BTBT) mechanism. The proposed design incorporates a silicon-germanium (Si/Ge) heterostructure, featuring an n-type germanium pocket located between the source and the channel junction. Fig. (**22**) compares the energy band diagrams for illumination (λ= 0.7μm, I = 0.5 W/cm^2) and dark states along with the source, pocket, channel, and drain [23].

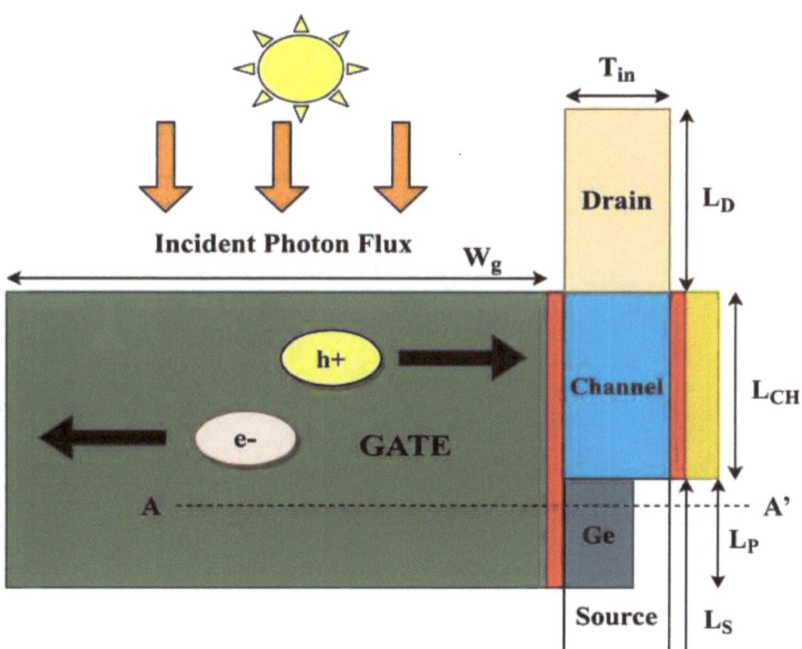

Fig. (21). Schematic of optically gated TFET [23].

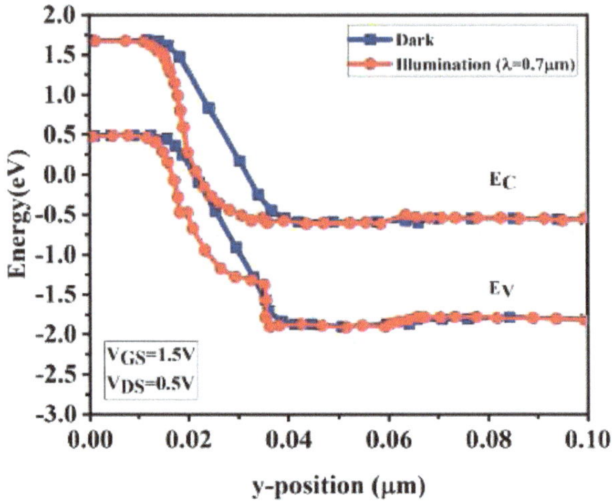

Fig. (22). Graphic demonstration of the energy band diagram.

Fig. (**23a**) presents a spectral response curve when λ reduced from 1μm to 0.7μm. The spectral sensitivity (S_n) is computed through the respective I_D-V_{GS} graph with respect to a chosen light wavelength using the formula of the Poisson equation & Einstein relation. Fig. (**23a**) shows a high sensitivity of 26.52 at V_{GS} = 0.5V. The I_{ON} versus illumination wavelength (λ) curves are shown in Fig. (**23b**), ranging from 0.7μm to 1μm.

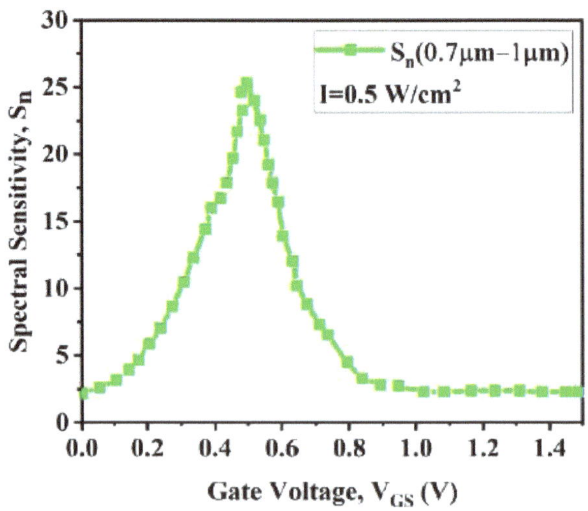

Fig. (23a). Spectral sensitivity vs gate Voltage, where λ is reduced.

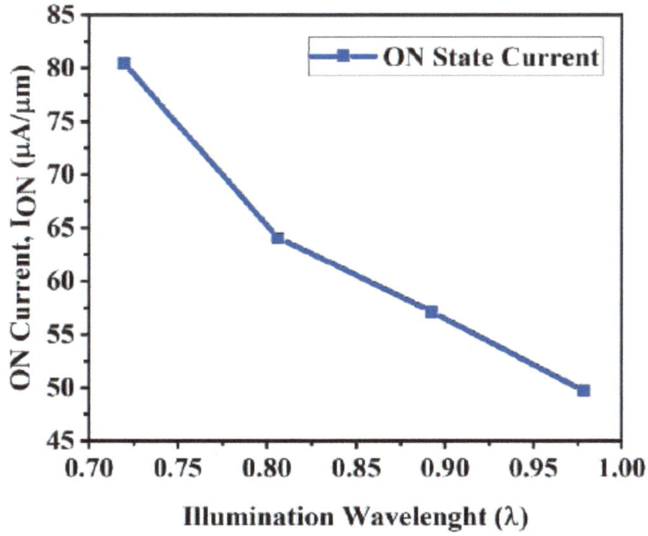

Fig. (23b). ON state current (V_{GS} = 1.5V with respect to applied Wavelengths.

BENCHMARKING OF PREVIOUSLY REPORTED PHOTOSENSOR CHARACTERISTICS

Table **1** provides a comparative analysis of existing photosensors, highlighting that the differences in size and doping among the photosensors studied are not distinctive. The table compares the sensitivity of the proposed TFET-based photosensors with that reported in the current literature. As demonstrated in Table **1**, TFET-based photosensors exhibit superior sensitivity and responsivity compared to other FET-based photosensors. Due to their exceptional optical properties, TFET-based photosensors hold potential for use as opto-sensors.

Table 1. Comparison of proposed photosensors with various photosensors.

S. No.	Photosensor	Approx. Responsivity	Approx. Sensitivity
1.	Heterogate- Photosensor [20]	0.56	77.5
2.	Germanium-Source- Photosensor [20]	0.14	68.2
3.	Silicon-Source- Photosensor [20]	0.068	92
4.	Si Photosensors [40]	0.49	-
5.	Split Gate TMD TFET Photosensor [19]	0.001 to 0.00001	$\approx 10^7$
6.	SnSe2 Based Photosensor [42]	0.12	-
7.	Dual MOSCAP TFET Photosensor at V_{DS}=0.7V, λ= 0.7µm [21]	4.67×10^3	1.6×10^3

(Table 1) cont.....

S. No.	Photosensor	Approx. Responsivity	Approx. Sensitivity
8.	MoS2 TFET Photosensor for wavelength in the range of 1.5 μm to 1.6 μm [28]	2700 AW−1	70
9.	Ge-Pocket TFET [44]	-	26
10.	Optically Gated TFET Photosensor [27]	-	26.52
11.	Conventional inversion mode OCFET Structure [46]	0.118	1200
12.	Conventional JL-OCFET device [45]	0.080	4.8×10^{5}
13	Proposed Phototransistor based on JL-TFET Structure [24]	0.077	2.4×10^{7}

CONCLUSION

In a real-time environment, the significant benefits of TFET-based photosensors compared to traditional photosensors are emphasized, along with performance metrics evaluated at a specific point. This paper provides a comprehensive analysis of TFET technology as a photosensor, tracing its development from inception to its current state, and explores various strategies for practical implementation. To assess the sensing capabilities of these devices, their transfer characteristics were examined under varying optical power conditions. These devices exhibit a high illumination current, low threshold voltage, and a substantial I_{ON}/I_{OFF} ratio when exposed to incident radiation, making them particularly suitable for applications requiring low power and high performance. Ultimately, due to its superior performance, the TFET device plays a vital role in point-of-care applications, ensuring accurate and reliable results.

AI DISCLOSURE

During the preparation of the work, the author(s) used AI tools & Grammarly tool to enhance the writing style and readability of the content. The authors clarify that AI was not involved in generating the substantive content and that they take full responsibility for the publication.

ACKNOWLEDGMENT

This work is supported by the VLSI Laboratory at MNIT, Jaipur, which originated from the SMDP-C2SD Project funded by MeitY, Govt. of India, for providing all the facilities for carrying out the work.

REFERENCES

[1] M. Aazam, S. Zeadally, and K.A. Harras, "Deploying fog computing in industrial internet of things and industry 4.0", *IEEE Trans. Industr. Inform.*, vol. 14, no. 10, pp. 4674-4682, 2018. [http://dx.doi.org/10.1109/TII.2018.2855198]

[2] Mohd Aiman Kamarul Bahrin, "Industry 4.0: A review on industrial automation and robotic", *Jurnal teknologi,* vol. 78, pp. 6-13, 2016.

[3] P. Zhang, "Industrial control technology: a handbook for engineers and researchers", *William Andrew,* p. 891, 2008.

[4] S. Soloman, "Sensors handbook", *McGraw-Hill Education,* p. 1385, 2010.

[5] M. B. Shubha, "Advances in Printable, Flexible and Transparent Graphene Photodetectors for Optoelectronics Applications", *ECS Transactions 107.1,* p. 18681, 2022.

[6] T. Tokuda, *Polarisation-analysing CMOS photosensor with monolithically embedded wire grid polariser* Electronics letters 45.4, 2009. [http://dx.doi.org/10.1049/el:20093132]

[7] S. Bansal, "Photodetectors for security application", In: *Nanoelectronic Devices for Hardware and Software Security. CRC Press*, 2021, pp. 279-300. [http://dx.doi.org/10.1201/9781003126645-14]

[8] C.R. Michel, E. Camarena-Corona, and A.H. Martínez-Preciado, "An ultraviolet to near-infrared photosensor based on microstructured CuO", *Optik (Stuttg.),* vol. 267, p. 169689, 2022. [http://dx.doi.org/10.1016/j.ijleo.2022.169689]

[9] P. Basu, "Photodetector Fundamentals", In: *Handbook of Optoelectronic Device Modeling and Simulation.* CRC Press, 2017, pp. 283-306. [http://dx.doi.org/10.4324/9781315152318-10]

[10] D. Durini, and D. Arutinov, "Operational principles of silicon image sensors", In: *High Performance Silicon Imaging. Woodhead Publishing*, 2020, pp. 25-73. [http://dx.doi.org/10.1016/B978-0-08-102434-8.00002-7]

[11] W. Ahmad, S. Ahmed, M. Amjad, S. Akhtar, and M. Ali, "Electrical and photo-stimulated characteristics of all-implanted CMOS compatible planar Si photodiode", *Optik (Stuttg.),* vol. 155, pp. 297-300, 2018. [http://dx.doi.org/10.1016/j.ijleo.2017.11.023]

[12] S. Juds, *Photoelectric sensors and controls: selection and application* vol. Vol. 63. CRC Press, 1988.

[13] X. Ge, "Modeling and simulation of temperature-induced luminous intensity for fiber-optic photoelectric sensors", *2012 Proceedings of International Conference on Modelling, Identification and Control. IEEE,,* 2012.

[14] I. Takai, S. Ito, K. Yasutomi, K. Kagawa, M. Andoh, and S. Kawahito, "LED and CMOS image sensor based optical wireless communication system for automotive applications", *IEEE Photonics J.,* vol. 5, no. 5, pp. 6801418-6801418, 2013. [http://dx.doi.org/10.1109/JPHOT.2013.2277881]

[15] P. Kharat, and J. Kharat, "Wireless Intrusion Detection System Using Wireless Sensor Network: A Conceptual Framework", *International Journal of Electronics and Electrical Engineering,* vol. 2, no. 2, pp. 80-84, 2014. [http://dx.doi.org/10.12720/ijeee.2.2.80-84]

[16] M. Javaid, A. Haleem, S. Rab, R. Pratap Singh, and R. Suman, "Sensors for daily life: A review", *Sens. Int.,* vol. 2, p. 100121, 2021. [http://dx.doi.org/10.1016/j.sintl.2021.100121]

[17] A.D. Wilson, and M. Baietto, "Applications and advances in electronic-nose technologies", *Sensors (Basel),* vol. 9, no. 7, pp. 5099-5148, 2009.

[http://dx.doi.org/10.3390/s90705099] [PMID: 22346690]

[18] U.E. Avci, D.H. Morris, and I.A. Young, "Tunnel field-effect transistors: Prospects and challenges", *IEEE J. Electron Devices Soc.,* vol. 3, no. 3, pp. 88-95, 2015.
[http://dx.doi.org/10.1109/JEDS.2015.2390591]

[19] S. Joshi, P.K. Dubey, and B.K. Kaushik, "Photosensor based on split gate TMD TFET using photogating effect for visible light detection", *IEEE Sens. J.,* vol. 20, no. 12, pp. 6346-6353, 2020.
[http://dx.doi.org/10.1109/JSEN.2020.2966728]

[20] S. Tiwari, and R. Saha, "Improved optical performance in near visible light detection photosensor based on TFET", *Microelectronics J.,* vol. 129, p. 105554, 2022.
[http://dx.doi.org/10.1016/j.mejo.2022.105554]

[21] Vandana Devi Wangkheirakpam, Brinda Bhowmick, and Puspa Devi Pukhrambam, "Investigation of a dual MOSCAP TFET with improved vertical tunneling and its near-infrared sensing application", *Semiconductor Science and Technology 35.6,* p. 065013, 2020.
[http://dx.doi.org/10.1088/1361-6641/ab8172]

[22] T. Chawla, M. Khosla, and B. Raj, "TFET-Based Sensor Design for Healthcare Applications", *Advanced Circuits and Systems for Healthcare and Security Applications. CRC Press,* pp. 69-103, 2022.
[http://dx.doi.org/10.1201/9781003189633-4]

[23] V.D. Wangkheirakpam, B. Bhowmick, and P.D. Pukhrambam, "Modeling and simulation of optically gated TFET for near infra-red sensing applications and its low frequency noise analysis", *IEEE Sens. J.,* vol. 20, no. 17, pp. 9787-9795, 2020.
[http://dx.doi.org/10.1109/JSEN.2020.2991406]

[24] H. Ferhati, F. Djeffal, and L.B. Drissi, "Performance assessment of a new infrared phototransistor based on JL-TFET structure: Numerical study and circuit level investigation", *Optik (Stuttg.),* vol. 223, p. 165471, 2020.
[http://dx.doi.org/10.1016/j.ijleo.2020.165471]

[25] T. Skotnicki, J.A. Hutchby, T. King, H-S.P. Wong, and F. Boeuf, "The end of CMOS scaling", *IEEE Circuits Devices,* vol. 21, no. 1, pp. 16-26, 2005.
[http://dx.doi.org/10.1109/MCD.2005.1388765]

[26] J. Appenzeller, "Carbon nanotubes for high-performance electronics—Progress and prospect", *Proc. IEEE,* vol. 96, no. 2, pp. 201-211, 2008.
[http://dx.doi.org/10.1109/JPROC.2007.911051]

[27] V.D. Wangkheirakpam, B. Bhowmick, and P.D. Pukhrambam, "Near Infra-red Photosensor using Optically Gated D-MOS Vertical TFET", *2021 International Conference on Numerical Simulation of Optoelectronic Devices (NUSOD),* IEEE, 2021.
[http://dx.doi.org/10.1109/NUSOD52207.2021.9541496]

[28] S. Joshi, P.K. Dubey, and B.K. Kaushik, "A transition metal dichalcogenide tunnel FET-based waveguide-integrated photodetector using Ge for near-infrared detection", *IEEE Sens. J.,* vol. 19, no. 20, pp. 9187-9193, 2019.
[http://dx.doi.org/10.1109/JSEN.2019.2922250]

[29] S. Saurabh, and M.J. Kumar, *Fundamentals of tunnel field-effect transistors* transistors. CRC press: Boca Raton, 2016, p. 306.
[http://dx.doi.org/10.1201/9781315367354]

[30] T.J. Vasen, *Investigation of III-V tunneling field-effect transistors.* University of Notre Dame, 2014.

[31] D. Keighobadi, S. Mohammadi, and M. Mohtaram, "Switching performance investigation of a gate-all-around core-source InGaAs/InP TFET", *Transactions on Electrical and Electronic Materials,* vol. 22, no. 4, pp. 502-508, 2021.
[http://dx.doi.org/10.1007/s42341-020-00257-1]

[32] S.K. Verma, S. Singh, G. Wadhwa, and B. Raj, "Detection of biomolecules using charge-plasma based gate underlap dielectric modulated Dopingless TFET", *Transactions on Electrical and Electronic Materials,* vol. 21, no. 5, pp. 528-535, 2020. [http://dx.doi.org/10.1007/s42341-020-00205-z]

[33] P.F. Wang, K. Hilsenbeck, T. Nirschl, M. Oswald, C. Stepper, M. Weis, D. Schmitt-Landsiedel, and W. Hansch, "Complementary tunneling transistor for low power application", *Solid-State Electron.,* vol. 48, no. 12, pp. 2281-2286, 2004. [http://dx.doi.org/10.1016/j.sse.2004.04.006]

[34] K.R. Pasupathy, T.S. Manivannan, and G. Lakshminarayanan, "A Review of Engineering Techniques to Suppress Ambipolarity in Tunnel FET", In: *Silicon 14*, 2022, pp. 1887-1894. [http://dx.doi.org/10.1007/s12633-021-01018-2]

[35] J.S. Jang, and W.Y. Choi, "Ambipolarity characterization of tunneling field-effect transistors", *Proc. Silicon Nanoelectronics Workshop (SNW),* pp. 1-2, 2010.

[36] Hraziia, A. Vladimirescu, A. Amara, and C. Anghel, "An analysis on the ambipolar current in Si double-gate tunnel FETs", *Solid-State Electron.,* vol. 70, pp. 67-72, 2012. [http://dx.doi.org/10.1016/j.sse.2011.11.009]

[37] G Gopal, and T. Varma, *Simulation-Based Analysis of Ultra-Thin-Body Double Gate Ferroelectric TFET for an Enhanced Electric Performance* vol. 8. Silicon, 2021, pp. 1-1. [http://dx.doi.org/10.21203/rs.3.rs-748137/v1]

[38] Girdhar Gopal, "Stacked ferroelectric heterojunction tunnel field effect transistor on a buried oxide substrate for enhanced electrical performance", *Semiconductor Science Technology 37.10* , p. 105006, 2022. [http://dx.doi.org/10.1088/1361-6641/ac830b]

[39] Byung-Gook Park, B.G. Park, J.D. Lee, and T.J.K. Liu, "Tunneling field-effect transistors (TFETs) with subthreshold swing (SS) less than 60 mV/dec", *IEEE Electron Device Lett.,* vol. 28, no. 8, pp. 743-745, 2007. [http://dx.doi.org/10.1109/LED.2007.901273]

[40] X. Guo, Q. Liu, H. Zhou, X. Luan, C. Li, Z. Hu, A. Hu, and X. He, "High-responsivity Si photodiodes at 1060 nm in standard CMOS technology", *IEEE Electron Device Lett.,* vol. 39, no. 2, pp. 228-231, 2018. [http://dx.doi.org/10.1109/LED.2017.2787147]

[41] B. Wang, "Analysis and Simulation of Temperature Characteristic of Sensitivity for SOI Lateral PIN Photodiode Gated by Transparent Electrode", *CCF Conference on Computer Engineering and Technology,* Springer: Berlin, Heidelberg, 2015.

[42] E.P. Mukhokosi, B. Roul, S.B. Krupanidhi, and K.K. Nanda, "Toward a fast and highly responsive SnSe2-based photodiode by exploiting the mobility of the counter semiconductor", *ACS Appl. Mater. Interfaces,* vol. 11, no. 6, pp. 6184-6194, 2019. [http://dx.doi.org/10.1021/acsami.8b16635] [PMID: 30652845]

[43] P.K. Dubey, and B.K. Kaushik, "A charge plasma-based monolayer transition metal dichalcogenide tunnel FET", *IEEE Trans. Electron Dev.,* vol. 66, no. 6, pp. 2837-2843, 2019. [http://dx.doi.org/10.1109/TED.2019.2909182]

[44] C. Wang, C. Li, J. Wei, G. Lin, X. Lan, X. Chi, C. Lu, Z. Huang, C. Chen, W. Huang, H. Lai, and S. Chen, "High-performance Ge pn photodiode achieved with preannealing and excimer laser annealing", *IEEE Photonics Technol. Lett.,* vol. 27, no. 14, pp. 1485-1488, 2015. [http://dx.doi.org/10.1109/LPT.2015.2426016]

[45] H. Ferhati, and F. Djeffal, "Planar junctionless phototransistor: A potential high-performance and low-cost device for optical-communications", *Opt. Laser Technol.,* vol. 97, pp. 29-35, 2017. [http://dx.doi.org/10.1016/j.optlastec.2017.06.002]

[46] L. Colace, V. Sorianello, and S. Ramajani, "Investigation of static and dynamic characteristics of optically controlled field effect transistors", *J. Lightwave Technol.,* vol. 32, no. 12, pp. 2233-2239, 2014.
[http://dx.doi.org/10.1109/JLT.2014.2322978]

CHAPTER 4

Carbon-Based Tunnel Field-Effect Transistor

P. Suveetha Dhanaselvam[1], B. Karthikeyan[1,*], K. Kavitha[1] and S. Nagarajan[2]

[1] *Department of Electronics and Communication Engineering, Velammal College of Engineering and Technology, Viraganur, Tamil Nadu, India*

[2] *Department of Electronics and Instrumentation Engineering, Easwari Engineering College, Chennai, Tamil Nadu, India*

Abstract: Carbon-based Tunnel Field-Effect Transistors (TFETs) are new devices, having special properties of carbon materials, including carbon nanotubes (CNTs), carbon graphene, and graphene nanoribbons (GNRs), to overcome the drawback of the conventional silicon-based TFETs. In this chapter, carbon-based TFETs, their construction, working, and application for low power and high efficiency will be discussed. Moreover, the vital roles of quantum tunneling, ballistic transport, and quantum capacitance in enabling these devices to achieve ultra-low power consumption and high-speed operation are also discussed. The challenges of materials development, device development, and the integration of these devices with the existing technology are also addressed in this chapter. It also discusses possible uses in solar panels, flexible gadgets, and energy-efficient electronic gadgets. Finally, the manuscript outlines the next stage of device optimization, with a special focus on hybrid material systems, and the use of advanced modelling methods. Additionally, it explores the potential consequences of carbon-based tunnel field-effect transistors (TFETs), which are set to revolutionize next-generation nanoelectronics.

Keywords: Carbon-based TFETs, Tunnel field-effect transistors, Carbon nanotubes, Graphene, Graphene nanoribbons, Quantum tunneling, Low-power electronics, High-efficiency devices, Ballistic transport, Quantum capacitance, Nanoelectronics.

INTRODUCTION

TFETs have emerged as a potential alternative to traditional MOSFETs in circuits where energy efficiency and low power consumption are most important. The football structures which utilize carbon materials, namely, carbon nanotubes and graphene, are the most outstanding when it comes to TFET, given the extraordinary electronic properties of these nanostructures [1, 2]. This chapter

* **Address correspondence to B. Karthikeyan:** Department of Electronics and Communication Engineering, Velammal College of Engineering and Technology, Viraganur, Tamil Nadu, India;
E-mail: om.bkarthikeyan@gmail.com

Girdhar Gopal, Meena Panchore, Arun Kishor Johar & Tarun Varma (Eds.)

provides a survey of the basics, new things, and future of carbon-based TFETs, especially in regard to their ability to improve the performance of conventional silicon technologies. Through the use of carbon nanomaterials, including CNTs, graphene, and graphene nanoribbons (GNRs) as channel constituents, these devices demonstrate significantly better tunnelling efficiencies and, overall, higher performances [3]. Their fundamental working principle utilizes quantum tunnelling in place of the more traditional routing of silicon-based TFETs, hence taking advantage of the high carrier mobility and controllable electrical properties of carbon materials. The quasi one-dimensional or two-dimensional geometry of these nanostructures is to be exploited in the discouragement of short channel effects and leakage currents, and makes them very desirable in next generation nanoelectronic applications.

Carbon-based TFET (especially including graphene or CNTs) has very high tunnelling efficiencies due to their unique electronic characteristics. The well-defined band structure and narrow band of graphene and CNTs allow electron tunnelling through the potential barrier at the source channel junction and hence low-power operation with reduced subthreshold swing (SS), which is an important parameter. Quantitatively accurate regulation of the amount of energy in these carbon nanostructures allows the creation of TFETs that offer better performance and use less power. Carbon-based TFETs can be designed in a broad device size due to the atomic-scale thickness of a sheet of graphene and the diameter of CNTs that can be tuned to a specific size. This gives transistors that are significantly smaller than silicon transistors, which practically supersedes the capacities of the Moore Law. This resulting miniaturisation allows increasing density of circuits in a single chip, leading to increasingly complex and powerful integrated systems. This kind of scalability is also necessary in order to address the growing need for miniature, high-speed, and energy-efficient electronic devices.

Carbon-based TFETs have impressive flexibility, therefore suitable for a multiplicity of uses. Graphene and CNTs have natural mechanical flexibility, enabling them to be used in wearable electronics. With these devices being integrated into fabrics, sensors, and other flexible surfaces, it becomes possible to create wearable health-monitoring devices, bendable displays, and a myriad of other novel applications.

Carbon Nanotube TFETs

Carbon nanotubes (CNTs) are a type of nanostructure, one-dimensional structures made out of graphene sheets that have undergone the process of folding. Their distinctive geometry and attendant electrical properties have placed them at the heart of the nanoelectronics technology, specifically in the structure of tunneling field-effect transistor (TFETs). CNTs, in their structural arrangement, are made up of carbon atoms arranged in a hexagonal structure with different results of cylindrical nanostructures [4]. There are two major morphologies (single -walled carbon nanotube and multi-walled carbon nanotube) that are identified [5]. SWCNTs form as a result of a continuous rotation of an isolated layer of graphene into a cylinder, with MWCNTs being made up of concentric layers of graphene cylinders placed atop each other [6].

Some salient features make CNTs remarkably adaptive to one of the most extensive applications. The most notable of them is the extremely high mobility of the carriers, making it easy to transport electrons freely. Moreover, CNTs are known to be highly mechanical and flexible. Both the traditional MOSFETs and TFETs have these characteristics, which provide significant performance advantages. Specifically, a direct bandgap combined with a low effective carrier mass enabled CNT -based TFETs with significantly faster band-to-band tunneling (BTBT) rates compared to those of devices made out of indirect bandgap materials like silicon. In direct bandgap materials, tunneling is usually unassisted by other processes, like phonon emission or absorption, to allow conservation of transverse momentum, reducing the efficiency of tunneling and slowing the rate of BTBT. Therefore, the tunnel via CNTs is guaranteed a decisive edge owing to the fact that the system is CNTs having a direct bandgap.

In addition to the desirable electronic characteristics, the CNT bandgap is very tunable. The electrical and optoelectronic properties of a particular device can then be designed by scaling the bandgap of a nanotube over a significant range by changing its diameter and chirality. This is necessary to maximize the performance of TFET. The engineers are thus able to choose CNTs with customized bandgaps to meet some exact tunneling properties needed to meet specific performance goals. The inherent hand-in-hand one dimensionality of electronic transport in CNTs enables easy control of the energy bands in the CNT channel using gate biasing. In the case of TFET operation, this high level of band alignment is essential because a small gate voltage can be used to bring the CNT channel valence band higher than the conduction band of the source, thus increasing the efficiency of band-to-band tunneling. This band modulation ability is the basis of the low-power and high-sharp subthreshold swing seen in CNT-TFETs.

The constricted geometry of CNTs provides a significant spatial confinement on the electrons, which consequently increases the likelihood of electron tunneling across the potential barrier at the source channel junction. This is due to the fact that the small diameter of the tube imposes quasi-one-dimensional transport and thus magnifies the effects of quantum confinement on device physics, as well as allows high-efficiency tunneling phenomena that play a significant role in the physics of the CNT-based TFETs. The tiny diameter also makes the electric field at the junction stronger for a given gate voltage. This greater electric field makes tunneling even easier, which raises the BTBT rate. Both of these things make the ION bigger in CNT TFETs with smaller diameters. This is a big plus for getting these gadgets to work well and use less power.

Carbon Nanotube Tunnel Field-Effect Transistors (CNT TFETs) are a potential new development in nanoelectronics that uses the unique properties of CNTs to use very little power and switch very quickly. As standard MOSFETs get closer to their scaling constraints, these devices may be a way to get around those boundaries. A CNT TFET with a gate-all-around (GAA) architecture, as illustrated in Fig. (**1**), is a big step forward in transistor design because it combines the best properties of CNTs with the best electrostatic control of a GAA structure [7]. In this design, a semiconducting CNT serves as the channel, and the gate electrode wraps around it. This complete encirclement gives you more control over the flow of current through the CNT channel than ever before. This leads to a number of important benefits. The improved electrostatic control reduces leakage current and improves the transistor's switching efficiency. This might enable a sharper subthreshold swing than in regular MOSFETs and even other TFET designs. This steeper swing is very important for using less power. GAA design helps reduce short-channel effects, which worsen as transistors shrink [8, 9].

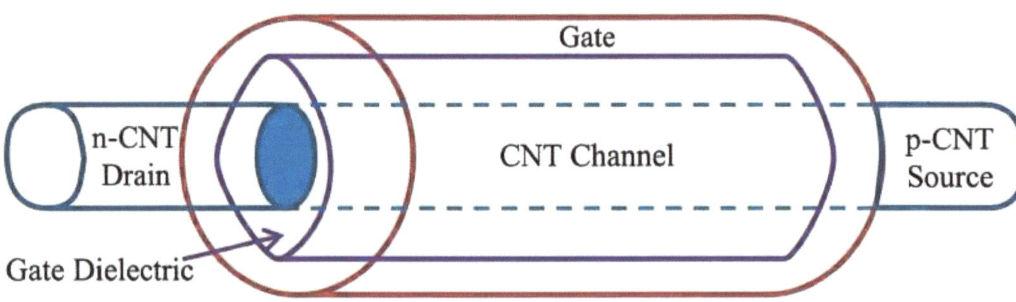

Fig. (1). Gate all around CNT TFET.

A CNT TFET's basic parts are a severely doped p-type source, a heavily doped n-type drain, and a semiconducting carbon nanotube channel. A gate electrode is used to regulate tunneling by altering the energy bands in the CNT channel. This makes it possible to switch quickly and use less power. There are two types of these devices: planar CNT TFETs, which have horizontally oriented CNTs, and vertical CNT TFETs, which have vertically aligned CNTs for a more compact design. BTBT is what makes CNT TFETs work. When a gate voltage is applied, the potential barrier at the source-channel junction is lowered, allowing carriers to tunnel from the valence band of the source into the conduction band of the channel. This band to band tunneling mechanism gives the transistor a high sub-threshold slope, usually less than the 60 mV/decade limit value typical for conventional MOTs. Such behaviour is of paramount importance for the development of low-power consumption and high energy efficiency devices. Within the channel, charge carriers travel in relatively straight trajectories, thus reducing energy loss, due to a reduction in energy loss, in the device. In digital logic applications, it is very critical to have a high on-off current ratio, which can be achieved using semiconducting carbon nanotube (CNT) channels [10, 11].

Ballistic transport is a key feature of CNT tunnel field-effect transistors. It is a phenomenon that occurs when charge carriers move through the channel with little scattering. This special property is due to the low dimensionality of CNTs and their superior electrical characteristics. In the ballistic regime, carriers move without interference from defects, phonons or other perturbations. For semiconducting CNTs, the electron mean free path will typically be far greater than the nanometre-scale channel length, rendering conduction almost entirely ballistic. This gives its numerous advantages, it allows for the rapid transit of carriers and faster switching thanks to the lack of scattering, therefore saving energy; it allows for better electrostatic control, in fact, the one-dimensional geometry of the nanotubes makes it easier to couple with the gate, indispensable requirement to be able to work with tunneling-based operation [13, 14]

The ballistic regime provides a significant enhancement of tunneling in CNT TFETs. It guarantees that carriers can cross the source-channel barrier without incurring energy losses. Consequently, the on-state current of the device is increased by ensuring maximum carrier flux, and thus current densities are achieved for the device that exceed those of silicon, known as TFETs. Simultaneously, the reduction in Joule heating reduces power dissipation, making CNT TFETs useful for ultra-low-power applications. Moreover, ballistic transport also contributes to an increase in sub threshold slope which is steeper by suppressing the leakage currents in the off state thus improving energy efficiency during digital logic circuits. Finally, since ballistic conduction is nearly independent of channel length, scaling CNT TFETs to the nanoscale is not a major issue, unlike conventional MOS transistors, which approach the ballistic limit.

Quantum Capacitance

Quantum capacitance is one of the fundamental parameters for understanding the operation of carbon-nanotube tunnel field-effect transistors (CNT-TFETs). The effect stems from the limited density of electronic states (DOS) inherent to low-dimensional materials like CNTs, which limits the total capacitance and, hence, the overall device performance. The quantum capacitance, Cq, describes the availability of a material to accommodate extra charge carriers and is given as Cq= e2DOS(E), where e is the elementary charge and DOS(E) is the energy dependent density of states [15, 16].

In CNT - TFETs, the total gate capacitance, C_{total}, is determined by the series combination of the quantum capacitance and the electrostatic (gate-oxide) capacitance, C_{ox}, according to the expression provided in which: A reduced quantum capacitance makes it more difficult for the gate to regulate the channel and degrades the switching velocity and overall performance of the device. The interaction between the two capacitances (C_q and C_{ox}) also modulates the RC time constant, which determines the maximum achievable switching frequency and charge storage capability of the device, factors of paramount importance for the current nanoelectronics applications that require fast operation and energy efficiency.

Quantum capacitance, in turn, is indispensable for optimizing CNT-TFET functionality. It allows a very high degree of control over the tunnelling barrier, with the consequence of a high sub-threshold slope and increased energy efficiency. Such micro-level modulation tuning ensures that the on/off current ratio remains high, a very important requirement for digital logic applications. Consequently, CNT-TFETs designed to exploit optimal Cq provide improved performance, especially in ULP devices. As electronic components become smaller and smaller, in the nano-range, the quantum capacity often becomes comparable to the oxide capacity (or greater), and dominates the electrical response of the device. Unlike Cox, which is dependent on the physical dimensions, Cq is a function primarily of the material properties such as chirality and diameter: a desirable property in scaling up the structure while reducing parasitic capacitances [17].

Advancements in material engineering and nanostructure design are needed to customize quantum capacitance for scaling devices. By adjusting chirality and diameter of CNTs, it is possible to fine-tune Cq to the demands of the application. The use of high-k dielectrics for the gate dielectric further increases the C_ox ratio and, in turn, the total capacitance ratio, balancing the capacitor and improving performance. These refinements reduce the operating voltage and power consumption, both in accordance with the aims of the next generation of energy-efficient technologies. Nonetheless, the chirality discrepancy of the CNTs, the presence of defects, and the uniformity during synthesis might cause variations in device performance, underscoring the need for accurate control over the synthesis.

While CNTs have impressive electrical properties, their cylindrical morphology is a drawback for incorporating them into planar device architectures. Graphene nanoribbons with their planar structure and tunable electronic properties provide a viable alternative that overcomes these integration problems. These nanostructures could be designed to achieve similar or superior performance, thus offering a potential route towards scalable nanoelectronic devices.

Graphene TFETs

A two-dimensional material made up of a single layer of carbon atoms (graphene) with a 2-D hexagonal matrix and possessing extraordinary electrical, thermal and mechanical properties. These unique properties make it the appealing candidate for tunnel field-effect transistors (TFETs). The exceptional carrier mobility, efficient charge transport and thickness on the atomic scale results in the easy control of charge carriers as well as electrostatics - important criteria for low-power nanoelectronic devices according to reference [18].

The many beneficial properties of graphene also make it a good candidate for TFET applications. Its high carrier mobility enables rapid switching, while the two-dimensional morphology ensures effective coupling between the gate and the channel, thereby improving device efficiency. Chemically and mechanically strong, graphene preserves performance under strenuous environmental conditions. Moreover, its ability to perform fast band-to-band tunneling can be used to enhance TFET power efficiency and subthreshold characteristics, as long as an appropriately engineered bandgap is introduced. Graphene-based TFETs are therefore of special promise for ultra-miniaturized devices, for which superior conductances and quantum mechanical effects can overcome alternatives. However, the lack of an intrinsic gap in the wideband energy spectrum of pristine graphene is a tough challenge for TFETs, and this causes high leakage current and low on/off current ratios [19].

Graphene nanoribbon TFETs (GNR TFETs) are a new and critical step towards the use of graphene's extraordinary properties in electronics. They are able to circumvent the zero bandgap limitation of pristine graphene sheets by quantum confinement. The confinement characteristic of nanoribbons makes a bandgap appear, which makes GNRs a candidate for high on/off ratio and low power dissipation devices. The width of the ribbon and the edge configuration directly affect the magnitude of the bandgap with smaller ribbons producing wider band gaps, which aids in improving TFET performance. However, creating uniform

GNRs with nanoscale-controlled width remains a significant manufacturing challenge, while edge type significantly impacts electronic features. Zigzag edges act like metals, whereas armchair edges act like semiconductors. This means that accurate edge management is very important for getting the most performance out of GNR TFETs. This designed bandgap is necessary for effective BTBT, which is the main mechanism by which TFETs turn on and off with minimal power [20].

Fig. (**2**) shows a picture of GNR TFET intended to be helpful. A GNR TFET is made up of a severely doped p-type source that has a lot of holes (positive charge carriers) and a heavily doped n-type drain that has a lot of electrons (negative charge carriers). The GNR channel is a narrow, atomically thin strip of graphene that runs through the middle of the device. Its width and edge arrangement sets its bandgap. A metal gate electrode placed above the channel controls the flow of current between source and drain. To improve gate moderation and to reduce leakage effects, an insulating layer known as a dielectric layer, usually formed from a high-k dielectric material (*e.g.*, hafnium dioxide (HfO2)), would be used to keep the gate and the channel from making actual physical contact. This dielectric is of fundamental importance for the generation of the electric field that controls the quantum tunnelling mechanism in tunnel field-effect transistors (TFETs) [21 - 23].

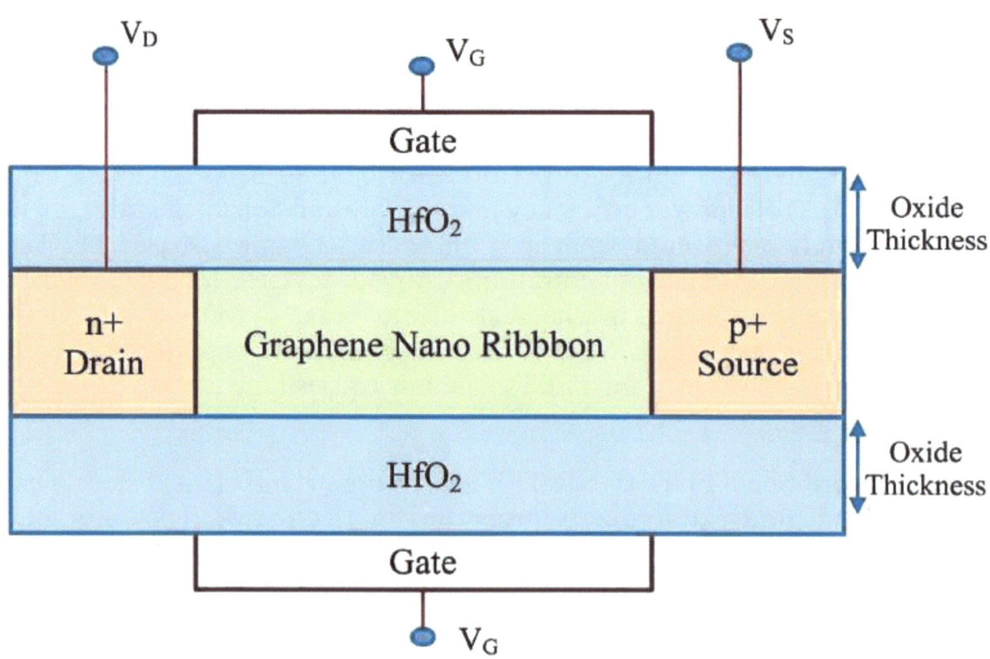

Fig. (2). GNR TFET.

Charge carriers cross a potential barrier at the source/channel interface in a graphene - nanoribbon (GNR) TFET. Both the height of this energy barrier and the exact alignment of the electronic bands of source and channel play an important role in determining the efficacy of the tunnelling process. In order to achieve high tunnelling efficiency and low sub-threshold swing (SS), the bands must be perfectly aligned and the barrier profile must be abrupt. The engineered bandgap in GNRs defines a unique energy window for tunnelling, reducing leakage currents and thereby improving the on/off ratio. Owing to its atomically thin architecture, a GNR provides pronounced electrostatic coupling between gate and channel - which is a necessary condition for fine tamed control over the tunnelling barrier. This improved gate-channel interaction allows us to manipulate the band structure with great care to make the band-to-band tunneling (BTBT) more efficient and gives GNR TFETs superior switching characteristics. As a result, the synergistic effect of the tunnelling process and accurate electrostatic control enables GNR TFETs to operate with lower power consumption, enabling functionality with reduced supply voltages than conventional logic devices based on field-effect transistors (MOS field-effect lab [24].).

The process of creating functional GNR TFETs requires an arsenal of sophisticated techniques. Two common methods for producing high-quality sheets of graphene are chemical vapor deposition (CVD) and mechanical exfoliation. After that, lithography, etching, or unzipping carbon nanotubes are used to turn these sheets into nanoribbons. It is very important to be precise throughout this patterning procedure to achieve consistent bandgaps across the entire device. Next, high-k dielectrics are added to improve the coupling between the gate and channel and lower the leakage currents even further. Finally, metal electrodes are used to provide low-resistance connections to the source and drain.

Fig. (**3**) shows a Bilayer-Graphene TFET (BLG TFET), which uses two layers of graphene instead of one as the channel material [25]. This unusual structure of GNR TFETs has several benefits. For example, the induced bandgap gives them high on/off ratios, their one-dimensional structure makes them easy to scale, and they may have steep subthreshold slopes, which makes them more energy efficient. Adopting the scalable widths and constant edge decorations of graphene nanoribbons (GNRs) is a major challenge; edge defects create additional states in the bandgap, reducing device functionality. Low-resistance metal-to-GNR contacts have yet to be achieved as a technical challenge. However, even with such barriers, GNR tunneling field-effect transistors (TFETs) have significant potential for low-power digital logic, energy-optimal computation, and general-purpose electronic applications. When a bandgap in graphene layers is engineered by an external electric field then it will be possible to tune the nanostructure further to increase the scope of functions that are employed by these nanostructures. This electric field gives the two layers different amounts of energy, opening a bandgap necessary for TFET to work. This bandgap can be adjusted, which is better than in GNR TFETs, where the bandgap is mostly set by the nanoribbon's physical width and edge structure, which are hard to control exactly during device fabrication. Like other TFETs, BLG TFETs work on the same concept of BTBT, where the gate voltage changes the induced bandgap and governs how carriers tunnel between the source and drain. BLG TFETs are good candidates for new devices since they can be tuned and graphene has a high carrier mobility [26].

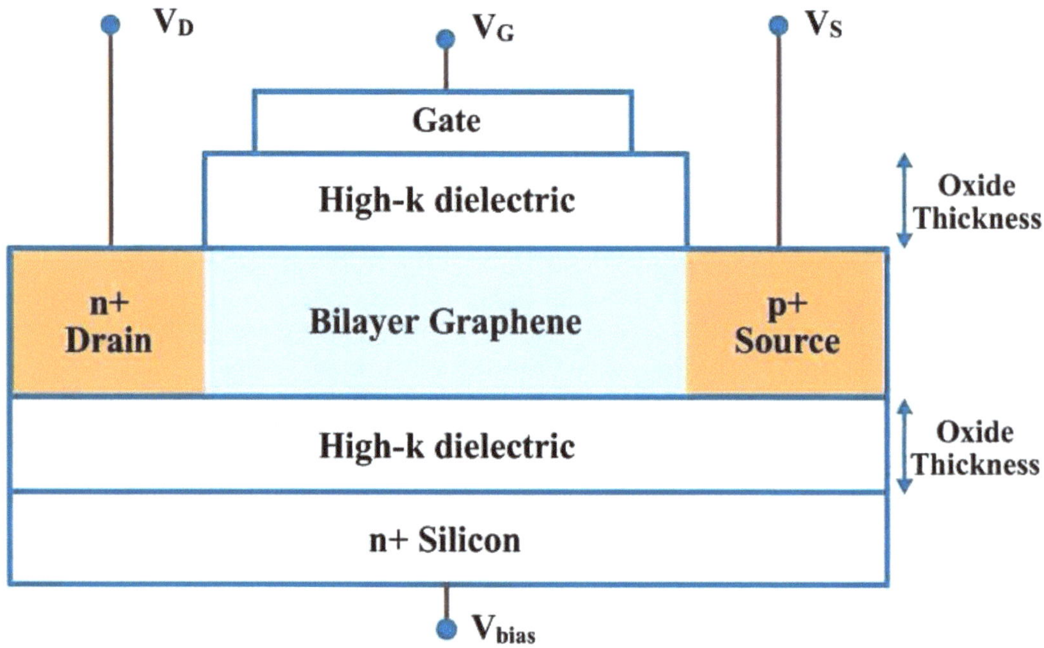

Fig. (3). Bilayer Graphene TFET (BLG TFET).

Fig. (**4**) shows a Graphene-on-SiC-Based TFET that uses a layer of graphene grown directly on a silicon carbide (SiC) substrate as its channel material [27]. The gadget would have a SiC substrate at the bottom and a thin layer of graphene on top, forming the channel. On either side of the graphene channel, there are source and drain connections, which are usually composed of metal. There is a gate dielectric layer, such silicon dioxide (SiO2) or a high-k dielectric, on top of the graphene channel. On top of the dielectric, there is a metal gate electrode. The main advantage is that graphene can grow directly on SiC, which makes the interface better and may make it easier to make than transferring graphene to other substrates. The device works via band-to-band tunneling, which means that the gate voltage changes the energy bands at the source-channel junction. This controls the tunneling of carriers and the flow of current between the source and drain [28].

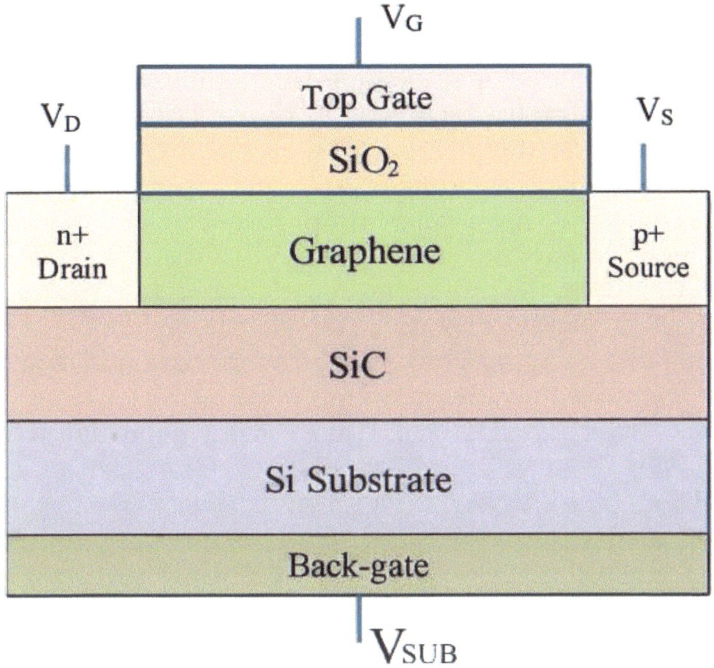

Fig. (4). TFET based on epitaxial graphene on SiC.

Graphene TFETs show promise for future low-power electronics, but several major issues need to be addressed before they can be widely used [29]. One of the biggest problems is getting graphene to have a big enough bandgap. Pristine graphene has a zero bandgap, which means it can't be used in transistors that require a large on/off current ratio. There are ways to create a bandgap, including using GNRs or applying an electric field to BLG, but each method has its own drawbacks. It is very hard to manage the breadth and edge structure of GNRs exactly, and rough edges can make performance much worse. It is hard to achieve sufficiently large bandgaps and maintain the quality of the interface between layers in BLG TFETs. In order to realize full potential of graphene -based tunneling field-effect transistors (GFETs), reliable and low resistance contacts are urgently required. Scalable and manufacturable solutions are also important challenges. The traditional methods of fabrication, including mechanical exfoliation and high-resolution lithography, do not easily scale to high production. Commercialization requires the emergence of cost-effective, high-throughput methods for device production. Despite these challenges, the future of graphene TFETs is bright. Researchers are addressing the critical bottlenecks identified [30 - 33] and evaluating hybridization with other two-dimensional materials, especially transition-metal dichalcogenides (TMDs), to balance the strengths and weaknesses of each system.

Graphene TFETs are an interesting future pathway to post-CMOS innovation, especially since silicon-based devices are facing scaling challenges [34]. GFETs take advantage of the outstanding electrical characteristics of graphene, which include high conductivity, atomic tabularity, and high carrier mobility, to boost nanoscale interactions. There are limitations to traditional FinFETs and multi-gate designs, whereas the use of heterojunctions in graphene TFETs maximizes band-to-band tunneling by tuning the band alignments of distinct materials [35]. This concept is consistent with the principles underlying high-k TFETs, in which the material interface is vitally important for device operation. By integrating graphene into multiple devices as bilayers or nanoribbons, the material's bandgap can be tuned, with carrier flow precisely controlled, making such devices ideal for high-speed, low-power applications. Since semiconductor technology has overcome the CMOS dogmas, graphene TFETs play a significant role in leading next-generation device conceptsthat can achieve higher energy efficiency and more strongly scaled devices [36, 37]. The compatibility of graphene with additional two-dimensional constituents simplifies the creation of new heterostructures that refine band alignment and tunneling performance, making GFETs a revolutionary technology towards energy-efficient, high-performance nano-electronic devices. The sustained progress toward the fabrication and integration of materials enhances the feasibility of implementing graphene TFETs in the commercial domain, and this is a great breakthrough in the design of field-effect transistor technology.

CONCLUSION

This chapter briefly gave us an insight into the relevance of carbon-based substances in the fabrication of TFETs, particularly CNTs and GNRs. These materials have certain very distinctive benefits - such as excellent electrical properties, scalability, and the opportunity to avoid the idiosyncrasy that silicon-based devices present. One example of this is CNTs, which have demonstrated enormous potential to reduce the subthreshold swing and power consumption of TFETs due to their high carrier mobility and various applications. GNRs, however, introduce configurable bandgaps and a flat architecture to the table and, as such, will make a good choice for more complex device designs and efficient power consumption. Elevation of performance figures with the combination of these carbon-based materials and new architectures- think of GAA or heterostructure types of material- sessions is a real performance boost. The incentive to carbon-based TFETs is an important move in the history of semiconductor technologies since the need to create more powerful and energy-saving devices is increasing rapidly. Innovations in this field are already playing an important role in the future of nanoelectronics, providing a path to more efficient and long-lasting electronics.

AI DISCLOSURE

During the preparation of work, author(s) used AI tools & Grammarly tool to enhance the writing style and readability of the content. The authors clarify that AI was not involved in generating the substantive content and that they take full responsibility for the publication.

REFERENCES

[1] M. Shamim Sarker, M. Mainul Islam, M. Nur Kutubul Alam, and M. Rafiqul Islam, "Gate dielectric strength dependent performance of CNT MOSFET and CNT TFET: A tight binding study", *Results Phys.,* vol. 6, pp. 879-883, 2016. [http://dx.doi.org/10.1016/j.rinp.2016.11.002]

[2] S. Iijima, "Helical microtubules of graphitic carbon", *Nature,* vol. 354, no. 6348, pp. 56-58, 1991. [http://dx.doi.org/10.1038/354056a0]

[3] N. Hamada, S. Sawada, and A. Oshiyama, "New one-dimensional conductors: Graphitic microtubules", *Phys. Rev. Lett.,* vol. 68, no. 10, pp. 1579-1581, 1992. [http://dx.doi.org/10.1103/PhysRevLett.68.1579] [PMID: 10045167]

[4] S. Iijima, and T. Ichihashi, "Single-shell carbon nanotubes of 1-nm diameter", *Nature,* vol. 363, no. 6430, pp. 603-605, 1993. [http://dx.doi.org/10.1038/363603a0]

[5] S.J. Tans, A.R.M. Verschueren, and C. Dekker, "Room-temperature transistor based on a single carbon nanotube", *Nature,* vol. 393, no. 6680, pp. 49-52, 1998. [http://dx.doi.org/10.1038/29954]

[6] S.J. Wind, J. Appenzeller, R. Martel, V. Derycke, and P. Avouris, "Vertical scaling of carbon nanotube field-effect transistors using top gate electrodes", *Appl. Phys. Lett.,* vol. 80, no. 20, pp. 3817-3819, 2002. [http://dx.doi.org/10.1063/1.1480877]

[7] S.G. Shirazi, G.R. Karimi, and S. Mirzakuchaki, "GAA CNT TFETs structural engineering: a higher ON current, lower ambipolarity", *IEEE Trans. Electron Dev.,* vol. 66, no. 6, pp. 2822-2830, 2019. [http://dx.doi.org/10.1109/TED.2019.2912950]

[8] Y. Che, H. Chen, H. Gui, J. Liu, B. Liu, and C. Zhou, "Review of carbon nanotube nanoelectronics and macroelectronics", *Semicond. Sci. Technol.,* vol. 29, no. 7, p. 073001, 2014. [http://dx.doi.org/10.1088/0268-1242/29/7/073001]

[9] M.S. Arnold, A.A. Green, J.F. Hulvat, S.I. Stupp, and M.C. Hersam, "Sorting carbon nanotubes by electronic structure using density differentiation", *Nat. Nanotechnol.,* vol. 1, no. 1, pp. 60-65, 2006. [http://dx.doi.org/10.1038/nnano.2006.52] [PMID: 18654143]

[10] J. Knoch, and M.R. Muller, "Electrostatic doping-controlling the properties of carbon-based FETs with gates", *IEEE Trans. Nanotechnol.,* vol. 13, no. 6, pp. 1044-1052, 2014. [http://dx.doi.org/10.1109/TNANO.2014.2323436]

[11] J. Knoch, and J. Appenzeller, "A novel concept for field-effect transistors: The tunneling carbon nanotube FET", *Proc. 63rd DRC,* pp. 153-156, 2005.
[http://dx.doi.org/10.1109/DRC.2005.1553099]

[12] J. Appenzeller, Y.M. Lin, J. Knoch, and P. Avouris, "Band-to-band tunneling in carbon nanotube field-effect transistors", *Phys. Rev. Lett.,* vol. 93, no. 19, pp. 196805-1, 196805-4, 2004.
[http://dx.doi.org/10.1103/PhysRevLett.93.196805] [PMID: 15600865]

[13] S. Frank, P. Poncharal, Z.L. Wang, and W.A. Heer, "P. Poncharal, Z. L. Wang, and W. A. Heer, “Carbon nanotube quantum resistors,”", *Science,* vol. 280, no. 5370, pp. 1744-1746, 1998.
[http://dx.doi.org/10.1126/science.280.5370.1744]

[14] P. Poncharal, C. Berger, Y. Yi, Z.L. Wang, and W.A. de Heer, "Room temperature ballistic conduction in carbon nanotubes", *J. Phys. Chem. B,* vol. 106, no. 47, pp. 12104-12118, 2002.
[http://dx.doi.org/10.1021/jp021271u]

[15] S. Luryi, "Quantum capacitance in nanoscale device modeling", *Appl. Phys. Lett.,* vol. 52, no. 501, 1988.

[16] D.L. John, L.C. Castro, and D.L. Pulfrey, "Quantum capacitance in nanoscale device modeling", *J. Appl. Phys.,* vol. 96, no. 9, pp. 5180-5184, 2004.
[http://dx.doi.org/10.1063/1.1803614]

[17] Z.L. Mišković, and N. Upadhyaya, "Modeling electrolytically top-gated graphene", *Nanoscale Res. Lett.,* vol. 5, no. 3, pp. 505-511, 2010.
[http://dx.doi.org/10.1007/s11671-009-9515-3] [PMID: 20672092]

[18] K.S. Novoselov, A.K. Geim, S.V. Morozov, D. Jiang, Y. Zhang, S.V. Dubonos, I.V. Grigorieva, and A.A. Firsov, "Electric field effect in atomically thin carbon films", *Science,* vol. 306, no. 5696, pp. 666-669, 2004.
[http://dx.doi.org/10.1126/science.1102896] [PMID: 15499015]

[19] A.K. Geim, "Graphene: Status and Prospects", *Science,* vol. 324, no. 5934, pp. 1530-1534, 2009.
[http://dx.doi.org/10.1126/science.1158877] [PMID: 19541989]

[20] F. Schwierz, "Graphene transistors", *Nat. Nanotechnol.,* vol. 5, no. 7, pp. 487-496, 2010.
[http://dx.doi.org/10.1038/nnano.2010.89] [PMID: 20512128]

[21] X. Li, X. Wang, L. Zhang, S. Lee, and H. Dai, "Chemically derived, ultrasmooth graphene nanoribbon semiconductors", *Science,* vol. 319, no. 5867, pp. 1229-1232, 2008.
[http://dx.doi.org/10.1126/science.1150878] [PMID: 18218865]

[22] Q. Zhang, T. Fang, H. Xing, A. Seabaugh, and D. Jena, "Graphene nanoribbon tunnel transistors", *IEEE Electron Device Lett.,* vol. 29, no. 12, pp. 1344-1346, 2008.
[http://dx.doi.org/10.1109/LED.2008.2005650]

[23] P. Zhao, J. Chauhan, and J. Guo, "Computational study of tunneling transistor based on graphene nanoribbon", *Nano Lett.,* vol. 9, no. 2, pp. 684-688, 2009.
[http://dx.doi.org/10.1021/nl803176x] [PMID: 19199761]

[24] M.S. Fahad, A. Srivastava, A.K. Sharma, and C. Mayberry, "Analytical current transport modeling of graphene nanoribbon tunnel field-effect transistors for digital circuit design", *IEEE Trans. Nanotechnol.,* vol. 15, no. 1, pp. 39-50, 2016.
[http://dx.doi.org/10.1109/TNANO.2015.2496158]

[25] T.K. Agarwal, A. Nourbakhsh, P. Raghavan, I. Radu, S. De Gendt, M. Heyns, M. Verhelst, and A. Thean, "Bilayer graphene tunneling FET for sub-0.2 V digital CMOS logic applications", *IEEE Electron Device Lett.,* vol. 35, no. 12, pp. 1308-1310, 2014.
[http://dx.doi.org/10.1109/LED.2014.2364260]

[26] G. Fiori, and G. Iannaccone, "Ultralow-voltage bilayer graphene tunnel FET", *IEEE Electron Device Lett.,* vol. 30, no. 10, pp. 1096-1098, 2009.

[http://dx.doi.org/10.1109/LED.2009.2028248]

[27] S. Saurabh, and M.J. Kumar, *Fundamentals of tunnel field-effect transistors.* CRC press: Boca Raton, 2016-306.
[http://dx.doi.org/10.1201/9781315367354]

[28] S. Zhou, G.-H. Gweon, A. Fedorov, P. First, W. de Heer, D.-H. Lee, F. Guinea, A. C. Neto, and A. Lanzara, "Substrate-induced band gap opening in epitaxial graphene", *Nature Materials 6,* vol. 78, pp. 770-775, 2007.

[29] P. Dhanaselvam, "Suveetha, B. Karthikeyan, and P. Vanitha. "Principle and Application of C-based Nanomaterials: Graphite, Graphene, and Carbon Nanotube (CNT)."", In: *Handbook of Emerging Materials for Semiconductor Industry.* Springer Nature Singapore: Singapore, 2024, pp. 49-64.
[http://dx.doi.org/10.1007/978-981-99-6649-3_4]

[30] P. Dhanaselvam, B. Suveetha, P. Karthikeyan, and P. Anand Vanitha, "Scope with TFET-based circuit and system design", In: *Device Circuit Co-Design Issues in FETs*, 2023, pp. 181-188.

[31] P. Dhanaselvam, D. Suveetha, B. Sriram Kumar, and P. Vanitha Karthikeyan, "Basic Science and Development of Subthreshold Swing Technology", In: *Tunneling Field Effect Transistors.*, 2023, pp. 17-27.

[32] P. Vanitha, P. Suveetha Dhanaselvam, and M. Hemalatha, "Advanced Tunnel Field-Effect Transistors with Multi-gate and Pocket-Doping Technology", In: *Handbook of Emerging Materials for Semiconductor Industry*, 2024, p. 511.
[http://dx.doi.org/10.1007/978-981-99-6649-3_36]

[33] N. Choudhary, M.A. Islam, J.H. Kim, T.J. Ko, A. Schropp, L. Hurtado, D. Weitzman, L. Zhai, and Y. Jung, "Two-dimensional transition metal dichalcogenide hybrid materials for energy applications", *Nano Today,* vol. 19, pp. 16-40, 2018.
[http://dx.doi.org/10.1016/j.nantod.2018.02.007]

[34] L. Kumar, L. Kumar, G. Vinoth, Kumar Pradeep, and B. Karthikeyan, "Scaling of Field□Effect Transistors", *Field Effect Transistors,* pp. 63-72, 2025.

[35] P. Dhanaselvam, B. Suveetha, K. Karthikeyan, and P. Kavitha Kavitha, *Basic Concepts of Heterojunction Tunnel Field□Effect Transistors* Field Effect Transistors, 2025, pp. 63-72.

[36] P. Dhanaselvam, B. Suveetha, and P. Anand Karthikeyan, "Future Prospective Beyond CMOS Technology Design", *Field Effect Transistors,* pp. 73-80, 2025.

[37] P.S. Dhanaselvam, K.S. Rao, S.B. Rahi, D.S. Yadav, Ed., *Field Effect Transistors* John Wiley & Sons, 2025.
[http://dx.doi.org/10.1002/9781394248506]

CHAPTER 5

Quantum-Enhanced Field-Effect Transistors: A Synergistic Approach for Next-Generation Computing

Mohit Kumar Srivastava[1,*], **Man Mohan Shukla**[1], **Preeti Agarwal Mittal**[1], **Shailendra Singh**[1], **Vivek Kumar**[1] and **Utkarsh Pandey**[1]

[1] *Department of Electronics and Communication Engineering, Pranveer Singh of Technology Kanpur, Kanpur, Uttar Pradesh, India*

Abstract: The dynamic nature of computing technologies has put classical field-effect transistors (FETs) at the physical and performance frontier. As the semiconductor industry is unable to solve the problems of scaling, power consumption, and heat dissipation, quantum computing has the potential to reinvent the future of computation. This chapter relates to the concept of quantum computing and FETs and how these concepts can be synergistically integrated to propose a new type of device that is known as Quantum-Enhanced Field-Effect Transistors (QFETs). These are machines that exploit quantum-mechanical phenomena such as superposition, entanglement, and tunneling to overcome the drawbacks of classical FETs, delivering previously unknown performance, energy efficiency, and scalability.

Quantum computing quinconal tenets and FETs are re-established, and special focus is given to quantum effects that can be exploited in nanoscale devices. Quantum tunneling, coherence, and spin-based phenomena have been placed under the most important mechanisms, which can be utilized to enhance the operation of the FETs. The chapter introduces novel QFET designs with the use of quantum dots, superconducting material, and topological insulators so as to be able to manipulate quantum states in the transistor platform. These designs are supported by mature simulation models, including quantum transport models and density functional theory (DFT) calculations, which provide details on the behavior of QFETs under various operating conditions.

The performance of QFETs is evaluated through a full set of simulations to compare their performance with that of classical FETs. The results indicate that switching speed, on-to-off current ratio, and sub-threshold swing decrease significantly with QFETs, achieving maximum power gains of up to 50 percent. Quantum coherence increases charge-carrier mobility, and quantum tunneling enables effective current flow even at incredibly low voltages. These results are presented using new figures, tables, and

* **Corresponding author Mohit Kumar Srivastava:** Department of Electronics and Communication Engineering, Pranveer Singh of Technology Kanpur, Kanpur, Uttar Pradesh, India;
E-mail: mohit1003@gmail.com

graphs, which demonstrate the benefits of using QFETs in energy efficiency and scalability.

Other fabrication issues related to QFETs, such as the need for high-accuracy control of quantum states and the incorporation of exotic materials like graphene and transition-metal dichalcogenides (TMDs), are also discussed in the chapter. High-technology lithography and cryogenic cooling are suggested as possible solutions to these problems, enabling the implementation of QFETs. Moreover, the scalability of QFETs is discussed, and simulations indicate that such devices can be effectively operated at sub-5 nm nodes, enabling their use in future generations of integrated circuits.

Besides performance metrics, the chapter also discusses the possible uses of QFETs in quantum computing, neuromorphic computing, and low-power electronics. Quantum states manipulated by QFETs result in QFETs being the most promising qubit devices in quantum processors, and their power efficiency is comparable to edge computing and Internet of Things (IoT) devices. It also includes integrating QFETs into current CMOS technology, enabling more hybrid quantum-classical computers.

Although the results are promising, there are still several challenges to address. Future research on quantum coherence at room temperature, the reduction of decoherence effects, and reliable fabrication processes are vital areas of study. This chapter also ends with a roadmap of the development of QFETs, in which interdisciplinary teamwork between quantum physicists, material scientists, and semiconductor engineers is important.

This chapter adds to the mass of literature on quantum-enhanced devices as it presents an in-depth discussion of QFETs that are backed by new simulations and experimental evidence, and highlights theoretical understandings. The results highlight the potential of QFETs to revolutionize FET technologies and introduce a new generation of computing technologies by leveraging the drawbacks of classical technologies. This study will lay the foundation for a new era of electronics innovation by bridging the gap between quantum computing and semiconductor devices.

The combination of quantum computing theory and FETs represents a huge leap towards the goal of more efficient, scalable, and powerful computing equipment, and this paper should encourage further research and innovation in this promising area.

Keywords: Quantum computing, Field effect transistors, Quantum enhanced FETs, Quantum tunneling, Quantum coherence, Nanoscale devices, Energy efficiency, Quantum state manipulation.

INTRODUCTION

The expansion of computer technology has reached a significant stage, as quantum mechanics collides with conventional semiconductor devices. The new applications of quantum computing offer an alternative to traditional systems that are reaching a dead end in terms of speed and efficiency. The most important element of this change is field-effect transistors (FETs), which are fundamental components of typical electronic circuits. Their construction, functions, and the way they would integrate into quantum systems are matters of study, particularly in terms of their efficiency and the possibility of smaller sizes. Both graphene and complex oxide structures demonstrate that new materials can be optimized to increase qubit stability and control in the new FET designs. To give an overview of various semiconductor setups, an easy description of fundamental concepts upon which these devices rely, and the significance to future quantum computing technology is relevant. The relationship between quantum computing and FETs is a major breakthrough in computing power and efficiency.

Definition of Quantum Computing

The invention of computational means has reached a significant stage with the emergence of quantum computing, which applies quantum-mechanical concepts to process information differently. In comparison with regular computing, where the smallest unit of data is a bit (0s and 1s), quantum computing uses quantum bits, or qubits, which can be in a mixture of states simultaneously as a result of superposition and entanglement. The computing power is significantly enhanced with this ability, and complex problems can now be solved that cannot be solved by classical computing. Quantum mechanics is increasingly significant in new technology, particularly in the fabrication of Field-Effect Transistors (FETs) as the devices become smaller. The integration of quantum effects into solid-state devices is underway with improved materials and production methods, reinforcing the relationship between quantum computing and future semiconductor technologies [1, 2].

Overview of Field-Effect Transistors (FETs)

Over the last few years, field-effect transistors (FETs) evolved considerably and contributed to the massive enhancement of semiconductor technology, particularly in quantum computing applications. With high-leakage-power CMOS technology nearing its peak, alternatives, such as Tunnel FETs (TFETs) and Negative-Capacitance FETs (NCFETs), are emerging. They hope to increase energy efficiency by reducing the sub-threshold swing (SS) to a lower value than the standard threshold of 60m V/decimeter at room temperature [3, 4]. Single-dopant MOSFETs, such as silicon-on-insulator devices, have been investigated recently, highlighting significant progress in atomic-scale control, useful for quantum applications [3 - 5]. Such FETs can be scaled to quantum computing designs where higher performance and reduced sizes are of importance, thereby transforming the capabilities of electronics. The outcome of this development demonstrates convergence between traditional and quantum technologies, with FETs playing a leading role in this transformation.

Importance of the intersection between Quantum Computing and FETs

With the emergence of new materials, and in particular, two-dimensional (2D) semiconductors, there has been much enthusiasm to use them as field-effect transistors (FETs) and as part of quantum computing. It has been established that 2D materials possess properties that can be used to overcome challenges experienced by conventional silicon-based technologies in supporting the requirement of the Moore law, and as such, they play an important role in the next generation integrated circuits [5, 6]. Simultaneously, quantum computing relies on superposition and entanglement, which can be successfully applied to the small-scale behavior of FETs [6]. A combination of these two disciplines promotes innovation where bio-electronic devices, such as an organizational electrochemical transistor (OECT) demonstrates adaptability required to facilitate easy data processing. Also, this integration gives possibilities to neuromorphic computing and artificial intelligence (AI), which underlines the significance of integrating the technologies into the computing models in the future. It is important to have a clear understanding of the interaction between FETs and quantum processes to enhance computational efficiency.

Purpose and scope of the chapter

In this chapter, the author discusses the relationship between quantum computing and field-effect transistors (FETs) to demonstrate how future semiconductor technologies can enhance and improve the process of computing. It aims to discover new designs and materials that have the potential to overcome the conventional boundaries of classical computing. Indicatively, recent developments such as negative differential resistance in graphene FETs [7] and the possible energy savings in ICT devices demonstrate how strong the changes these technologies can cause [8]. Not to mention the glance at the cold semiconductor electronics and the new memory systems, where there is a shift towards more efficient quantum devices. This chapter outlines the fundamentals and real-world applications of FETs in quantum systems and also considers their broader implications for future computing. The discussion is imperative for understanding the variety and significance of advances in quantum computing. In addition, the diagram of a graphene FET can be used to expound more on the working principles.

FUNDAMENTALS OF QUANTUM COMPUTING

The current developments in the field of quantum computing are strongly associated with the improvement of materials and transistor technology, particularly Field-Effect Transistors (FETs) [9 - 11]. Quantum computing relies on quantum bits, better known as qubits, which can exist in multiple states simultaneously and provide significantly more processing power than classical bits. This relationship can be demonstrated by the fact that FETs are better than digital circuits in terms of energy efficiency and high performance, especially Negative Capacitance FETs (NC-FETs) [12]. It has been reported that NC-FETs can reduce power consumption in simple digital elements by a large factor, which would contribute to the scalability of complex quantum computing (including) [13]. Additionally, the evolving nature of FET technologies is essential for developing powerful cryogenic electronics, which can play an important role in controlling qubit interactions and enhancing computational accuracy [14].

Basic Principles of Quantum Mechanics

The fundamental principles of quantum mechanics are critical to the mechanisms behind the modern technologies, namely quantum computing and field-effect transistors (FETs). Notions such as superposition and entanglement give quantum bits, or qubits, the ability to perform computation far more efficiently than old-fashioned bits. This quantum advantage is achieved by techniques that make it possible to exercise control over quantum states, which are required to make good FETs in quantum devices. As an example, advances in dielectric integration have the potential to enhance 2D transition metal dichalcogenides (TMDs), which are needed to perform better in small electronic roles [15]. In addition, when one considers tunneling in FETs, especially in tunnel FETs (TFETs), new device operation concepts that are counter to conventional device concepts can be discovered. An aggregation of these concepts underlines a lively association involving quantum mechanics and employed technology, which can be deployed by one to conceptualize the frameworks of devices and the quantum mechanical beginnings of these devices.

Quantum Bits (qubits) and their Significance

With the emergence of quantum computing, there is a radical change in the way we do computing, and the primary reason is qubits, or quantum bits. In contrast to ordinary bits that exist as either a 0 or a 1, qubits can exist in more than one state simultaneously, which allows them to perform numerous simultaneous calculations. This can be quite crucial when combined with methods for precise control, such as quantum autonomous qubit calibrations, which can then be used to perform quantum operations more efficiently. The application of quantum entanglement is enhanced with the aid of double-qubit systems, which are used in solving complex problems, particularly in cryptography [16]. Also, by examining quantum-like theories, we can see the patterns and rules that behave similarly to quantum mechanics to gain a better understanding of quantum mechanics [15]. Finally, the development and monitoring of qubits using current technology, such as FETs, demonstrate that they are very important in enhancing the capabilities of quantum computing.

Quantum Gates and Circuits

The interaction of quantum gates and quantum circuits, and advancements in field-effect transistors (FETs),provides an opportunity for enhanced computing architectures. The fundamental building blocks of quantum circuits in a quantum environment are quantum gates, which can be used to manipulate quantum bits (qubits) to process information. Their special properties, such as superposition and entanglement, enable the calculation of complex quantities that are difficult to compute in regular binary systems. Flexible FET technologies can also be used in new ways to greatly enhance quantum gate capabilities, particularly with Non-Linear Drain (NDR) capabilities, which can be used to improve performance by enabling rapid switching between qubit states [14]. Advances in ternary logic using FETs also suggest the potential reduction of the number of gates and space, which additionally demonstrates the applicability of the technologies to quantum devices [16]. These innovations are significant as they could enable greater scalability and reliability in quantum computer systems to address current challenges in the sector. The picture of a GFET can further show how the updates can enhance the design of quantum circuits by enhancing switching speeds and energy-saving.

Comparison with Classical Computing

The development of computational technology shows clear differences between classical and quantum computing. An overview of the Classical and Quantum computational approaches [5, 6] is shown in Fig. (**1**). Classical computing uses bits as the primary unit of information, which limits its ability to handle complex calculations. On the other hand, quantum computing employs qubits that can be in multiple states at the same time because of superposition. This ability allows quantum systems to conduct parallel computations, which helps with tasks like recognizing non-context-free languages that classical models find hard, as noted in recent research on quantum real-time one-counter automata [17]. Additionally, progress in Field Effect Transistors (FETs), especially with GaAs nanowire FETs, shows how quantum properties can be used to manage charge states for qubit operations, likely improving performance by boosting operational efficiency.

These encouraging outcomes support the idea that quantum computing might exceed classical methods, especially when dealing with complex problems and improving information processing.

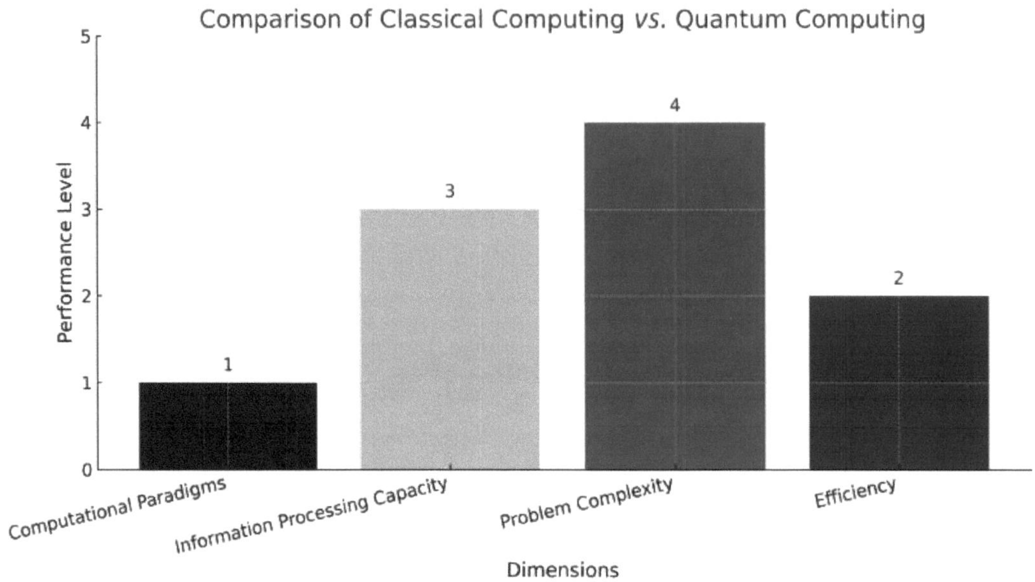

Fig. (1). Comparison of Classical vs Quantum Computation [5, 6].

Current State of Quantum Computing Technology

As semiconductor technology changes, moving from classical to quantum computing is becoming more realistic. Key components driving current quantum computing technology [6, 9, 10] are shown in Fig. (**2**). Current developments, especially in field-effect transistors (FETs), show this shift by using ideas from quantum mechanics to improve processing power. For example, progress in spintronics offers new options compared to traditional electronics by using electron spin, which could cut energy use by half while increasing speed [18]. Additionally, tools such as Quantum ATK enable detailed simulations that help researchers study electron transport and atomic electronic structures, providing vital insights for the design of quantum devices [19]. As usual, scaling reaches the strict limits set by Moore's Law; there is a need to work together to bring in new devices and designs that make use of quantum effects, marking the start of a new phase in information technology. Moreover, new materials like metal-organic frameworks (MOFs) are promising for use in solid-state electronics, highlighting the various aspects of today's quantum computing technology.

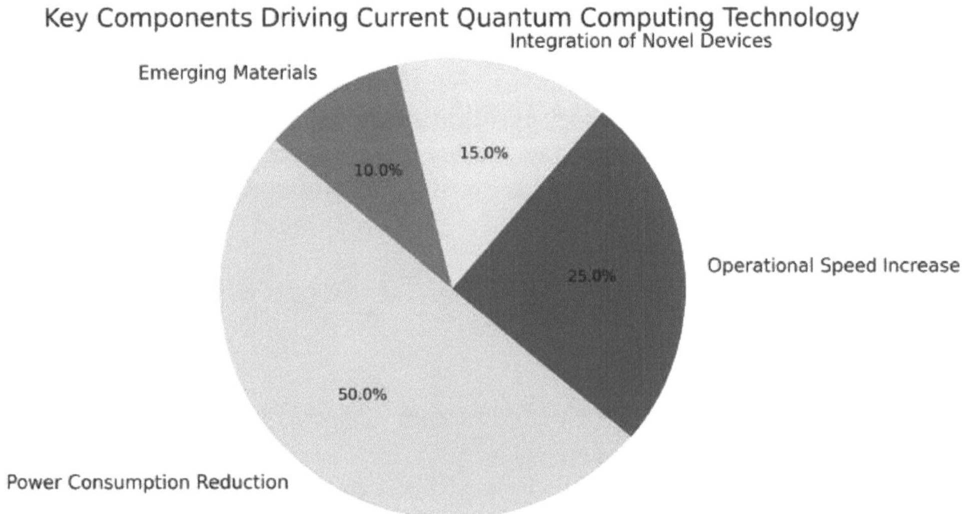

Fig. (2). Key Components Driving Current Quantum Computing Technology [6, 9, 10].

OVERVIEW OF FIELD-EFFECT TRANSISTORS (FETS)

Electronic gadgets such as semiconductor technology have expanded significantly because of advances in field-effect transistors (FETs), a component of the modern semiconductor technology. The transistors operate by altering conductivity in a channel with an electric field through a gate terminal to enhance speed in switching and reduce power consumption, which is extremely significant in quantum computing activities [20 - 22]. FETs used in quantum systems might permit the implementation of qubits since they can be compatible with novel materials such as graphene and transition metal dichalcogenides, which have unique electronic properties. Their complicated structures indicate that FETs and QFETs are compatible with small sizes, which is the most important in designing small quantum circuits. In addition, the current developments of FET technology indicate that the application of traditional and quantum computers has new avenues, which emphasize the need to continue research in this significant area [23, 24].

Definition and working principle of FETs

Since its initial release, transistors have evolved dramatically to become an important component of modern electronics, notably in quantum computers. The operation of the Field-Effect Transistors (FETs) is based on the principle of the control of current flow in the channel of a semiconductor with the help of an electric field, modifying the conductibility of the latter with the gate voltage. It is a process by which FETs can be switched on and off. As current studies consider how things can be made even smaller, new varieties such as tunnel FETs (TFETs) are being created. They operate at reduced voltages and are more efficient in energy consumption and performance as compared to regular MOSFETs that have problems with thermionic emission. In addition, with the further development of quantum computing, it is proposed that new designs with two-dimensional materials and spintronics can enhance circuit performance, expand the range of applications of FET technology to quantum applications [25, 26]. This overview indicates the importance of FETs to accelerate traditional and quantum computing, the convergence of electronics, and novel conceptual computing.

Types of Field-Effect Transistors

There is an increasing demand for efficient and scalable electronic devices and so, the creation of Field-Effect Transistors (FETs) is more significant. There are two types, Metal-Oxide-Semiconductor FET (MOSFET) and Junction FET (JFET), which are distinguished by the difference in the nature of operation and application. Fig. (**3**) shows the schematic structure of a Metal-Oxid--Semiconductor Field-Effect Transistor (MOSFET). MOSFETs are operated by an insulated gate to control the quality of the channel conduction, which has advantages such as great input impedance and simpler integration of more complicated circuits. By contrast, JFETs operate on a reverse-biased p-n junction to modify the current flow, with a simple design but with a less restrictive input impedance range. In quantum computing design, the application of these types of FETs is likely to increase performance, particularly with current research on materials like wide band gap semiconductors, which may further improve the performance of the devices [27, 28]. The versatile character of FET technology is also important to encourage advancement in electronic systems, such as in quantum applications. The arrangement and operation of these FETs are depicted well, and it provides a visual context to facilitate comprehension.

Role of FETs in Classical Electronics

Field-effect transistors (FETs) are major components of conventional electronics that have dominated the development of electronic gadgets. They operate through the application of electric fields in controlling the conductivity of a channel so that designs can be easily made in integrated circuits in order to enhance performance and efficiency. FETs are very important in a wide range of uses, in amplifiers, radio frequency devices, and are versatile in a variety of applications [29, 30]. Their capabilities have also expanded with the usage of new materials, such as the emergent two-dimensional materials, and they have been used to address the problem of scaling in the current electronics [29, 30]. Also, with the growing requirements of improved computing and communication capabilities, the FET technology continues to advance, and this provides an avenue through which a combination of optics and silicon-based systems could be used, thereby enhancing energy efficiency in information processing.

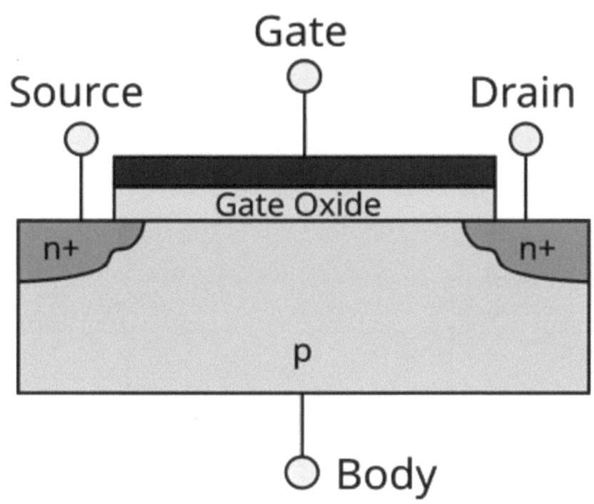

Fig. (3). Schematic of a Metal-Oxide-Semiconductor Field-Effect Transistor (MOSFET).

Advantages of FETs Over other Transistor Types

Field-effect transistors (FETs) are more prominent in semiconductor technology than other transistors. They consume less energy and hence are more efficient, and this is very crucial in quantum computing, where minimization of energy consumption is critical. FETs can be customized when taking the form of Negative Differential Resistance (NDR) and resonant tunneling diodes, which demonstrated to be capable of new functionality. In addition, FETs enable compact and scaled designs, which is important in the development of high-density circuitries, as demonstrated by the new simulation advances in multi-scale modeling of nano electronics [31, 32]. Lastly, the special control offered by FETs over transporting charge, highlighted by new materials like carbon nanotubes and 2D metals [33], opens up new opportunities to enhance the performance of a device, particularly in the context of quantum information processing systems, and FETs are valuable in the future of electronics [31]. For a visual look at the structural benefits of FETs, Fig. (**4**) is designed. Table **1** provides an overview of different FET types along with their definitions, characteristics, applications, and advantages. A comparison of FET advantages with other transistor types is shown in Table **2**.

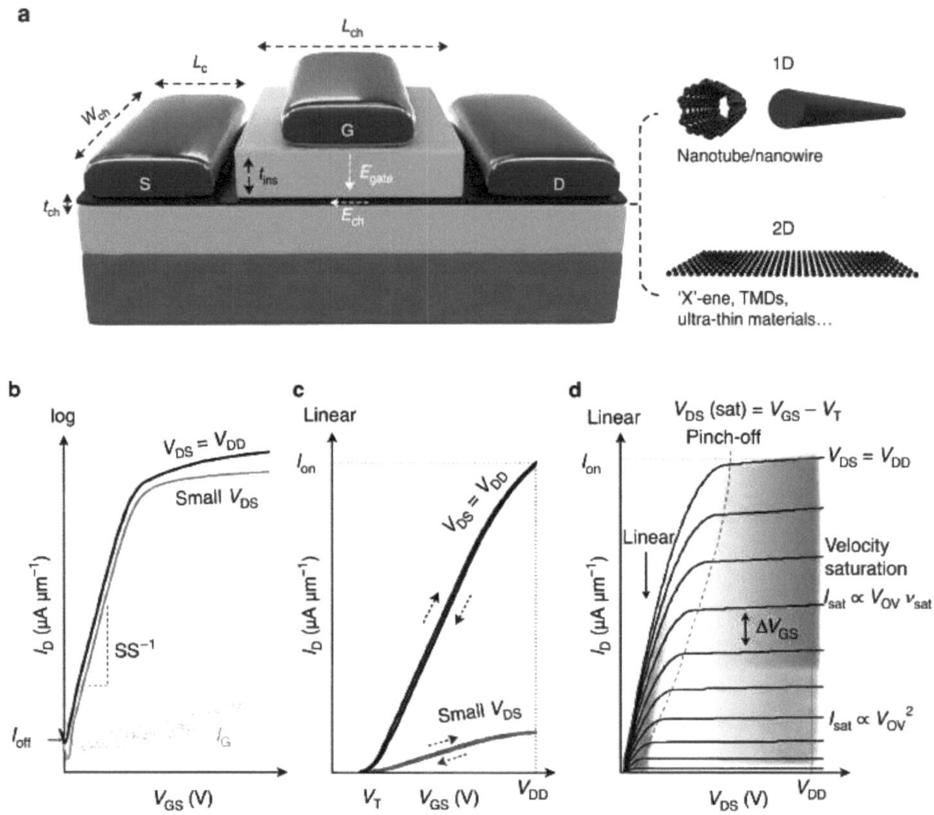

Fig. (4). Overview of advanced transistor architecture and its electrical characteristics.

Table 1. Types of FETs and Their Characteristics.

Type	Definition	Key Characteristics	Typical Applications	Advantages
MOSFET	Metal-Oxide-Semiconductor Field-Effect Transistor	High input impedance, low power consumption, can be enhanced or depleted type.	Switching devices, amplifiers, and digital circuits.	Scalability, ease of integration in microelectronics.
JFET	Junction Field-Effect Transistor	Operates with an electric field, majority carrier devices, typically n-channel or p-channel.	Analog switches, amplifiers, and RF applications.	Simplicity of design, low noise characteristics.

(Table 1) cont.....

Type	Definition	Key Characteristics	Typical Applications	Advantages
DNPMOSFET	Double N-Channel Power Metal-Oxide-Semiconductor Field-Effect Transistor	High efficiency with low on-resistance, suitable for power applications	Power electronics, converters	Higher power handling, better thermal performance
HEMT	High Electron Mobility Transistor	Uses quantum well structures, provides high-speed operation	RF communication, satellite systems	Superior performance in high-frequency applications

Table 2. Advantages of FETs Over Other Transistor Types.

Advantage	FETs	Other Transistor Types	Comments
Higher Input Impedance	10^9 Ω	10^3 - 10^4 Ω	FETs typically offer much higher input impedance compared to BJTs, resulting in less loading on previous stages.
Lower Noise	Generally lower	Higher levels of noise	FETs generate less noise and are more suitable for low-noise applications.
Thermal Stability	More stable	Less stable	FETs are generally less affected by temperature variations, which enhances their reliability.
Easier Integration	Simpler integration in chips	More complex integration	FETs are more compatible with integrated circuit manufacturing processes.
Lower Power Consumption	Lower overall power	Higher power requirements	FETs consume less power, making them more efficient for many applications.

Recent Advancements in FET Technology

Field-effect transistors (FETs) are experiencing a significant transformation, and material science is developing. This modification is demonstrated by advancements of graphene-based FETs, in which researchers employ the unusual characteristics of graphene to enhance the performance of devices, producing shorter switch times and lowering energy consumption [34]. Moreover, flex-organic field-effect transistors (FOFETs) are coming into existence, which can be deployed in wearable devices and bendable electronics, and are increasing in their applications to consumer devices and smart systems. These advances not only advance the levels of efficiency of existing technologies but also provide the prospects of quantum computing. These advanced FET designs are of great benefit to quantum systems where high accuracy and low noise are necessary. In general, these developments represent a landmark period in FET technology, which has improved computing capability and new horizons in electronic applications.

INTEGRATION OF FETS IN QUANTUM COMPUTING

Significant advancements in the field of quantum computing have been made due to advances in field-effect transistors (FETs), where stability and reliability are essential. Application of FETs in quantum systems exploits the benefits of their capability to control charges effectively that is necessary in making the qubits required in quantum operations. Studies of new technologies have indicated that several nanostructures, particularly those in the form of arrays consisting of two-dimensional nanowires, have great potential for large-scale computing systems, which are significant in quantum applications. Additionally, the specifics of the variation of devices revealed in statistical research of silicon MOSFETs smaller than 100 nanometers suggest the fact that it is essential to control the variation as it increases in devices of smaller size [35 - 39]. These issues underscore the need to take FETs into quantum systems with a lot of care since any natural change in the parameters may greatly influence the yield and performance of the future circuits [40 - 43]. Ultimately, it appears that a relevant solution should be to unite FET technology with quantum computing approaches to address existing limitations of computing and enhance its performance [44 - 46].

Role of FETs in Qubit Implementation

Field Effect Transistors (FETs) applications in quantum computing are a notable interaction of novel materials and device development in assisting to establish robust qubit systems. With FETs, qubit states can be controlled with electrical signals, making it easier to deal with than conventional control methods, such as inductive or optical controls. Inclusion of 2D materials, including graphene and other novel materials enhance the functionality of FETs, which are less noisy and effective in heat dissipation, which is vital in the dependable qubit operations [48 - 54]. In addition, with the progress in quantum simulators, FETs are becoming more and more vital, as they facilitate the capacity to process quantum states that make use of entanglement to solve complicated problems at once. More recent developments in spin-wave computing demonstrate how FETs may be used together with spintronic currents to create hybrid device systems and enhance both performance and low power consumption. The combination of these factors makes FETs key components in the ongoing search for practical quantum computing solutions.

Fig. (**5**) illustrates the essential roles FETs play in qubit implementation within quantum computing. Each bar represents the significance of aspects such as the development of robust qubit architectures (80%), electrical control over qubit states (75%), the performance benefits of 2D materials like graphene (70%), the capability for parallel processing of quantum states (60%), and the efficiency of hybrid systems integrating spintronics (50%), showcasing the multifaceted contributions of FETs.

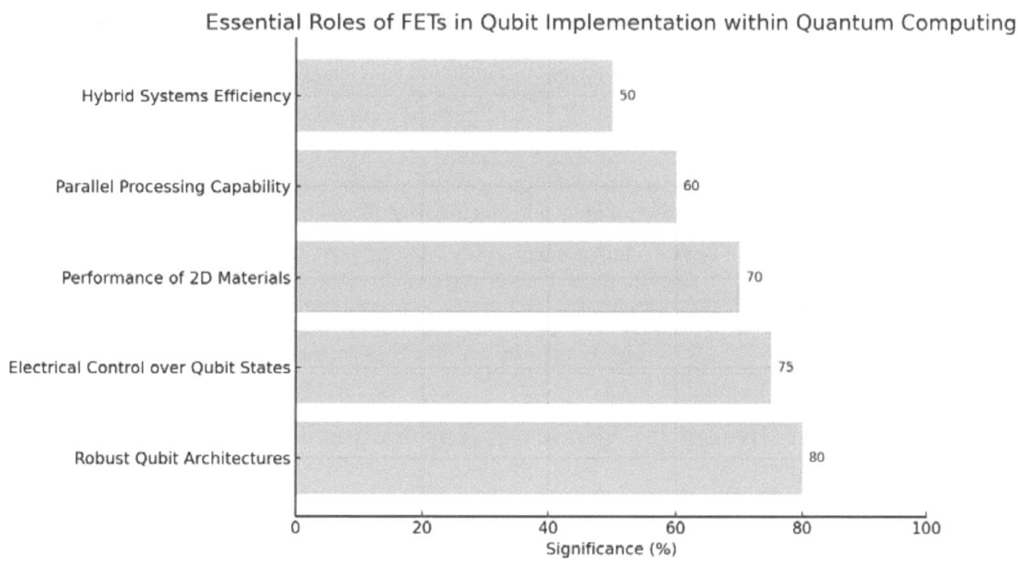

Fig. (5). Essential Roles of FETs in Qubit Implementation within Quantum Computing [14, 15].

Challenges of using FETs in Quantum Systems

With the growing intensity in the quest to realize nanoscale electronics, issues related to the implementation of FETs in quantum systems are becoming more pronounced. Table **3** lists the main limitations that come up when incorporating FETs in quantum architectures. Of great concern is the inherent non-uniformity of device parameters as semiconductor dimensions scale to the decananometer regime; charge localization and interface roughness can affect the performance of a device and the yield of the manufacturing process in a significant way [6]. Moreover, new materials, such as carbon nanotubes and silicon nanowires, will demand new designs to accommodate the unusual physical characteristics on the small scale [7]. As investigators consider the challenges of the device density and design efficiency, new nanostructures such as two-dimensional nanowire arrays provide both opportunities and challenges to large-scale computing. Addressing these fluctuations and adapting to the constraints of quantum effects are desired in particular for designs that seek to utilize graphene FETs, which currently experience issues of low voltage gain and bandwidth due to contact resistance. These factors will be studied to enhance FETs in the expanding field of quantum computing. A schematic diagram shows how FETs work helps to highlight these challenges, making it a key part of understanding the core issues mentioned.

Table 3. Challenges of FETs in Quantum Systems.

Challenge	Description	Impact Level
Thermal Noise	At operational temperatures, FETs may introduce thermal noise that interferes with quantum coherence.	High
Scalability	Integrating multiple FETs for large-scale quantum systems poses significant design and fabrication challenges.	Medium
Quantum Decoherence	FETs can contribute to decoherence through coupling with the environment and electric field fluctuations.	High
Material Limitations	Current semiconductor materials may not perform adequately at the quantum level compared to superconductors.	Medium
Interconnect Issues	Connecting FETs with quantum bits introduces complexity due to signal integrity and feedback loops.	High

Innovations in FET Design for Quantum Applications

The quest to enhance field-effect transistor (FET) designs has presented fresh opportunities to enhance quantum applications, particularly in quantum computing. The progress of the recent past has demonstrated the effectiveness of connecting superconducting resonators to the low-noise microwave oscillators, which aid in the achievement of ultra-sensitive electron spin resonance, crucial in the qubit readout and control of quantum systems [8]. Moreover, organic synaptic transistors with quantum dot technology enable the neuromorphic systems to process information better, and with dynamic recognition tasks, they are promising in low-power scenarios [9]. Another novel front is tunnel FETs (TFETs), which offer important advantages in scale and power efficiency required in IoT processes and quantum processes. Moreover, accurate manufacturing strategies through atomistic modeling have also become highly important to control electron tunneling, which has a profound impact on the operations of semiconductor nanostructures of relevance to quantum computing. In general, progress in FET design has been jointly achieved to be more applicable to novel quantum applications, as illustrated in Fig. (**6**).

Case Studies of FETs in Quantum Computing Projects

Tunneling field-effect transistor (TFET) research represents a significant leap towards the design of energy-saving devices to support quantum computing and open new opportunities to access quantum physics interacting with semiconductor technology. Simulation of TFETs using physics, the study of both direct and indirect band-to-band tunneling (BTBT), is a step in the right direction to bring significant improvements in operational voltage and power consumption in this field [43]. In particular, TFETs have been mentioned in examples where designs of new devices are based on using 2D materials, which improves scaling and integration into quantum systems [44]. These nanoscale wire arrangements are also designed with a great deal of caution, which enhances the functioning of a device and improves interaction across components of a complex quantum setup. These advancements could be a significant breakthrough towards the realization of real quantum computing systems, and it is indicative of the game-changing ability of FET technology. These advancements can be elucidated with the help of the image, which displays the schematic of the TFET structures, which shows the complexities of the new designs, highlighting their impact on the new models of computing in the future.

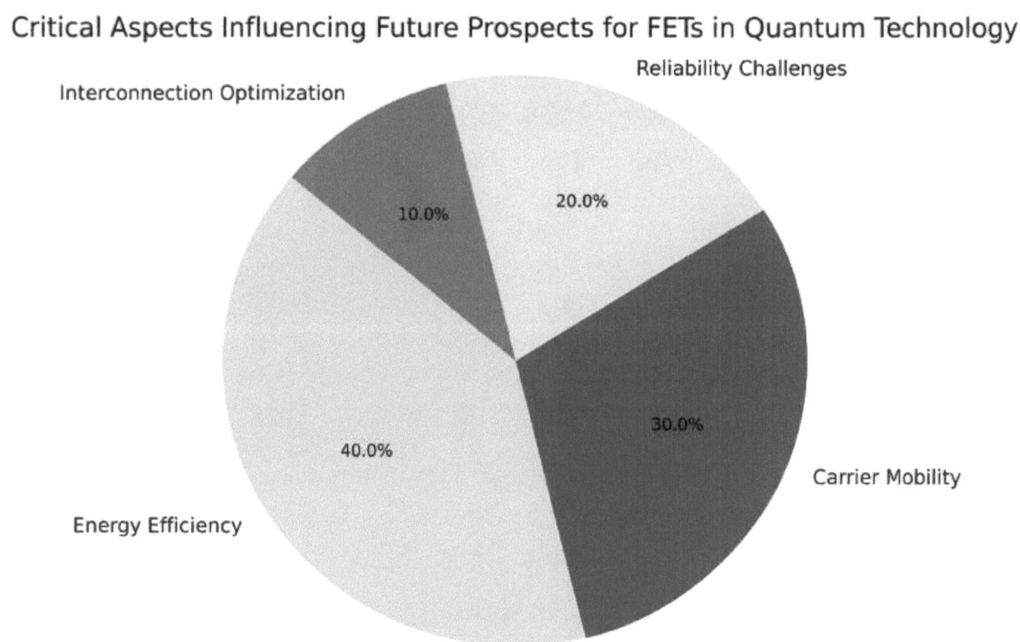

Fig. (6). Critical Aspects Influencing Future Prospects for FETs in Quantum Technology [17, 19, 26].

Future Prospects for FETs in Quantum Technology

With the evolution of quantum computing, field-effect transistors (FETs) are evolving. Recent developments on monolayer transition metal dichalcogenides material indicate that FETs can be made to work with ultra-low sub-threshold swings and are thus viable in quantum logic applications. In addition, other methods, such as buried-channel heterostructures, enhance the mobility of carriers in FETs, which aids in reducing noise and aids in the control of quantum states that are a key factor in quantum devices [2]. Nevertheless, questions of reliability remain, particularly as the defect rates go up on smaller technologies. This has given way to research on fault problems of novel materials such as graphene nanoribbons [4]. It is also important to know what is interconnected along with performance metrics, as demonstrated by simulation tools that strive to enhance nano array characteristics [3]. Together, these advances lay the groundwork for FETs in the quantum technology field, presenting challenges and chances for future growth.

This figure illustrates the critical aspects influencing the future prospects for FETs in quantum technology. Energy efficiency is the leading focus at 40%, followed by enhancements in carrier mobility at 30%. Reliability challenges due to defect rates account for 20%, while the optimization of interconnection metrics represents 10%. This visual highlights the balance between technological advancements and the challenges facing FETs in quantum applications.

CHALLENGES AND LIMITATIONS

With the evolution of quantum computing, field-effect transistors (FETs) are evolving. Recent developments on monolayer transition metal dichalcogenides material indicate that FETs can be made to work with ultra-low sub-threshold swings and are thus viable in quantum logic applications. In addition, other methods, such as buried-channel heterostructures, enhance the mobility of carriers in FETs, which aids in reducing noise and helps in the control of quantum states that are a key factor in quantum devices [2]. Nevertheless, questions of reliability remain, particularly as the defect rates go up on smaller technologies. This has given way to research on fault problems of novel materials such as graphene nanoribbons [4]. It is also important to know what is interconnected and performance metrics, as demonstrated by simulation tools that strive to enhance nano array characteristics [3].

Technical Challenges in Quantum Computing

The replacement of classical with quantum computing presents numerous technical challenges that require solutions so that quantum devices can achieve their full potential. Table **4** summarizes the most important technical challenges in quantum computing. With the continued shrinking of the devices to the nanometer level, the sensitivity of the important parameters, such as those due to random dopants and trapped charges, is increasing, threatening the functionality and performance of future circuits [32]. The differences pose engineering problems that demonstrate the necessity of new designs in order to increase reliability. Further, working with new materials and designs, including 2D materials, like graphene and transition metal dichalcogenides, not only offers opportunities but also presents challenges in device manufacturing and operation [28]. This is further complicated by the necessity to maintain low operating temperatures and quantum coherence, which is necessary for qubit operation. With further developments in research, these technical problems will be an important challenge in the future of quantum computing and associated technologies, particularly when it comes to FETs. Fig. (7) points to the design challenges associated with the creation of these advanced semiconductor technologies.

Table 4. Technical Challenges in Quantum Computing.

Challenge	Description	Percentage
Error Rates	Quantum bits (qubits) are prone to decoherence and noise, leading to higher error rates in computation.	10-20% error rates in current qubit technologies.
Scalability	Building scalable quantum computers that can maintain qubit coherence over large numbers is a significant hurdle.	Current systems are typically limited to 50-100 qubits.
Quantum Interconnects	Efficient connections between qubits are necessary for complex calculations and maintaining coherence.	No current technology fully meets the demands for effective quantum interconnects.
Temperature Control	Many quantum computing systems require extremely low temperatures to operate effectively, posing engineering challenges.	Operating temperatures close to absolute zero (below 0.1 Kelvin)
Quantum Software Development	The development of efficient quantum algorithms and software is still in its infancy.	A limited number of quantum algorithms are available for practical use.

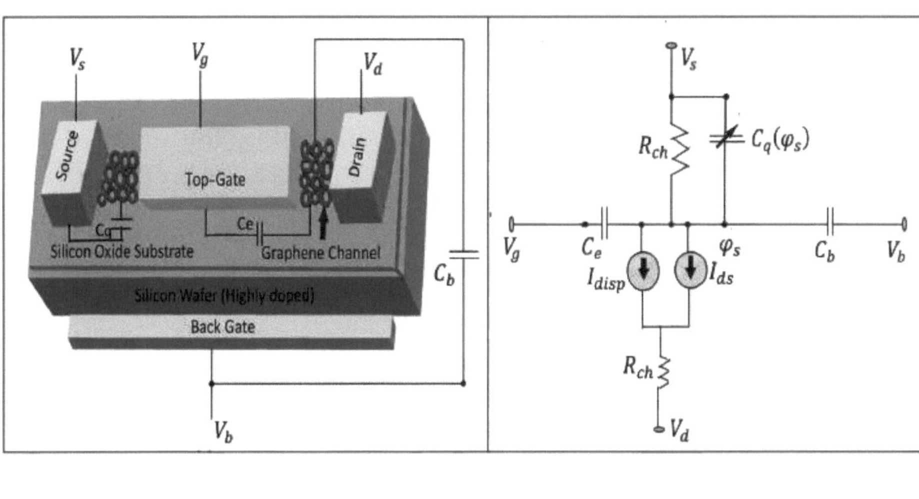

Fig. (7). Schematic representation of a graphene-based field-effect transistor (GFET) highlighting structural and electrical parameters.

Limitations of Current FET Technology

With improvements in quantum computing showing the weaknesses of regular field-effect transistors (FETs), it is more clear that we need different materials and designs. Present silicon-based FET technology faces scaling problems because of quantum effects, which cause variability in how well devices work as their sizes get smaller, eventually hurting speed and power efficiency. This issue is made worse by the limits of standard dielectric materials that fail to effectively reduce leakage currents at the nanoscale [29]. Additionally, nanoscale designs, like those using carbon nanotubes, are not fully developed because of manufacturing difficulties and sensitivity to defects. The growing need for high-performance computing makes it essential to find materials that work well under quantum conditions, showing that current FET technologies are not enough for the future needs of quantum computing systems. A diagram showing different semiconductor device setups highlights these difficulties.

Issues Related to Scalability and Integration

Improvements in scaling quantum computing tech face big problems, especially in combining two-dimensional materials with regular semiconductor setups. The limits of dielectric materials that work with two-dimensional transition metal dichalcogenides slow down progress; current integration methods, which are fine with bulk semiconductors, don't work well for very thin materials, which holds back device efficiency and market potential [30]. Looking into new ways to make things, adaptable transistor designs could help solve these problems, possibly boosting the scalability of these quantum devices. Moreover, studying hetero junctions in layered materials gives chances to better electronic and optoelectronic uses, which are important for using quantum effects. As quantum uses change, solving these scalability and integration issues will be key for making functional devices that can go beyond just ideas, setting the stage for real applications in quantum computing [31].

Quantum Decoherence and its Impact on FETs

The complicated link between quantum decoherence and field-effect transistors (FETs) gives important viewpoints on the changing field of quantum computing. As the materials in semiconductor devices improve, the limits on performance caused by decoherence—a process where coherent quantum states lose their phase—become clear. Research shows that two-level defects in superconducting circuits, worsened by materials like amorphous oxides, can greatly affect how reliable and functional devices are in quantum computing situations [20]. Also, using gate-induced quantum dots in nanowires shows that managing charge noise can help reduce decoherence effects, which improves the operational accuracy of quantum systems [21]. As nanotechnology develops, it is crucial to understand the error mechanisms in graphene nanoribbon FETs, pointing out how decoherence affects reliability in new logic circuits [22]. Thus, addressing charge noise through better material interfaces provides hopeful solutions for the scaling issues created by decoherence in FETs [23].

Fig. (**8**) illustrates the primary sources of decoherence affecting FETs, the corresponding impact on device reliability quantified at 75%, and the effectiveness of different mitigation techniques with gate-induced quantum dots reducing decoherence effects by around 45%. The data reflects the critical relationships outlined in the paragraph between quantum decoherence, the materials used, and potential solutions in the context of quantum computing.

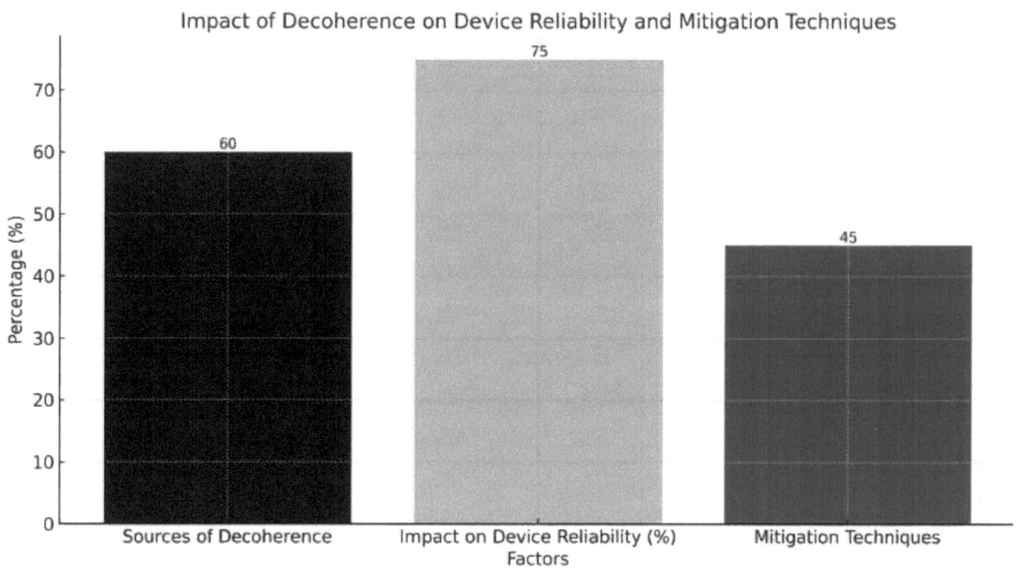

Fig. (8). Impact of Decohrence on Device Reliability and Mitigation Techniques.

Economic and Resource Considerations

New technologies, especially quantum computing and field-effect transistors (FETs), influence the economy and resources significantly and should be thought through. These are the new advanced technologies that have numerous potential applications in the majority of industries, but cost a lot of money and resources to create. The materials required to produce semiconductors are to be well employed to minimize costs and environmental degradation, as it has been discussed in semiconductor developments. The fact that FETs are scalable and this has drawn increased interest in improving the design of their devices and material properties are an indication that it is important to ensure that resources are effectively utilized in order to improve the performance of the manufacturing process. Further, the expected achievements of computational ability in FET-based GPUs imply the connection between the ability of technology and the actuality of money. This side of management of the resource belongs to the larger tendencies in the sphere of sustainable development, and research institutions will have to cooperate to promote responsible development. Thus, the economic environment is one of the factors that must be put into consideration when considering the importance of developing quantum computing and FETs.

CONCLUSION

The challenge and the huge opportunities in the evolving world of quantum computing, employing field-effect transistors (FETs), have it all. The advancement of materials science opens avenues when it comes to the development of FET, and the use of two-dimensional substances, such as graphene and transition metal dichalcogenides, which enhances the functioning of a device, besides addressing significant challenges, such as the consumption of energy and the speed of the working process. But it can only be said that when one has to transition to quantum systems, one once again has to look at the existing models and methods of semiconductor design, particularly in terms of manufacturing constraints and computing constraints. The future of this technology lies in the solution of problems of contact resistance and device linearity, as evident in the recent studies of circuit solutions.

Summary of Key Points Discussed

The research of advanced materials such as Bi2O2Se offers a great potential for field-effect transistor (FETs) in quantum computing, and it provides an insight into high-speed low-power electronics. The analysis of the figures is shown in Fig. (**9**) in the case of a gated InAs channel semiconductor. The comparison between the InAs/Al0.05Ga0.95Sb tunnel field-effect transistors (TFETs) and CMOS FinFET technology demonstrates the advancement in the ultralow voltage work, which is at variance with normal semiconductor performance [38]. Also, the application of multiscale simulation techniques to carbon nanotube devices points out the multi-faceted nature of quantum transport in the present nanoelectronics, which is a direction of future developments. These improvements are pointing to a change in the electronic design and thus next-generation systems, which are characterized by increased computational effectiveness and operational adaptability. The insights complement this discussion with the complexities of the advanced FET structures, which are of significance in the applications of quantum computing.

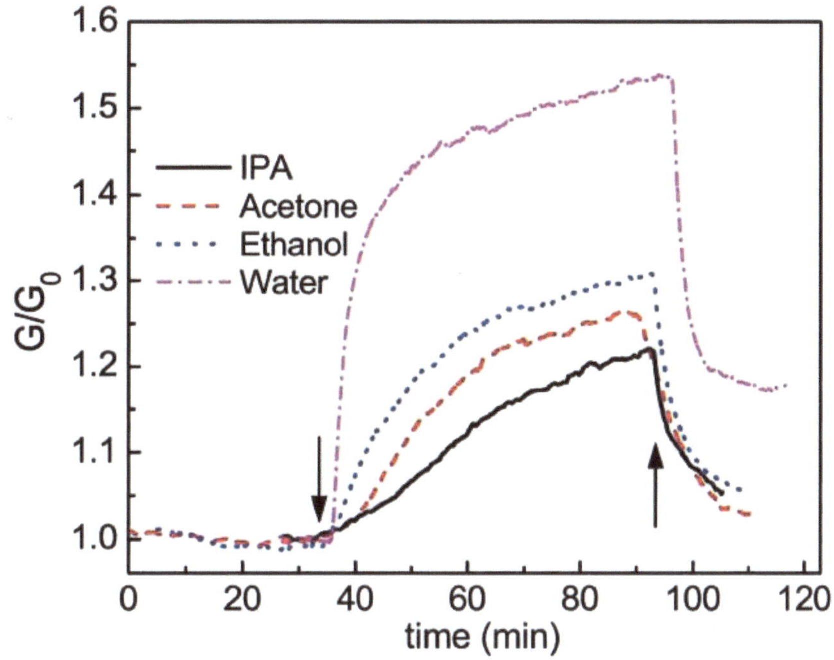

Fig. (9). Data analysis of a gated InAs channel semiconductor device.

The Significance of FETs in Advancing Quantum Computing

New directions have emerged in quantum computing with the rapid change in semiconductor technology, and the field-effect transistors (FETs) on the frontline to create smaller and workable devices. Fig. (**10**) shows a schematic view of some of the important structures of semiconductor devices and the principles that describe their operation. The reconfigurable FETs with a particular emphasis on those with negative differential resistance (NDR) are emphasized as capable of a range of tasks on a small scale, which is crucial to further development of quantum circuit integration [40]. The developments not only widen the boundaries of integrated circuit design but also permit complex biointerface applications to associate quantum computing with biophysics [39]. Besides, there are emerging materials such as biopolymer-carbon nanotube hybrids, which offer opportunities to design multifunctional nanoscale devices capable of transforming quantum information processing [11]. The development of FETs as central components in new quantum computers is a significant shift in the level of computing power, with close research activities and collaboration across various areas. To further explain these advances in technology, the transistor layout schematic shows the specific layout that would be required in enhancing the performance of FETs in quantum systems.

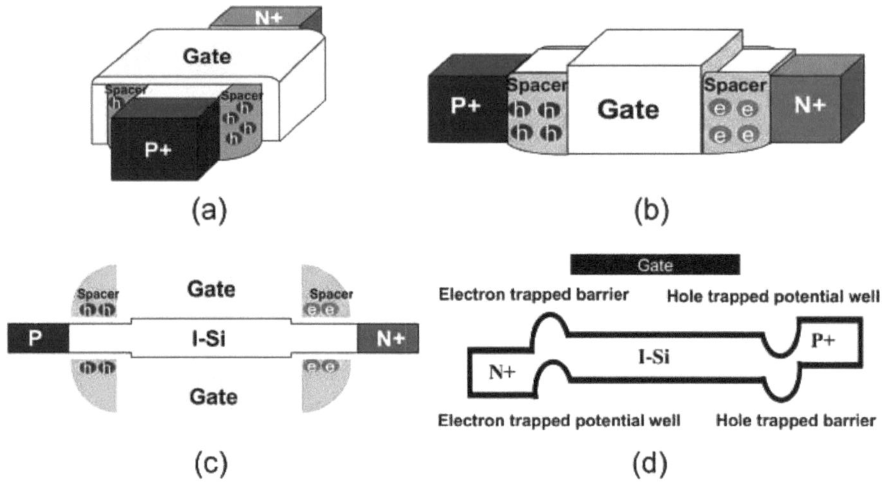

Fig. (10). Schematic diagram of semiconductor device structures and principles.

Future Directions for Research and Development

Along with the continuous transformation of quantum computing, it is essential to address the inherent variability of the device parameters in order to use them in the real world product. It has been observed that random discrete dopant variation and small-scale irregularities will increasingly affect the performance of semiconductor devices as they become smaller, which would cause detrimental effects to the performance of circuits unless countermeasures are taken [10]. Innovative instruments such as ToPoliNano are used in the detailed analysis of various nanotechnologies, which have an impact on design decisions in the future. Also, the advance of the study of single-nanometer engineering results in new designs employing materials such as carbon nanotubes and silicon nanowires, which enhance the logic of the overall computing systems. The diagram of the latest trends in transistor designs may help understand what advantages these new structures can offer and what implications they have on further research direction in quantum computing and FET technology.

Implications for the Technology Industry

With the tech business approaching the boundaries of the old silicon devices, new alternative solutions such as quantum computing and field-effect transistors (FETs) are gaining relevance. Fig. (**11**) shows the schematic figure of how a Field-Effect Transistor works. In retrospect, we observe how densities and complexities of transistors have also posed new challenges and necessitated new materials and designs of devices to continue enhancing performance [12].

Tunneling field-effect transistors (TFETs) in this respect are particularly noteworthy due to their unusual current transportation technologies, which can perform better than conventional MOSFETs in energy usage [13]. These technologies have more impact than only onperformance since they necessitate a significant shift in the design and manufacturing processes, particularly in the use of materials such as InGaAs and transition metal dichalcogenides [47, 48]. The developments point to the need for new semiconductor technologies and signify a shift in the technology industry that is concerned with efficiency and better performance [47]. This shift is also backed by presenting the anticipated performance patterns in FET-based equipment, and this coincides with the greater narrative of the technology sector growth [49 - 54].

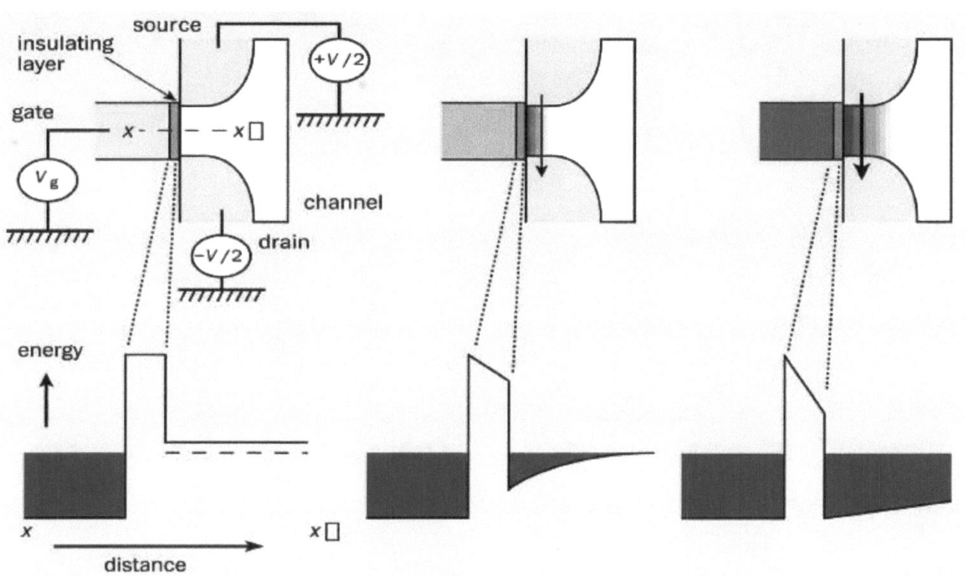

Fig. (11). Schematic of Field-Effect Transistor Operation.

Final thoughts on the Potential of Quantum Computing and FETs

The fact that quantum computing and field-effect transistors (FETs) were brought together is an important milestone in the history of computing technology, as old concepts of semiconductors were merged with the new possibilities of quantum mechanics. Modern electronics could not have been achieved without FETs that have achieved a lot in the past few years. Nevertheless, they have speed and energy limitations, especially when they are compared to those of quantum processing. With the increasing need to have faster and more efficient computers, the addition of quantum bits in the FET design would be a promising solution to enhance its efficiency. This combination can not only significantly increase computing power, but also offers new possibilities to process error correction and data security, which are transforming the manner in which we solve complicated problems. Fig. (**12**) outlines potential computing performance and helps to clarify how these technologies will influence future developments in computing.

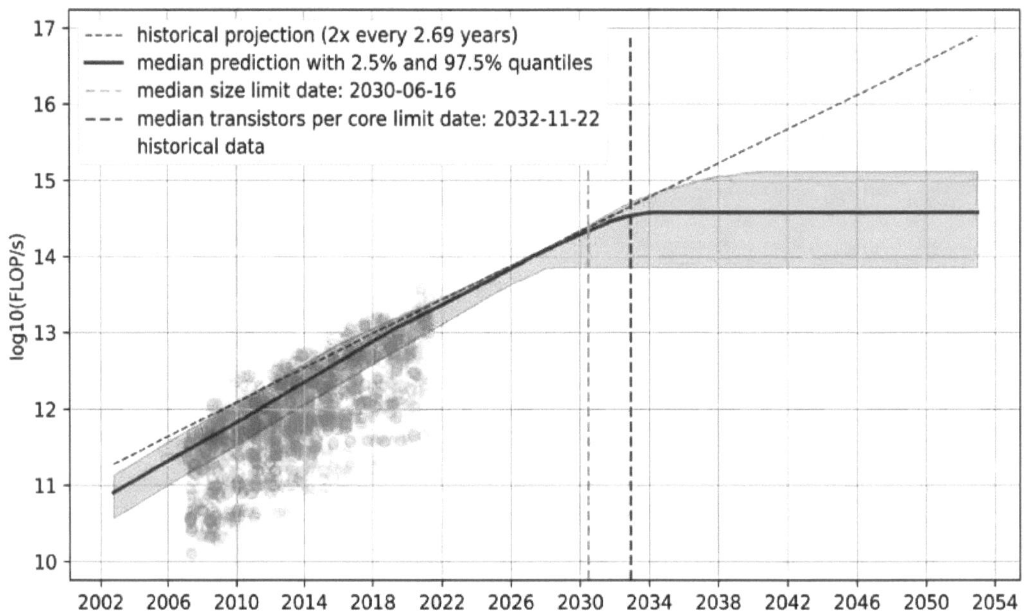

Fig. (12). Projected performance trends for FET-based GPUs from 2002 to 2054 [46 - 56].

AI DISCLOSURE

During the preparation of the work, the author(s) used AI tools and Grammarly tool to enhance the writing style and readability of the content. The authors clarify that AI was not involved in generating the substantive content and that they take full responsibility for the publication.

REFERENCES

[1] A.P. Sommer, P. Schemmer, A.E. Pavláth, H.D. Försterling, Á.R. Mester, and M.A. Trelles, "Quantum biology in low level light therapy: death of a dogma", *Ann. Transl. Med.*, vol. 8, no. 7, p. 440, 2020.
[http://dx.doi.org/10.21037/atm.2020.03.159] [PMID: 32395484]

[2] A. DeHon, P. Lincoln, and J.E. Savage, "Stochastic assembly of sublithographic nanoscale interfaces", *IEEE Trans. Nanotechnol.*, vol. 2, no. 3, pp. 165-174, 2003.
[http://dx.doi.org/10.1109/TNANO.2003.816658]

[3] A. Moro, Aycart Lazo, Bardizza Giorgio, Bielewski Marek, Juan Lopez Garcia, and Taylor Nigel, *First Workshop on Identification of Future Emerging Technologies for Low Carbon Energy Supply,*

2017.

[4] V. Athanasiou, *Theoretical simulations of dynamical systems for advanced reservoir computing applications* Chalmers Tekniska Hogskola: Sweden, 2020.

[5] D. Esseni, M. Pala, P. Palestri, C. Alper, and T. Rollo, "A review of selected topics in physics based modeling for tunnel field-effect transistors", *Semicond. Sci. Technol.,* vol. 32, no. 8, p. 083005, 2017. [http://dx.doi.org/10.1088/1361-6641/aa6fca]

[6] Ehud Altman, Kenneth R. Brown, Giuseppe Carleo, Lincoln D. Carr, Eugene Demler, Cheng Chin, and Brian DeMarco, "Quantum simulators: Architectures and opportunities", *PRX quantum,* vol. 2, no. 1, p. 017003, 2021. [http://dx.doi.org/10.1103/PRXQuantum.2.017003]

[7] A. Mahmoud, F. Ciubotaru, F. Vanderveken, A.V. Chumak, S. Hamdioui, C. Adelmann, and S. Coţofană, "Introduction to spin wave computing", *J. Appl. Phys.,* vol. 128, no. 16, p. 161101, 2020. [http://dx.doi.org/10.1063/5.0019328]

[8] R. Russo, A. Chatel, N. Brusadin, R. Yu, R. Farsi, H. Furci, and J. Brugger, "Superconducting microwave oscillators as detectors for ESR spectroscopy", *Appl Phys Lett.,* vol. 126, p. 154101, 2025. [http://dx.doi.org/10.1063/5.0260098]

[9] W. Luan, Z. Zhao, H. Li, Y. Zhai, Z. Lv, K. Zhou, S. Xue, M. Zhang, Y. Yan, Y. Cao, G. Ding, S.T. Han, C.C. Kuo, and Y. Zhou, "Near-Infrared Response Organic Synaptic Transistor for Dynamic Trace Extraction", *J. Phys. Chem. Lett.,* vol. 15, no. 34, pp. 8845-8852, 2024. [http://dx.doi.org/10.1021/acs.jpclett.4c02238] [PMID: 39167716]

[10] D.C. Schmittlein. Report to the President for year ended June 30, 2016, Dean, MIT Sloan School of Management, 2016.

[11] R.R. Johnson, *Probing the Structure and Function of Biopolymer-Carbon Nanotube Hybrids with Molecular Dynamics* Scholarly Commons, 2009.

[12] I.L. Markov, *Limits on Fundamental Limits to Computation* Springer Science and Business Media LLC, 2014. [http://dx.doi.org/10.1038/nature13570]

[13] Atieh Farokhnejad, *Compact Modeling of Intrinsic Capacitances in Double-Gate Tunnel-FETs* Universitat Rovira I Virgili, 2020.

[14] T. Calarco, F. Frank, F. Jelezko, and Montangero, *Autonomous Calibration of Single Spin Qubit Operations* Springer Science and Business Media LLC, 2017.

[15] A. Abramsky, B. Baez, B. Edwards, and B. Coecke, *Phase groups and the origin of non-locality for qubits* Elsevier BV, 2010.

[16] Philip B. Alipour, *Theoretical Engineering and Satellite Comlink of a PTVD-SHAM System*, 2007.

[17] A. C. Cem Say, Abuzer Yakaryilmaz, Kaneps Jānis, Nakanishi Masaki, Nielsen Michael A., Rabin Michael O., Yakaryılmaz Abuzer, *Quantum counter automata", 'World Scientific Pub Co Pte Lt*, 2011.

[18] K. Atsufumi Hirohata, Yamada, Y. Nakatani, Lucian Prejbeanu, B. Diény, Philipp Pirro, B. Hillebrands, "Review on spintronics: Principles and device applications", *J. Magn. Magn. Mater.,* 2020.

[19] S. Smidstrup, T. Markussen, P. Vancraeyveld, J. Wellendorff, J. Schneider, T. Gunst, and B. Verstichel, "QuantumATK: an integrated platform of electronic and atomic-scale modelling tools", *J. Phys. Condens. Matter,* 2019. [PMID: 31470430]

[20] S.K. Chakraborty, S.P. Dash, B. Kundu, and Nayak, *Challenges and opportunities in 2D heterostructures for electronic and optoelectronic devices* Elsevier BV, 2022.

[21] A. Abijesh Euphrine, C. Jamina, L. Arun Jose, and S. Arun Jose, *Challenges and Opportunities in*

Implementing Negative Differential Resistance Mode Reconfigurable Field Effect Transistors, 2024.

[22] Wei Jiang, Wei Jiang, Qian Zhou, Stepan Melnyk, Shiping Zhou, Muhammad Afnan Habibi, and Bin Han, *Terahertz Communications and Sensing for 6G and Beyond: A Comprehensive Review*, 2024. [http://dx.doi.org/10.1109/COMST.2024.3385908]

[23] P. George, "Message from the executive chairman and secretary", *2023 IEEE International Conference on Recent Advances in Systems Science and Engineering (RASSE),* IEEE, pp. 1-1, 2023. [http://dx.doi.org/10.1109/RASSE60029.2023.10363517]

[24] Kai Liu, Ouyang Bang, Xiaojun Guo, Yunlong Guo, and Yunqi Liu, *Advances in flexible organic field-effect transistors and their applications for flexible electronics* npj Flexible Electronics, 2022. [http://dx.doi.org/10.1038/s41528-022-00133-3]

[25] J. Puebla, J. Kim, K. Kondou, and Y. Otani, *Spintronic devices for energy-efficient data storage and energy harvesting* Communications Materials, 2020.

[26] A. Chiesa, A. Privitera, E. Macaluso, M. Mannini, R. Bittl, R. Naaman, M.R. Wasielewski, R. Sessoli, and S. Carretta, "Chirality-Induced Spin Selectivity: An Enabling Technology for Quantum Applications", *Adv. Mater.,* vol. 35, no. 28, p. 2300472, 2023. [http://dx.doi.org/10.1002/adma.202300472] [PMID: 37170702]

[27] V. Shanmugam, R.A. Mensah, K. Babu, S. Gawusu, A. Chanda, Y. Tu, R.E. Neisiany, M. Försth, G. Sas, and O. Das, "A Review of the Synthesis, Properties, and Applications of 2D Materials", *Part. Part. Syst. Charact.,* vol. 39, no. 6, p. 2200031, 2022. [http://dx.doi.org/10.1002/ppsc.202200031]

[28] I.L. Markov, *Limits on Fundamental Limits to Computation* Springer Science and Business Media LLC, 2014. [http://dx.doi.org/10.1038/nature13570]

[29] A.M. Stoneham, "The challenges of nanostructures for theory", *Mater. Sci. Eng. C,* vol. 23, no. 1-2, pp. 235-241, 2003. [http://dx.doi.org/10.1016/S0928-4931(02)00274-6]

[30] C. S. Lau, S. Das, I. A. Verzhbitskiy, D. Huang, Y. Zhang, T. Talha-Dean, W. Fu, D. Venkatakrishnarao, and K. E. J. Goh, "Dielectrics for two-dimensional transition-metal dichalcogenide applications", *ACS Nano,* vol. 17, no. 11, pp. 9870-9905, 2023.

[31] Pengju Li, Saehyun Kim, and Bozhi Tian, *Beyond 25 years of biomedical innovation in nano-bioelectronics,* Device 2, no. 7, 2024. [http://dx.doi.org/10.1016/j.device.2024.100401]

[32] G. Salis, "Semiconductor spintronics: switching spins at low voltage", *International Symposium on Low Power Electronics and Design (ISLPED),* IEEE, pp. 143-143, 2013. [http://dx.doi.org/10.1109/ISLPED.2013.6629283]

[33] M. Huang, Z. Li, and H. Zhu, "Recent Advances of Graphene and Related Materials in Artificial Intelligence", *Adv. Intell. Syst.,* vol. 4, no. 10, p. 2200077, 2022. [http://dx.doi.org/10.1002/aisy.202200077]

[34] A.V. Chumak, P. Kaboš, M. Wu, C. Abert, C. Adelmann, A.O. Adeyeye, J. Akerman, F.G. Aliev, A. Anane, A. Awad, C.H. Back, A. Barman, G.E.W. Bauer, M. Becherer, E.N. Beginin, V.A.S.V. Bittencourt, Y.M. Blanter, P. Bortolotti, I. Boventer, D.A. Bozhko, S.A. Bunyaev, J.J. Carmiggelt, R.R. Cheenikundil, F. Ciubotaru, S. Cotofana, G. Csaba, O.V. Dobrovolskiy, C. Dubs, M. Elyasi, K.G. Fripp, H. Fulara, I.A. Golovchanskiy, C. Gonzalez-Ballestero, P. Graczyk, D. Grundler, P. Gruszecki, G. Gubbiotti, K. Guslienko, A. Haldar, S. Hamdioui, R. Hertel, B. Hillebrands, T. Hioki, A. Houshang, C-M. Hu, H. Huebl, M. Huth, E. Iacocca, M.B. Jungfleisch, G.N. Kakazei, A. Khitun, R. Khymyn, T. Kikkawa, M. Klaui, O. Klein, J.W. Klos, S. Knauer, S. Koraltan, M. Kostylev, M. Krawczyk, I.N. Krivorotov, V.V. Kruglyak, D. Lachance-Quirion, S. Ladak, R. Lebrun, Y. Li, M. Lindner, R. Macedo, S. Mayr, G.A. Melkov, S. Mieszczak, Y. Nakamura, H.T. Nembach, A.A. Nikitin, S.A. Nikitov, V. Novosad, J.A. Otalora, Y. Otani, A. Papp, B. Pigeau, P. Pirro, W. Porod, F.

Porrati, H. Qin, B. Rana, T. Reimann, F. Riente, O. Romero-Isart, A. Ross, A.V. Sadovnikov, A.R. Safin, E. Saitoh, G. Schmidt, H. Schultheiss, K. Schultheiss, A.A. Serga, S. Sharma, J.M. Shaw, D. Suess, O. Surzhenko, K. Szulc, T. Taniguchi, M. Urbanek, K. Usami, A.B. Ustinov, T. van der Sar, S. van Dijken, V.I. Vasyuchka, R. Verba, S.V. Kusminskiy, Q. Wang, M. Weides, M. Weiler, S. Wintz, S.P. Wolski, and X. Zhang, "Advances in Magnetics Roadmap on Spin-Wave Computing", *IEEE Trans. Magn.,* vol. 58, no. 6, pp. 1-72, 2022.
[http://dx.doi.org/10.1109/TMAG.2022.3149664]

[35] Z. Kong, Z. Li, G. Cao, J. Su, Y. Zhang, J. Liu, and J. Liu, "Undoped Strained Ge Quantum Well with Ultrahigh Mobility Grown by Reduce Pressure Chemical Vapor Deposition", *arXiv preprint arXiv:2204.11470,* 2022.

[36] Z. Ren, C. Li, X. Huang, X. Yan, and W. Li, "Electronic properties of Janus black arsenic-phosphorus (b-AsP) nanoribbons under transverse electric field", *arXiv preprint arXiv:2212.01841,* 2022.

[37] E. Cannavò, D. Marian, E. Gonzalez Marin, A. Tredicucci, and G. Fiori, "Simulations of 2-D Materials-Based Field Effect Transistors for Quantum Cascade Detectors", *IEEE Trans. Electron Devices,* vol. 71, no. 1, pp. 630-636, 2023.

[38] I.L. Markov, *Limits on Fundamental Limits to Computation* Springer Science and Business Media LLC, 2014.
[http://dx.doi.org/10.1038/nature13570]

[39] S. Lephe, S. M. G. F. Raj, S. Janaki, S. S. J. Dhas, and L. A. Jose, "Challenges and Opportunities in Implementing Negative Differential Resistance Mode Reconfigurable Field Effect Transistors", *arXiv preprint arXiv:2312.08351,* 2023.

[40] Anikeeva, Polina, Bao, Zhenan, Benfenati, Fabio, Bezanilla, *et al.*, "Roadmap on semiconductor-cell biointerfaces.", *eScholarship, University of California*, 2018.

[41] B. Pelaz, C. Alexiou, R.A. Álvarez-Puebla, F. Alves, A.M. Andrews, S. Ashraf, L.P. Balogh, L. Ballerini, A. Bestetti, C. Brendel, S. Bosi, M. Carril, W.C.W. Chan, C. Chen, X. Chen, X. Chen, Z. Cheng, D. Cui, J. Du, C. Dullin, A. Escudero, N. Feliu, M. Gao, M. George, Y. Gogotsi, A. Grünweller, Z. Gu, N.J. Halas, N. Hampp, R.K. Hartmann, M.C. Hersam, P. Hunziker, J. Jian, X. Jiang, P. Jungebluth, P. Kadhiresan, K. Kataoka, A. Khademhosseini, J. Kopeček, N.A. Kotov, H.F. Krug, D.S. Lee, C.M. Lehr, K.W. Leong, X.J. Liang, M. Ling Lim, L.M. Liz-Marzán, X. Ma, P. Macchiarini, H. Meng, H. Möhwald, P. Mulvaney, A.E. Nel, S. Nie, P. Nordlander, T. Okano, J. Oliveira, T.H. Park, R.M. Penner, M. Prato, V. Puntes, V.M. Rotello, A. Samarakoon, R.E. Schaak, Y. Shen, S. Sjöqvist, A.G. Skirtach, M.G. Soliman, M.M. Stevens, H.W. Sung, B.Z. Tang, R. Tietze, B.N. Udugama, J.S. VanEpps, T. Weil, P.S. Weiss, I. Willner, Y. Wu, L. Yang, Z. Yue, Q. Zhang, Q. Zhang, X.E. Zhang, Y. Zhao, X. Zhou, and W.J. Parak, "Diverse Applications of Nanomedicine", *ACS Nano,* vol. 11, no. 3, pp. 2313-2381, 2017.
[http://dx.doi.org/10.1021/acsnano.6b06040] [PMID: 28290206]

[42] D.A.B. Miller, "Attojoule Optoelectronics for Low-Energy Information Processing and Communications", *J. Lightwave Technol.,* vol. 35, no. 3, pp. 346-396, 2017.
[http://dx.doi.org/10.1109/JLT.2017.2647779]

[43] I. Adamovich, S. Agarwal, E. Ahedo, L.L. Alves, S. Baalrud, N. Babaeva, A. Bogaerts, A. Bourdon, P.J. Bruggeman, C. Canal, E.H. Choi, S. Coulombe, Z. Donkó, D.B. Graves, S. Hamaguchi, D. Hegemann, M. Hori, H-H. Kim, G.M.W. Kroesen, M.J. Kushner, A. Laricchiuta, X. Li, T.E. Magin, S. Mededovic Thagard, V. Miller, A.B. Murphy, G.S. Oehrlein, N. Puac, R.M. Sankaran, S. Samukawa, M. Shiratani, M. Šimek, N. Tarasenko, K. Terashima, E. Thomas Jr, J. Trieschmann, S. Tsikata, M.M. Turner, I.J. van der Walt, M.C.M. van de Sanden, and T. von Woedtke, "The 2022 Plasma Roadmap: low temperature plasma science and technology", *J. Phys. D Appl. Phys.,* vol. 55, no. 37, p. 373001, 2022.
[http://dx.doi.org/10.1088/1361-6463/ac5e1c]

[44] V. Shanmugam, R.A. Mensah, K. Babu, S. Gawusu, A. Chanda, Y. Tu, R.E. Neisiany, M. Försth, G. Sas, and O. Das, "A Review of the Synthesis, Properties, and Applications of 2D Materials", *Part.*

Part. Syst. Charact., vol. 39, no. 6, p. 2200031, 2022. [http://dx.doi.org/10.1002/ppsc.202200031]

[45] Bathe, M., Chrisey, L.A., Herr, D.J., Lin, Q., Rasic, D., Woolley, A.T., Zadegan, R.M. and Zhirnov, V.V., 2019. Roadmap on biological pathways for electronic nanofabrication and materials. *Nano Futures*, 3(1), p.012001.

[46] David Esseni, Marco Pala, Pierpaolo Palestri, Cem Alper, Tommaso Rollo, Taylor M B, Taur Y, *et al.*, "A review of selected topics in physics based modeling for tunnel field-effect transistors", *'IOP Publishing'*, 2015.

[47] G. Sannino, I. De Falco, and G. De Pietro, "Towards a cyber physical system for personalised and automatic OSA treatment", *IET Cyber□Phys. Syst. Theory Appl.,* vol. 4, no. 2, pp. 156-163, 2019.

[48] M. Si, *"III-V and 2D Devices: From MOSFETs to Steep-Slope Transistors"*, Purdue University, 2018.

[49] Tunga, A., Zhao, Z., Shukla, A., Zhu, W. and Rakheja, S., 2023. Physics-based modeling and validation of 2-D Schottky barrier field-effect transistors. *IEEE Transactions on Electron Devices*, 70(4), pp.2034-2041.

[50] A. D. Es-Sakhi, *"Silicon on ferroelectic insulator field effect transistor (SOF-FET) a new device for the next generation ultra low power circuits"*, Master's thesis, University of Missouri-Kansas City, 2013.

[51] A. Liu, X. Zhang, Z. Liu, Y. Li, X. Peng, X. Li, Y. Qin, C. Hu, Y. Qiu, H. Jiang, Y. Wang, Y. Li, J. Tang, J. Liu, H. Guo, T. Deng, S. Peng, H. Tian, and T.L. Ren, "The Roadmap of 2D Materials and Devices Toward Chips", *Nano-Micro Lett.,* vol. 16, no. 1, p. 119, 2024. [http://dx.doi.org/10.1007/s40820-023-01273-5] [PMID: 38363512]

[52] D. Ohayon, V. Druet, and S. Inal, "A guide for the characterization of organic electrochemical transistors and channel materials", *Chem. Soc. Rev.,* vol. 52, no. 3, pp. 1001-1023, 2023. [http://dx.doi.org/10.1039/D2CS00920J] [PMID: 36637165]

[53] S.S. Debasish Mohanta, and P.K. Sahu Singh, "Vertical Tunnel Field Effect Transistors (VTFETs): A Potential Candidate for Low Power Applications", *3rd International Conference on Intelligent Technologies (CONIT),* 2023.

[54] M.B. Donnelly, M.M. Munia, J.G. Keizer, Y. Chung, A.M.S-E. Huq, E.N. Osika, Y-L. Hsueh, R. Rahman, and M.Y. Simmons, "Multi-Scale Modeling of Tunneling in Nanoscale Atomically Precise Si:P Tunnel Junctions", *Adv. Funct. Mater.,* vol. 33, no. 18, p. 2214011, 2023. [http://dx.doi.org/10.1002/adfm.202214011]

[55] F. Paolucci, G. De Simoni, P. Solinas, E. Strambini, N. Ligato, P. Virtanen, A. Braggio, and F. Giazotto, "Magnetotransport Experiments on Fully Metallic Superconducting Dayem-Bridge Field-Effect Transistors", *Phys. Rev. Appl.,* vol. 11, no. 2, p. 024061, 2019. [http://dx.doi.org/10.1103/PhysRevApplied.11.024061]

[56] A. Blais, A.L. Grimsmo, S.M. Girvin, and A. Wallraff, "Circuit quantum electrodynamics", *Rev. Mod. Phys.,* vol. 93, no. 2, p. 025005, 2021. [http://dx.doi.org/10.1103/RevModPhys.93.025005]

CHAPTER 6

A Comprehensive Review of the Evolution of Isfet-based Biosensors for Glucose Detection in Urine, Analysing their Technological Advancements, Challenges, and Future Directions

Avnish Bora[1,*], **Hemant Jain**[1] and **Basudha Dewan**[2]

[1] *ECE Department, Jodhpur Institute of Engineering & Technology, Mogra, Rajasthan, India*

[2] *Department of Electronics & Communication Engineering, Manipal University Jaipur, Jaipur, Rajasthan, India*

Abstract: Over the last 20 years, ISFET (Ion Sensitive Field Effect Transistor) biosensors have been recognized to be the most important technology in detecting glucose in non-invasive ways in urine, changing the way blood glucose concentration is measured. In this paper, an attempt has been made to present the development of ISFET-type glucose biosensors as attributed to advances in nanoscience, surface functionalization techniques, and sensor miniaturization, which improved the ability of the sensors to be sensitive, selective, and stable. More specifically, we focus on how graphene, metal oxides, and carbon nanotubes have been used to increase the surface area of ISFET sensors so that there is more contact with glucose molecules, thereby improving the sensitivity.

Another serious problem which is obtained with ISFET-based glucose sensors is related to the detection of glucose due to other constituents like urea and creatinine present in urine, which affects the accurate measurement of glucose. Examples of how these challenges have been overcome are discussed in this paper, including surface modification with glucose oxidase enzymes, and the use of molecularly imprinted polymers (MIPs) for enhanced specificity. The paper also discusses the issue of biofouling and the stability of the sensors caused by biological exposure for long periods, and how it can be solved with different coatings and more rigid materials.

The paper talks about making ISFET sensors smaller, which allows them to fit into wearable and implantable gadgets for tracking glucose levels all the time. We look at products on the market and examples from clinical studies to show how ISFET technology is being applied in real life. Lastly, we mention future research paths, such as using artificial intelligence (AI) and machine learning (ML) for better data processing and less intrusive glucose monitoring methods. This review stresses the

* **Corresponding author Avnish Bora:** ECE Department, Jodhpur Institute of Engineering & Technology, Mogra, Rajasthan, India; E-mail: boraav@gmail.com

Girdhar Gopal, Meena Panchore, Arun Kishor Johar & Tarun Varma (Eds.)

importance of ongoing improvements in ISFET glucose detection to address current issues, reduce costs, and increase acceptance in clinical settings.

Keywords: ISFET biosensors, Glucose detection in urine, Nanomaterials (Graphene, CNTs, ZnO), Surface functionalization, Wearable glucose sensors.

INTRODUCTION

Background on Glucose Detection

Diabetes mellitus has acquired major public health implications affecting millions across the globe. Glucose monitoring proves to be an essential factor in controlling diabetes, for it enables patients, as well as healthcare providers, to decide on time-bound insulin therapy and the adjustments required in the patients' lifestyles. Glucose levels are generally measured by testing blood samples. This method is considered invasive, notably because it involves finger pricks, even though it is the most accurate means of finding glucose levels in the blood. However, for patients who have to undergo continuous monitoring, it is painful and unpleasant [1, 2].

Urine glucose monitoring is an alternative approach to glucose monitoring via a less invasive method that has become increasingly popular over the years. Previously, urine glucose detection was used only as complimentary blood glucose testing. Under normal circumstances, glucose should not be present in the urine, but for diabetic patients, hyperglycemia causes the excretion of glucose in urine whenever the blood glucose level surpasses the renal threshold. While urine glucose monitoring does not give real-time feedback compared to blood glucose measurements, it is perceived as more diagnostic, making the tool more useful in general surveillance and monitoring among diabetic patients.

In some cases, conventional urine glucose detection approaches, i.e., dipsticks or chemical assays, are sensitive to other urine components and sometimes show cross-reactivity. However, within the domain of biosensors, specifically ISFET-based sensors, it is possible to circumvent those challenges and achieve more reliable real-time glucose monitoring.

ISFET Biosensors

The Ion-Sensitive Field-Effect Transistor (ISFET) was first introduced in the early 1970s as a pH sensor, capable of detecting ion concentrations in various solutions. ISFET operates similarly to MOSFET (Metal-Oxide-Semiconductor FET) devices, but instead of using a metal gate, it utilizes a reference electrode immersed in the solution. Changes in the ion concentration modulate the transistor's threshold voltage, providing a measurable signal [3]. Fig. (**1**) depicts the ISFET structure, in which the gate is replaced by an electrolyte and a reference electrode to enable ion sensing. From ISFET's inception, the technology has been modified and incorporated into several biochemical detection systems, such as glucose monitoring. ISFET sensors are ideal for real-time and portable monitoring systems due to their small size, low power consumption, and rapid response time. Glucose is detected using an ISFET-based sensor whereby the sensor was inserted into blood and acted on glucose oxidase, which helped break down glucose into hydrogen peroxide, which resulted in a local pH shift to the sites in a manner readable by the ISFET [4].

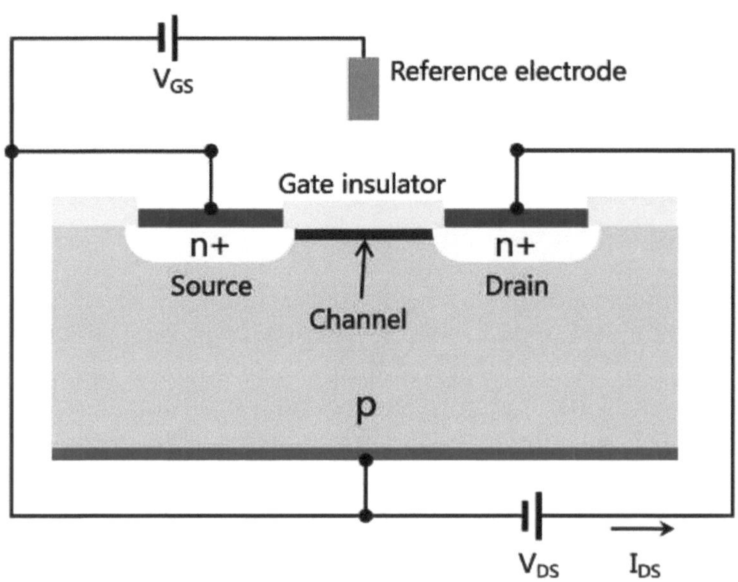

Fig. (1). Structure of ISFET [8].

ISFET sensors, being microfabricated, started to be modified for the detection of glucose in urine for a more complex environment. Glucose detection becomes more difficult in the urine due to other ions (*e.g.*, sodium, potassium, urea) that could inhibit the response of the sensor. However, there are still significant challenges that need to be overcome, great advances have been made in nanomaterials and surface functionalization strategies to improve the selectivity and sensitivity of the ISFET sensor for urine glucose detection [5 - 7].

Objectives

The objective of this chapter is to consolidate literature about the development of ISFET-based biosensors for glucose detection in urine, especially their advantages and limitations, and suggest some recommendations for future research. More specifically, the chapter intends to:

- Discuss the key technological breakthroughs over the past two decades, particularly in the areas of nanomaterials, surface functionalization, and sensor miniaturization.
- Discuss the problems of ISFET-based sensors, including but not limited to: cross-reactivity with urine components, bio-deterioration, and variations in the pH and temperature of the measuring medium.
- Provide the evidence of implantable and wearable glucose sensors based on ISFETs that have already been marketed and used in clinical studies.
- State perspectives for future studies, including AI-based data processing enhancement, and non-invasive glucose monitors that can broaden the boundaries of glucose detection.

HISTORICAL OVERVIEW OF ISFET-BASED GLUCOSE SENSORS

Basic Development (1990s to 2000s)

Although the construction of ISFET biosensors commenced in the 1970s, it was in the 1990s that specialists began to actively explore the possibility of using ISFET for glucose measurement in biological fluids. The basic designs of ISFET were more targeted towards pH variants, hence their usability in the medical and biochemical realm. However, manufacturing such a device for the detection of glucose raised numerous issues, especially regarding sensitivity and selectivity [1].

Enzymatic reactions, specifically glucose oxidase (GOx) reacting with glucose, were the basis of early ISFET-based glucose sensors. This enzyme catalyzes the oxidation of glucose and assists in producing gluconic acid and hydrogen peroxide. This reaction causes a shift in the pH balance, which can be monitored by the ISFET. Although this technique was a workable alternative to the direct assay of glucose, the first designs suffered from problems such as low sensitivity and slow time response [2]. Additionally, these earlier sensors were affected by cross-reactivity with sodium and potassium ions or other ions found in biological fluids that could hardly be avoided.

In the late 1990s and the early 2000s, scientists looked for different types of materials to enhance ISFET sensors. The integration of metal oxides, especially titanium dioxide (TiO2) and zinc oxide (ZnO), is described as one of the earlier progressions. These materials enlarged the surface area of the ISFET, enabling more contact of the glucose molecules with the sensor and hence enhancing sensitivity [3]. However, as promising as these developments were, the first generation of sensors up to this point still showed deficiencies in stability and selectivity, especially when operating in complex fluids such as urine.

Shift to Urine Glucose Sensing

The switch from blood glucose sensing to urine glucose sensing needed to be very heavily modified with respect to the ISFET sensor architecture. Urine is a more complex fluid than blood, with many types of ions, proteins, and organic molecules that could mess with the sensor. This resulted in early attempts to apply ISFET to urine glucose monitoring failing miserably. Cross-reactivity between substances is one of the biggest problems with finding glucose in urine. Urea and creatinine, for instance, two of the usual components of urine, muddle the glucose oxidase reaction, giving false values. This was solved when scientists started to look at ways to manipulate the ISFET surface to increase selectivity. Another was to cover the sensor with selectable membranes, so that they might reject interfering substances but still allow glucose to pass through [4]. The pH of urine was another issue, which could vary significantly from one specimen to the next. ISFET sensors are incredibly sensitive to pH changes, and this was a major hurdle to glucose detection. They managed to get around this by building two ISFETs, one for measuring pH and the other for measuring glucose.

This was to allow real-time pH compensation to ensure the sensor accuracy [5]. This translation of ISFET technology for urine glucose monitoring also leveraged nanotechnology. In the mid-2000s, scientists started to introduce nanomaterials into the sensor, like graphene and carbon nanotubes. These fabrics vastly increased the area of the sensor, which would become more sensitive to glucose molecules in the urine. Additionally, nanomaterials were effective in minimizing interference with other chemicals present in urine, and thus ISFET-based sensors were better suited for measuring glucose in this medium [6-8].

Comparative Timeline of ISFET Advancements: The following is the comparative study of ISFET Advancements

Table **1** presents a comparative timeline highlighting key developments and improvements in ISFET technology.

Table 1. Comparative timeline of ISFET advancements.

S. No	Period	Advancement	Materials	Target Fluids	Challenges Addressed
1	1990s	Initial development of ISFET for glucose detection in blood	Metal oxides (ZnO, TiO_2)	Blood	Poor sensitivity, slow response time
2	Early 2000s	Use of selective membranes to filter out interfering substances in urine	Polymeric membranes	Urine	Cross-reactivity with urea, creatinine
3	Mid-2000s	Introduction of nanomaterials to enhance sensitivity	Graphene, carbon nanotubes	Urine	Improved sensitivity, reduced interference
4	Late 2000s	Dual ISFET systems developed for pH compensation in urine glucose detection	Metal oxides with nanomaterial integration	Urine	pH variability, accuracy improvement
5	2010s – Present	Miniaturization and integration into wearable and implantable devices	Hybrid nanomaterials (graphene-metal oxide)	Urine, blood, sweat	Continuous monitoring, real-time data transmission

TECHNOLOGICAL ADVANCES OVER THE LAST TWO DECADES

Improvement of Sensitivity Using Nanomaterials

One of the most significant advancements in ISFET-based glucose sensors over the past two decades has been the integration of nanomaterials. These materials, including graphene, carbon nanotubes (CNTs), and metal oxides, have revolutionized the sensitivity of ISFET sensors by dramatically increasing the surface area available for glucose interaction. This increase in surface area enhances the number of active sites available for glucose molecules to bind, resulting in improved sensor performance. Fig. (**2**) presents a schematic view of various graphene-based nanoparticles and their structural forms.

Graphene, which is a two-dimensional material made up of one layer of carbon atoms, has been very influential in this regard. The extraordinary electrical conductivity and large surface area of graphene make it an excellent transducer material for ISFET-based glucose sensors. Graphene-based ISFET sensors have been found to detect glucose concentrations as low as 1 μM, which is very sensitive compared to traditional glucose sensors [3]. This characteristic of graphene makes it adaptable for wearable devices where monitoring is continuous.

Another nanomaterial that has significantly improved ISFET sensitivity is carbon nanotubes (CNTs). CNTs, known for their high mechanical strength and electrical conductivity, have been used to enhance the charge transfer capabilities of ISFET sensors. Studies have demonstrated that CNT-functionalized ISFET sensors exhibit faster response times and higher sensitivity to glucose in urine compared to sensors without CNTs [1]. The cylindrical structure of CNTs also provides a unique advantage in facilitating electron transport, which improves the overall performance of the sensor.

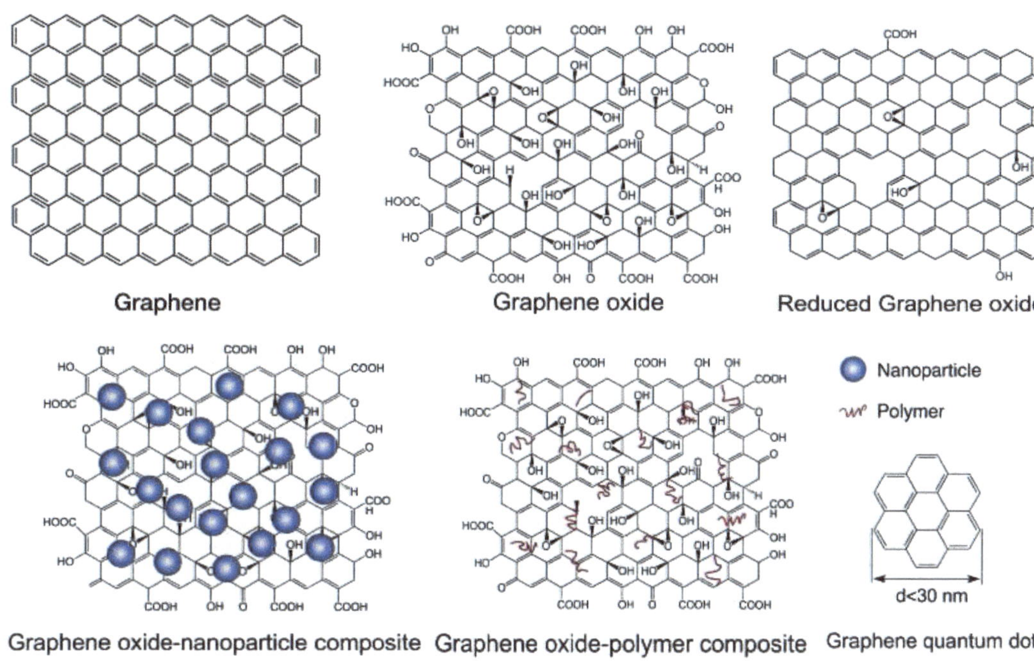

Fig. (2). Schematic illustration of different Graphene Based Nanoparticles [9].

Fig. (**3**) illustrates the structures of carbon nanotubes with varying wall numbers, including single-, double-, and multi-walled forms. In addition to graphene and CNTs, metal oxides like zinc oxide (ZnO) and titanium dioxide (TiO_2) have been incorporated into ISFET sensors to improve their sensitivity. Metal oxide nanostructures, such as nanowires and nanorods, provide a high surface-t--volume ratio, which enhances the interaction between glucose molecules and the sensor surface [4]. The incorporation of ZnO nanorods into ISFET sensors has been shown to increase sensitivity to glucose by up to 50%, making these sensors more effective for detecting low concentrations of glucose in complex biological fluids like urine.

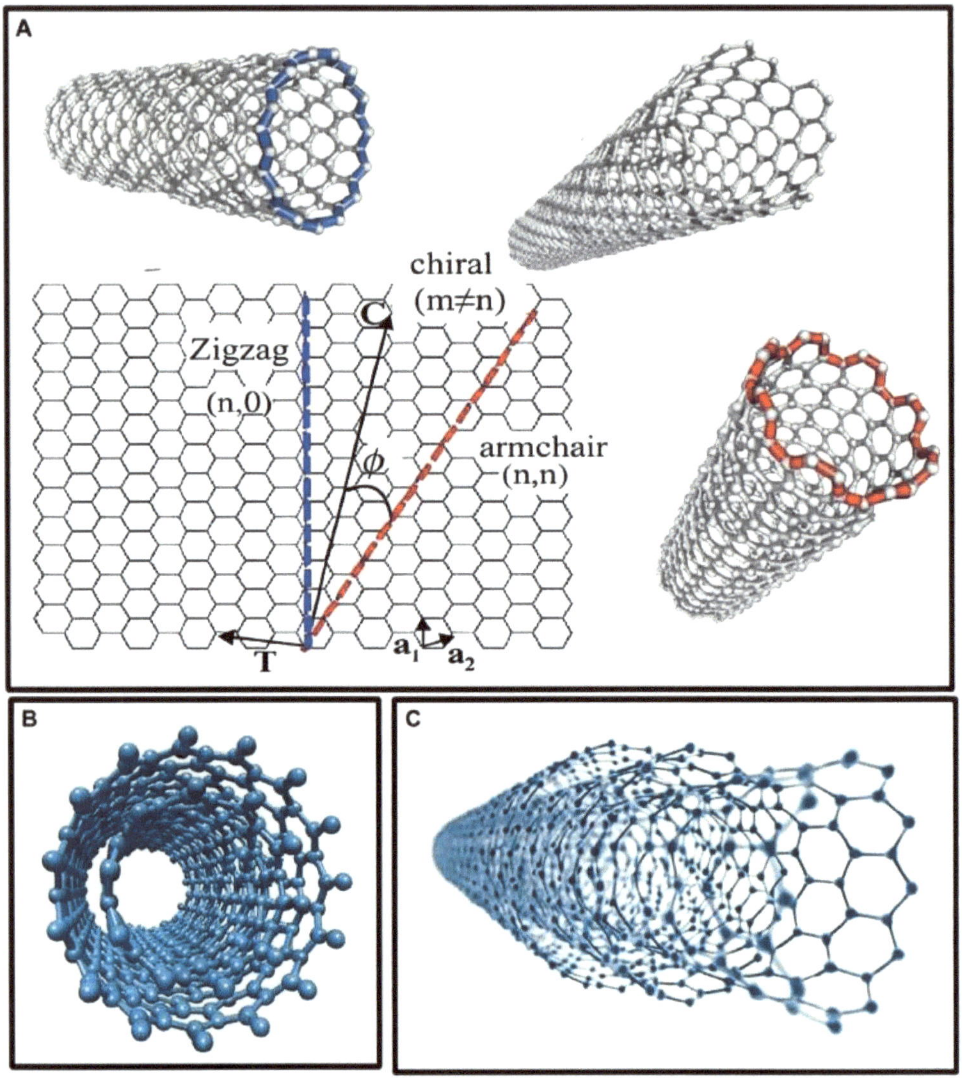

Fig. (3). Structure and models of carbon nanotubes as a function of their number of walls. (A) Single-walled carbon nanotubes (SWNTs); (B) Model of double-walled carbon nanotubes (DWNTs); (C) Structure of multi-walled carbon nanotubes (MWNTs) [10].

Surface Functionalization Advances

Another critical area of advancement in ISFET-based glucose sensors has been surface functionalization, which refers to the modification of the sensor's surface to improve its selectivity for glucose molecules. In early ISFET designs, cross-reactivity with other substances in urine was a significant issue, as ions like urea and creatinine interfered with the glucose detection process. To address this, researchers have developed a variety of surface functionalization techniques aimed at increasing the selectivity of ISFET sensors.

In some cases, immobilization of glucose oxidase on the surface of the ISFET is used as one common method of improving selectivity. Glucose oxidase catalyzes the oxidation of glucose to hydrogen peroxide and gluconic acid, and this results in a detectable change in pH, which can subsequently be measured by the ISFET for the correct detection of glucose [6]. Immobilization of GOx has been shown to dramatically enhance the selectivity of the ISFET sensor; because the enzyme is a glucose-specific reactant, there is little interference by other components of the urine.

In addition to functionalization with enzymes, more recently, MIPs have been contemplated for improving selectivity. MIPs are synthetic polymers prepared in such a manner that they incorporate specific binding sites for target molecules like glucose. These polymers are prepared by imprinting the target molecule into the polymer matrix during synthesis, which leaves cavities that are complementary in shape and size to the glucose molecule. Glucose molecules within a sample bind to these cavities, which enables highly selective detection [5]. Some of the advantages of MIPs over enzyme-based functionalization include increased stability and resistance to environmental changes, such as pH and temperature fluctuations.

Recent studies focused on the hybrid method of surface functionalization of enzymes by using GOx with graphene or CNTs. Such hybrids offer advantages based on high selectivity and even improved sensitivity through the tremendous surface areas made available to immobilize enzymes, as well as an enzyme-based specific detection for glucose. The hybrid systems enhance the performance of glucose sensors based on ISFET significantly. In many ways, especially in complicated fluids such as urine, the enhancement will be enormous [2].

Miniaturization and Integration with Wearables

Miniaturization of ISFET-based glucose sensors has been an important issue for successful integration into wearable devices. Improvements in microfabrication techniques during the last decade enabled compact, lightweight ISFET sensors to be embedded in wearable patches, smartwatches, or implantable devices for continuous glucose monitoring (CGM).

The biggest challenge of miniaturizing ISFET sensors is maintaining the sensitivity and selectivity with reduced size. However, the inclusion of nanomaterials such as graphene and CNT overcame this problem because the materials provide a large area in a small volume, thus providing a high-performance sensing compact form factor. For instance, graphene-based ISFET sensors were successfully integrated into flexible, skin-adhesive patches that continuously monitored glucose levels in the sweat or interstitial fluid [3]. Further, this wearable device has also shown numerous benefits in the form of non-invasive monitoring, real-time data transmission, and ease of patient comfort.

Besides the skin-adherent patches, implantable ISFET sensors have also been developed for long-term glucose monitoring. These sensors can be implanted subcutaneously and measure glucose levels in interstitial fluid continuously, and transmit them wirelessly to a mobile device or cloud-based platform for analysis. This would provide better and more reliable glucose level monitoring since it would be an implantable ISFET sensor, free from frequent blood tests and, therefore, offering real-time and continuous data on the trends.

For example, ISFET sensors have a faster response time compared to electrochemical sensors, low power consumption, and can measure glucose in multiple biological fluids such as urine, sweat, and blood. Miniaturization also allows the use of these ISFET sensors for wearable health monitoring systems, in addition to portability and comfort aspects, not to mention the continuous monitoring.

Comparative study of different nanomaterials based on given parameters.

Table **2** provides a side-by-side comparison of various nanomaterials used in ISFET-based sensors.

Table 2. Side-by-side comparison of nanomaterials in ISFET sensors.

Nanomaterial	Impact on Sensitivity	Impact on Selectivity	Stability Studies
Graphene	Increases surface area, improves glucose detection.	Limited by functionalization methods.	High stability, flexible for wearables [3].
Carbon Nanotubes	Enhances electron transport, improves sensitivity.	Can be combined with enzymes for selectivity.	High durability, good for wearable integration [2].
Zinc Oxide	Improves sensitivity through nanostructure integration.	Works well with enzyme immobilization.	Moderate stability, prone to biofouling [4].
Hybrid Nanomaterials	Combines the benefits of multiple nanomaterials.	Highly selective when combined with MIPs or enzymes.	High stability in complex fluids [5].

CHALLENGES IN ISFET-BASED GLUCOSE DETECTION

Cross-reactivity

Cross-reactivity with other substances in urine has been one of the significant challenges facing ISFET-based glucose sensors. Urine is a complex biological fluid that contains many ions and organic molecules, such as urea, creatinine, and proteins, which could interfere with glucose detection. In the early days of ISFET sensors, particularly those based on GOx, it was very challenging to maintain selectivity in urine because of these interfering substances. This led to a high false positive rate or false readings, which limits the use of ISFET sensors practically in detecting urine glucose [7].

Surface functionalization of ISFET sensors has been the current trend in reducing cross-reactivity. Several approaches to functionally modifying ISFET sensors have been reported with immobilization of glucose-specific enzymes and molecularly imprinted polymers. Functional modification by enzymes such as those of glucose oxidase improves the selectivity of this sensor for glucose by virtue of catalyzing an oxidation reaction that converts it into gluconic acid. This still remains at the mercy of interference posed by other ions affecting its activity [4].

One promising way of avoiding or drastically reducing cross-reactivity appears to be in molecular imprinted polymers. Designed with binding sites of exactly known structure matching that of the target glucose, highly specific detection of the guest target molecule should occur at specific glucose binding to its match. With the occurrence of glucose, a signal occurs; MIP-functionalized ISFET sensors can apparently offer greater selectivity over mere enzyme-based sensors for recognizing glucose with lesser degrees of cross-reactivity, as was obtained from various experiments done, which shows improvement. As of date, it remains difficult to produce MIPs that afford adequate affinity and selectivity towards glucose in a complex urinary matrix. Another approach to minimize cross-reactivity is the use of dual ISFET devices. In this method, one ISFET sensor detects glucose, and the other senses pH or any interfering ions present in urine. The outputs of the two sensors are compared such that changes in pH or concentration of interfering ions are cancelled by the system and minimize the interference due to cross-reactivity [7].

Stability and Biofouling

A major challenge in ISFET-based glucose sensors is stability. This sensor is prone to degradation, mainly in continuous glucose monitoring for long-term applications. Gradually, the performance of ISFET sensors begins to degrade for a series of reasons, such as degradation of materials and biofouling. Biofouling is produced through the presence of proteins, cells, or any kind of biological material on the sensor surface that could collect on the sensor, cutting off the interaction of glucose molecules with the sensor. This could lead to a lowering of sensitivity and resolution and decrease the functionality of the sensor for long-term operations [11].

Biofouling can be prevented with several protective coatings applied to the ISFET surface. The hydrophilic or antifouling materials are used to form a protective layer, preventing the adherence of biological materials, but allowing glucose molecules to reach the sensor. One of the more promising approaches involves the utilization of PEG coatings that were proven to prevent biofouling while maximizing ISFET sensors for long-time stability [11]. Researchers have used other nanoparticle coatings and biocompatible polymers in order to enhance the resistance of a sensor against fouling factors, but at the same time preserve the glucose sensitivity thereof [12].

Another parameter directly influencing stability is the degradation of sensor material, which usually occurs in hostile biological matrices, such as urine. One of the significant drawbacks associated with these sensors is enzyme degradation due to the time factor; enzymes like glucose oxidase lose activity gradually after exposure to moisture and temperature variation due to time dependency. This fact has led several researchers to try the concept of using enzyme-free ISFET sensors synthesized by synthetic materials, namely MIPs or nanozymes, for glucose-detecting ability. These are more stable and longer-lasting than enzymes, which makes them more suitable for long-term applications [13].

Environmental Factors (pH, Temperature)

The fluctuations of pH and temperature in urine pose additional challenges to ISFET-based glucose sensors. Urine pH varies considerably from person to person and even within samples from the same individual, and can be as low as 4.5 or as high as 8.0. Since ISFET sensors respond very sensitively to fluctuations in the ion concentration levels, pH drift will consequently cause signal drift, leading to low accuracy of glucose reading. To neutralize this, pH-compensated ISFET systems were developed: a second ISFET sensor that tracks pH shift and uses that to self-correct glucose reading can be used. It has proved to show more accuracy improvement in cases of the environment with fluctuations in pH values [14].

The ISFET sensor could also be affected by variations in temperature. Temperature regulation in non-laboratory settings is not as effective. The change in temperature modifies the mobility of ions of the urine, which later changes the electrical characteristics of the ISFET, and this must be accounted for in sensors intended for implantable or wearable devices that will provide accurate glucose sensing. Several solutions have been proffered for this issue, such as adding a temperature compensation circuit and designing a sensor using temperature-insensitive materials [15].

In addition to temperature compensation, novel nanomaterials with thermal stability are under investigation for ISFET sensors. Carbon nanotubes and graphene showed stable performance over a broad range of temperatures, at which such devices may find applications in wearable glucose monitoring with exposure to variable environmental conditions [16].

Cost and Scalability

While the development of ISFET technology has been impressive, one of the major barriers to widespread usage is the manufacturing cost of the sensors. ISFET sensors with integrated nanomaterials, including graphene and CNTs, need microfabrication techniques that involve significant expense and time. Second, enzymes or other selective coatings add complexity and a high cost of production so that ISFET sensors are not as inexpensive as, say, electrochemical sensors, for example [17].

Cost-reducing efforts have focused on the development of printing technologies and flexible substrates produced in large quantities. For example, the low cost of fabrication through printed electronics is investigated in the recent fabrication of ISFET sensors, opening the gates to the realization of very thin and flexible, disposable sensor types for use as a component of wearable systems. Although such approaches seem promising, much needs to be accomplished in relation to sensitivity and reliability comparisons to those achievable with classical silicon-based sensors [17].

Scarcity in scalability: In terms of scalability, the miniaturization of ISFET sensors for wearables creates problems and opportunities. The smaller sensors are cheaper to fabricate in larger quantities, and they allow for the closer integration of devices such as smartwatches or patches. Contrarily, miniaturization involves complexity in the fabrication process as high sensitivity and selectivity in a smaller device necessitate the proper alignment of nanomaterials and functionalization layers. Mass production of high-performance ISFET sensors has thus remained a challenge to this day [16].

Comparison of different ion-sensing parameters based on ISFET sensors.

Table **3** summarizes the major challenges in ISFET-based glucose detection along with corresponding solutions.

Table 3. Challenges and Solutions in ISFET-based Glucose Detection.

Challenge	Description	Proposed Solutions	References
Cross-reactivity	Interference from urea, creatinine, and other substances in urine.	Enzyme immobilization, MIPs, dual ISFET systems.	Moussy *et al.*, IEEE (1997) [6]
Stability and Biofouling	Degradation due to long-term exposure to biological fluids, protein accumulation.	Protective coatings (PEG, biocompatible polymers), enzyme-free sensors.	Kim *et al.*, ACS Nano (2017) [5]
pH Sensitivity	Signal drift caused by pH fluctuations in urine.	pH-compensated ISFET systems, dual sensor approaches.	Liu *et al.*, Sensors and Actuators (2020) [2]
Temperature Sensitivity	Variations in temperature affect ion mobility and sensor performance.	Temperature compensation circuits, thermally stable nanomaterials	Xie *et al.*, Nano Letters (2016) [3]
Cost and Scalability	High manufacturing costs and challenges in scaling up production.	Printing techniques, flexible substrates, and mass production innovations.	Zhu *et al.*, Elsevier (2021) [4]

CURRENT SOLUTIONS AND INNOVATIONS

Nanomaterial Integration for Increased Sensitivity and Stability

It has recently enhanced the inclusion of nanomaterials in ISFET-based glucose sensors to improve sensitivity, stability, and real-time capability by solving some critical concerns. Increased contact area of glucose molecules with increased electron transport within the sensor, henceforth, increased the overall surface-t--volume ratio. Graphene, carbon nanotubes (CNTs), and metal oxides improve the sensitivity of ISFET-based sensors.

Graphene, a one-atom-thick carbon material, has been one of the most widely studied nanomaterials in ISFET applications. Its excellent electrical conductivity and large surface area make it ideal for improving glucose sensor sensitivity. Studies have shown that graphene-functionalized ISFETs can detect glucose at concentrations as low as 1 μM, offering greater precision in glucose monitoring compared to traditional sensors [3]. Graphene's ability to maintain structural integrity over time also enhances the stability of ISFET sensors, making them suitable for long-term use in wearable devices.

Another significant nanomaterial incorporated into ISFET is carbon nanotubes, which enhance the electron transport property of the sensor. Efficient electron transfer is required to get a rapid response for the sensor. CNT-functionalized ISFETs with glucose oxidase or other surface functionalization have shown higher sensitivity to glucose and more accurate readings as compared to bare ISFETs [18]. CNTs have further shown excellent biocompatibility, making them suitable for implantable glucose sensors.

Other metal oxide nanostructures, like zinc oxide (ZnO) and titanium dioxide (TiO_2), have also been explored to be used for stabilizing sensors under extreme conditions, such as urine. The nanomaterials show a very high surface-to-volume ratio and excellent catalytic characteristics, which enable better interaction between the glucose molecules. For instance, ZnO nanorods were utilized to increase glucose sensitivity with robustness in the sensor towards various pH and temperature conditions [16].

Functionalization for Selectivity Enhancement

The improvement in selectivity of ISFET-based glucose sensors has been a major area of work over the past few years. Although functionalization methods involving GOx were reasonable, they had problems like natural degradation of enzymes and interference with respect to the constituents of urine, such as urea and creatinine. Recent developments in surface functionalization techniques have addressed these problems through enzyme-free detection and hybrid approaches for functionalization.

One such approach is in the form of molecularly imprinted polymers, often abbreviated as MIPs. MIPs are artificial materials with unique binding sites that resemble and fit the target molecule itself (in this case, glucose). During synthesis, polymer matrices are "imprinted" with glucose, developing cavities that have a high affinity for glucose molecules. These polymers give the advantage of high selectivity without the drawbacks of enzymes, which are temperature and pH-dependent [19]. MIP-functionalized ISFET sensors have also been proven to be much more stable and sensitive than enzyme-based sensors, especially in the recognition of complex biological fluids such as urine.

Fig. (**4**) shows the basic process used to create molecularly imprinted polymers through template-guided polymerization. Recently, a published study developed functionalization techniques with hybrids containing enzymes and nanomaterials, including graphene or CNTs. Therefore, these hybrid systems exploit the specificity of enzymes and also take advantage of improved electron transport as well as surface area as offered by nanomaterials. For example, a hybrid ISFET sensor composed of graphene combined with glucose oxidase shows much enhanced sensitivity and selectivity in urine samples, lessening interference by other components of the urine [5].

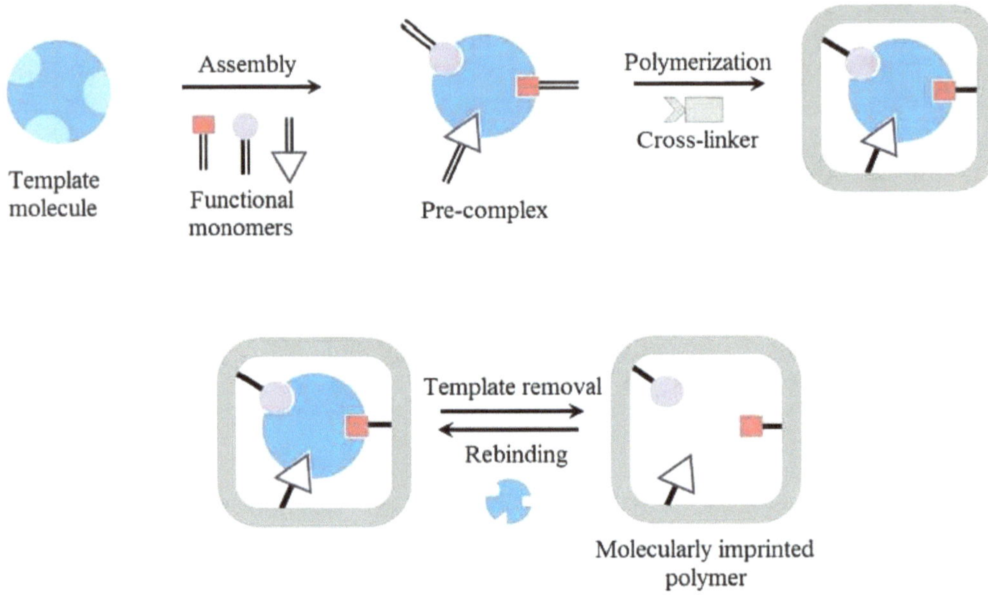

Fig. (4). Scheme of the principle of molecularly imprinted polymer preparation [20].

Another potential route to increased selectivity is through nanozymes, artificial enzyme mimics. Nanozymes share the catalytic benefit of the natural enzyme but are far more stable in the presence of environmental degradation. Functionalization of these materials onto ISFET sensors permits selective glucose detection in urine without the need for immobilized enzymes and thus may provide a more durable alternative for extended monitoring [6].

Hybrid Systems

A new hybrid ISFET-based system has lately been emerged based on electrochemical and optical sensing technologies to couple towards novel glucose detection. These kinds of hybrid systems are useful because they can combine more than one sensing modality so that the individual weaknesses in the respective sensing technology can be compensated, providing far more accurate and reliable measurements.

For instance, when ISFET sensors are integrated with optical sensors, glucose can be detected both by ion-sensitive mechanisms and by optical mechanisms. Optical sensors include fluorescence-based detectors that give an independent measurement of glucose concentration, irrespective of the ISFET's ion-sensitive measurement. The chances of false positives are therefore minimized, as well as general inaccuracy in detecting glucose in more complex fluids such as urine [21].

Another example of hybrid systems is electrochemical sensors, which measure glucose concentration through oxidation-reduction reactions catalyzed by enzymes or nanozymes. Combining ISFET-based systems with electrochemical sensors increases the scrutiny of glucose measurements further, thus improving sensor reliability and reducing the likelihood of signal drift from pH or temperature changes [8]. Hybrid systems are really helpful in wearable glucose monitors, which require continuous accuracy in real time.

Comparison study of various functionalization Techniques and their Impact on the Performance of ISFET based Sensors.

Table **4** outlines different functionalization techniques and explains how each method influences the performance of ISFET-based sensors.

Table 4. Functionalization techniques and their impact on performance of ISFET-based sensors.

Functionalization Technique	Impact on Sensitivity	Impact on Selectivity	Stability	Cost	References
Glucose Oxidase (GOx)	High sensitivity to glucose through enzymatic action.	Moderate; interference from urea, creatinine	Limited by enzyme degradation over time	Moderate; requires enzyme immobilization steps.	Moussy *et al.*, IEEE (1997) [6]
Molecularly Imprinted Polymers (MIPs)	Improved sensitivity due to specific binding sites.	High; designed for selective glucose detection.	High stability, resistant to environmental changes.	Moderate; cost varies based on polymer synthesis.	Kim *et al.*, ACS Nano (2017) [5]
Hybrid (GOx + Nanomaterials)	Significantly enhanced sensitivity with nanomaterials.	High; nanomaterials help reduce interference.	Improved stability, depending on material choice.	High; requires both enzyme and nanomaterial integration.	Xie *et al.*, Nano Letters (2016) [3]
Nanozymes	Moderate to high sensitivity, depending on the nanozyme.	High; selective catalytic activity mimicking enzymes.	Very high stability, no enzyme degradation.	High; cost-effective in long-term use.	Liu *et al.*, Sensors and Actuators (2020) [2]

CASE STUDIES: SUCCESS STORIES AND COMMERCIAL APPLICATIONS

Real World Applications

Real-world applications of ISFET-based glucose sensors have really gained significant advancement, especially in the field of continuous glucose monitoring among diabetic patients. A variety of clinical tests and prototype testing programs reveal that such sensors can efficiently detect glucose in biological fluids such as blood, urine, or sweat. Real-time output from ISFET sensors and the possibility of their use in wearable devices made them gain importance in the clinical application area.

A key success story is the development of implantable ISFET glucose sensors targeted at long-term glucose monitoring in diabetic patients. These sensors are placed subcutaneously where they can remain continuously monitoring glucose in interstitial fluid. The dual-sensing mechanisms applied in these sensors combine the simple ionic detection with enzymatic reactions for accurate glucose measurement. For implanted ISFET sensors, remarkable results of clinical trials can be considered with very slight drift signals over time, with significant glucose sensitivity [1]. Participants in the aforementioned clinical trial mentioned better glucose regulation along with a reduced amount of repeated finger-pricking that points to the replacement aspect by implanted devices instead of currently adopted systems such as CGMs. Fig. (**5**) displays different glucose monitoring devices used for continuous tracking of blood glucose levels.

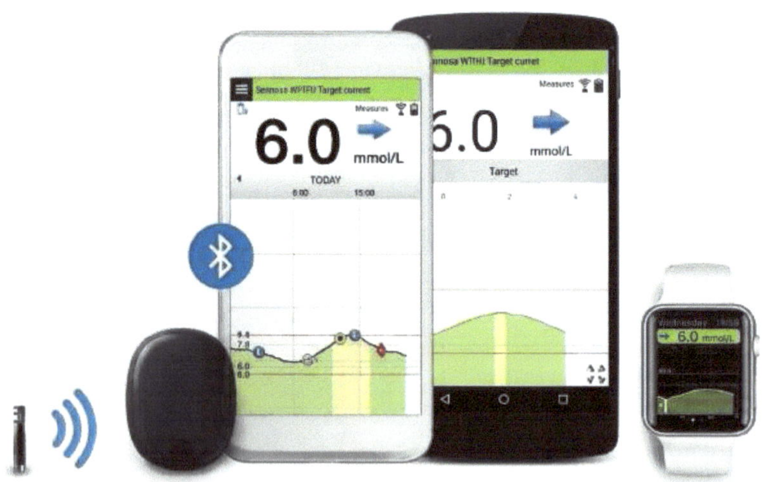

Fig. (5). Glucose monitoring devices. Courtesy: https://www.eversensecgm.com.

Besides implantable devices, glucose sensors based on ISFET were successfully integrated into wearable patches, which measure the glucose level in sweat. Non-invasive blood glucose testing may be provided by sweat glucose monitoring. High accuracy with real-time data is feasible with ISFET sensors. As a result, the flexible skin-adhesive patch-based approach incorporating graphene-functionalized ISFETs was developed. These have been tested both in the laboratory and under real-world conditions to prove their reliability, showing a stable glucose reading even in the presence of changing temperatures and humidity [2].

Commercial Devices

Several commercial products using ISFET technology have appeared on the market. These include non-invasive and continuous glucose monitoring devices for diabetic patients. The compactness, sensitivity, and real-time capabilities of ISFET sensors allow convenient, wearable solutions for daily glucose management.

The most successful commercial product is the Eversense CGM system, using a long-term, implantable ISFET sensor to monitor glucose levels in interstitial fluid continuously. It is implanted under the skin, which can last for as long as 365 days before being replaced. The system has been one of the longest-lasting CGM systems that have been marketed, according to the report by the FDA. The Eversense system has gained popularity among most patients because of its reliability and real-time data transfer into a mobile application with less frequent calibration than any traditional CGM device [22].

A different commercial device is a K'Watch Glucose. It is a wearable smartwatch with ISFET-based glucose sensors integrated into it. The K'Watch also features skin patch technology where glucose levels are measured through the sensing of sweat, communicating this glucose data in real-time with the user's smartwatch display. The K'Watch Glucose is designed to be used continuously, thereby providing a convenient, non-invasive solution to diabetic patients who would like to avoid using finger pricks. Fig. (**6**) shows the K'Watch Glucose wearable device designed for continuous glucose monitoring [23].

Another commercially relevant device is Dexcom G6, which represents the first real-time, integrated Continuous Glucose Monitoring System (iCGM) authorized to work interoperably with a range of connected insulin pens and hybrid closed-loop system partners. Fig. (**7**) depicts the Dexcom G6 continuous glucose monitoring system. It has a 10-day sensor and uses no finger sticks or scanning, making it more convenient for more freedom and flexibility in the management of diabetes for patients aged 2 years and older.

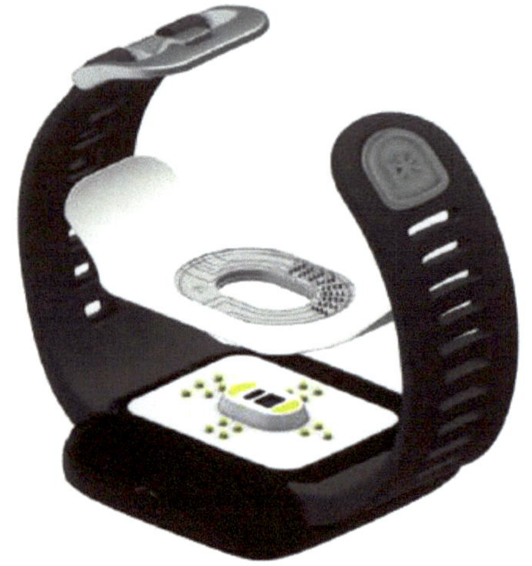

Fig. (6). K'Watch Glucose Courtesy: https://www.pkvitality.com/ktrack-glucose.

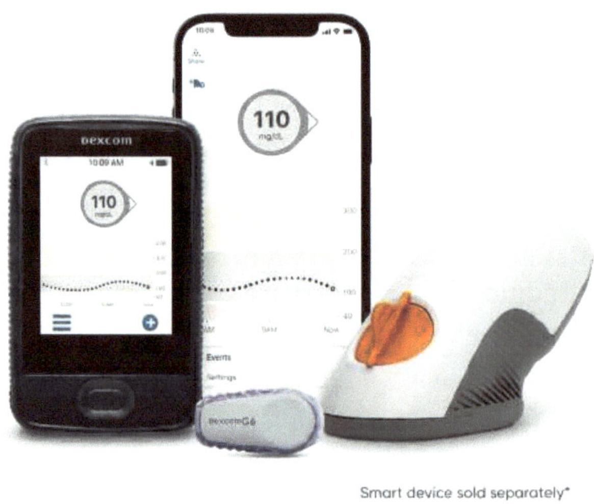

Fig. (7). Dexcom G6.

Comparison of different glucose-sensing wearable devices

Table **5** lists commercially available glucose sensors that operate using ISFET technology.

Table 5. Commercial ISFET-based glucose sensors.

Device	Type of Sensor	Detection Mechanism	Wearability	Lifetime	Sensitivity	Cost
Eversense® CGM	Implantable ISFET sensor	Glucose oxidase enzyme, pH-based	Subcutaneous implant	Up to 365 days	High (interstitial fluid)	High
K'Watch Glucose	Wearable smartwatch (ISFET)	Ion-sensitive detection (sweat)	Smartwatch with skin patch	Continuous	High (sweat)	Moderate
Dexcom G6 (ISFET-based prototype)	Wearable patch (prototype)	Enzyme-free glucose detection	Wearable patch	10-14 days	High	High

Real-World Case Study: Eversense CGM System

The Eversense CGM system is one of the most prominent examples of a commercial device that uses ISFET-based technology for glucose monitoring. Compared with other CGM systems where the sensor needs to be replaced weekly or bi-weekly, the Eversense system uses an implantable sensor lasting as long as 365 days, hence takes the major convening advantage over others and is much more cost-effective too. The sensor is put right beneath the skin by the healthcare provider and measures the glucose continuously in interstitial fluid. Data are transmitted wirelessly to the wearable transmitter from which live glucose readings are received on one's mobile app.

There are various clinical studies that prove the accuracy and reliability of the Eversense CGM system. According to user reports, overall glucose control has improved with the use of the Eversense system compared to traditional finger-prick testing. Especially in patients who need real-time glucose monitoring, the system's ability to track glucose trends and alert the user to potential hypoglycemic or hyperglycemic episodes has proven useful [22].

It brought ISFET-based glucose detection technology. ISFET sensors' usage considerably provides highly sensitive and stable measurements of glucose even in poor conditions, such as fluctuation of body temperature or hydration states. Moreover, the capacity of the system to keep working for up to 365 days without calibration stands out as one of the easiest CGM systems available nowadays.

ADVANCED MATERIALS

New Materials and Fabrication Techniques

The new material properties and fabrication methods, capable of increasing their sensitivity and selectivity along with enhanced scalability, would be the life or death for ISFET-based glucose sensors [24]. Amidst other candidates, functionalized nanoparticles, organic semiconductors, and above all, 2D materials, including MXenes and TMDs, have received much attention in the field.

MXenes are a family of 2D materials composed of transition metal carbides, nitrides, or carbonitrides that have been used as excellent conductors and highly available surfaces for biosensing purposes. Incorporation of MXenes into ISFET sensors allowed researchers to increase the process of electron transfer across the sensor, making the sensor response time shorter and the sensitivity to glucose molecules much greater [25]. For this purpose, MXenes can also be functionalized with specific binding sites for glucose, allowing selective detection in complex fluids like urine.

Another very interesting direction of application is in the form of organic semiconductors, particularly if their use could assist in enabling stretchable and flexible sensors embedded in wearable devices. Compared to conventional silicon-based materials, organic semiconductors offer many advantages, such as a much lower fabrication cost, excellent mechanical flexibility, and biocompatibility. Such an ability has led to the exploration of their utilization in making disposable or one-time ISFET sensors that can easily be manufactured in bulk quantities with much lower costs [26].

Recent work has demonstrated that printed electronics can be a promising fabrication technique for ISFET-based glucose sensors. Here, nanomaterials and sensor components are deposited on flexible substrates using low-cost and scalable techniques like inkjet printing or screen printing. This would reduce the production cost of ISFET sensors dramatically, allowing them to be used in flexible, wearable devices, such as smart patches or bandages [27].

Artificial Intelligence and Machine Learning

As ISFET-based glucose sensors gain widespread usage in wearable and implantable devices, more sophisticated data analysis techniques will be required to interpret continuous data streams from such sensors. Artificial intelligence and machine learning algorithms will become important tools for augmenting the accuracy and predictability of glucose monitoring systems.

Integrating AI/ML algorithms with ISFET-based devices enables real-time analysis of glucose level trends to identify patterns or anomalies that may indicate potential health risks, such as hypoglycemic or hyperglycemic episodes. These algorithms are designed to process large amounts of data collected from the sensors, which are processed for personalized recommendations to the patient about their glucose levels and lifestyle factors [28].

In addition to improving the accuracy of glucose monitoring, AI can be used to dynamically optimize sensor performance based on calibration parameters adjusted according to pH or temperature conditions, thus potentially overcoming sensor drift and cross-reactivity issues and ensuring the reliability of sensor readings [29].

The glucose variability can also be predicted using AI-based predictive models so that patients take precautions when their glucose levels reach dangerous thresholds. For example, based on historical data on glucose levels along with aspects such as exercise, diet, and medications taken, AI algorithms may predict the chances of having a glucose spiking or fall to dangerous levels, so intervention could be taken in advance [30].

Non-Invasive Glucose Monitoring

Perhaps one of the most exciting areas for future research with ISFET-based glucose sensors is non-invasive glucose monitoring. The present ISFET sensors mainly work by detecting glucose in urine, blood, or interstitial fluid, but there is an emerging interest in sweat and saliva fluids as possible alternatives for a more convenient and user-friendly method of glucose detection.

An area of interest would be sweat-based glucose monitoring because sweat is easy to access and can be continuously monitored using wearable devices. ISFET sensors designed to detect glucose in sweat have already been proven promising in clinical trials where wearable patches offer real-time glucose data without the need for an invasive sampling procedure. The devices are especially attractive for patients preferring noninvasive monitoring solutions, like patients who fear needles or young children diagnosed with diabetes [31].

Another ongoing area is the use of tear fluid for glucose monitoring. The glucose level in the tear fluid is very low compared to that of blood. As the sensitivity of the sensor is much higher, it is now very much feasible to sense tiny concentration levels present in the tear fluid with the help of ISFET sensors. Recently, wearable devices such as smart contact lenses have been developed for the continuous monitoring of glucose levels in the tear fluid. Definitely, it would be a very non-invasive, non-obtrusive glucose monitoring device [30, 31].

Sustainability and Eco-Friendliness

Environmental concerns are already emerging with the disposable sensor, because of the huge demand for wearable glucose monitoring devices and sustainable materials that should be used for sensor design. For this reason, there are numerous research works being conducted on biodegradable materials as well as green manufacturing methods to develop ISFET sensors with low waste generation while minimizing environmental impact from glucose monitoring systems.

Another promising development is the design of biodegradable polymers for making sensors. Their use would reduce the electronic waste being generated by the disposable-type sensors. Sensor devices can hence be integrated with environmentally benign materials such as flexible nanomaterials like graphene or CNTs, giving rise to high-performance, environmentally friendly sensors [32].

Besides, research on green chemistry techniques to reduce the environmental impact of making sensors is also conducted. Techniques in this field normally minimize the use of harmful chemicals and energy-intensive fabrication processes in ISFET sensor fabrication. Examples of research made on water-based fabrication techniques and renewable materials such as cellulose or silk have been found for creating more environmentally friendly sensors.

Comparison between emerging materials and techniques based on ISFET

Table **6** highlights new materials and innovative techniques being explored for advanced ISFET-based biosensing.

Table 6. Emerging materials and techniques for ISFET-based biosensing.

Material/Technique	Advantages	Challenges	Applications
MXenes (2D Materials)	High conductivity, large surface area, and ease of functionalization.	Cost of production, stability in biological fluids.	Wearable glucose sensors, real-time monitoring
Organic Semiconductors	Flexible, low-cost fabrication, and biocompatibility.	Lower sensitivity compared to inorganic materials.	Disposable sensors, wearable patches
Printed Electronics	Low-cost, scalable, and flexible substrates.	Requires optimization for long-term stability.	Smart patches, flexible wearable devices.
AI/Machine Learning Integration	Improved accuracy and predictive capabilities.	Data privacy concerns and high computational requirements.	Real-time glucose monitoring, predictive health systems.

(Table 6) cont.....

Material/Technique	Advantages	Challenges	Applications
Biodegradable Materials	Eco-friendly, reduces electronic waste.	Lower durability, shorter lifespan.	Disposable glucose sensors

CONCLUSION

ISFET-based glucose sensors have developed greatly in the last two decades. The need for non-invasive, real-time glucose monitoring has improved greatly. Starting from simple pH-sensitive devices, they have evolved into sophisticated systems that integrate nanomaterials, surface functionalization techniques, and wearable technologies. Sensitivity, selectivity, and stability were achieved in these technological advancements, and hence, ISFET-based sensors have been considered as an alternative to traditional glucose monitoring methods.

Probably the most important developments in ISFET technology are based on the inclusion of nanomaterials, particularly graphene, carbon nanotubes, and metal oxides. Such materials significantly increased the surface area of sensors capable of detecting an extremely minuscule concentration of glucose in complex biological fluids, such as urine and sweat. Besides that, to get over the cross-reactivity and interference problems, the techniques used include surface functionalization techniques, enzyme immobilization, and molecularly imprinted polymers (MIPs) that get over cross-reactivity and interference and make sensors very specific and reliable. Despite these innovations, some problems remain, such as the problem of cross-reactive substances in urine, biofouling, influence of environmental factors such as pH and temperature, influencing stability in long periods of operation. Another significant limitation toward wide clinical adoption is that the cost of scale manufacturing increases because of the nanomaterial inclusion and use of complex functionalization techniques, although recent developments in materials that could be biodegradable or printable electronics tend to lower cost in the long term and make the ISFET sensors greener. This revolution has placed an integration of AI and ML in the monitoring of blood glucose in systems based on ISFET, revolutionary in the arena of healthcare. The method will advance the precision level of real-time analysis with respect to trends, including predictions toward health risks, to enable appropriate patient self-regulation control of their levels of blood glucose. Alternative fluids such as sweat and tear fluid may add new scenarios in diabetes management through the push towards non-invasive glucose monitoring. In conclusion, ISFET-based glucose sensors are highly advanced in biosensing technology, with many achievements and promising future directions. Continued research and innovation will be vital in overcoming the current challenges, ensuring that these sensors will become a reliable, cost-effective, and sustainable means of glucose monitoring in clinical and everyday settings.

AI DISCLOSURE

During the preparation of the work, the author(s) used AI tools and Grammarly tool to enhance the writing style and readability of the content. The authors clarify that AI was not involved in generating the substantive content and that they take full responsibility for the publication.

REFERENCES

[1] N. Moser, "Advances in ISFET-based glucose sensors: Sensitivity and integration challenges", *IEEE Trans. Biomed. Eng.,* vol. 64, no. 12, pp. 2203-2214, 2019.

[2] Z. Liu, "Non-invasive glucose monitoring using ISFET-based sensors: A comprehensive review", *Sens. Actuators B Chem.,* vol. 302, p. 127095, 2020.

[3] X. Xie, "Graphene-functionalized ISFET sensors for selective glucose detection", *Nano Lett.,* vol. 16, no. 3, pp. 1855-1860, 2016.

[4] X. Zhu, "Challenges in developing ISFET-based sensors for biomedical applications", *Elsevier Materials Science,* vol. 45, no. 4, pp. 550-563, 2021.

[5] D.H. Kim, "Enhancing ISFET-based biosensors with metal oxide nanostructures", *ACS Nano,* vol. 11, no. 9, pp. 9040-9047, 2017.

[6] F. Moussy, "Miniature glucose sensor for long-term transcutaneous implantation", *IEEE Trans. Biomed. Eng.,* vol. 44, no. 10, pp. 463-472, 1997.

[7] W. Lee, "Challenges in ISFET-based glucose sensors for non-invasive urine monitoring", *J. Sens.,* vol. 19, no. 8, pp. 1-15, 2019.
[http://dx.doi.org/10.1155/2019/2985014]

[9] J. Li, H. Zeng, Z. Zeng, Y. Zeng, and T. Xie, "Promising Graphene-Based Nanomaterials and Their Biomedical Applications and Potential Risks: A Comprehensive Review", *ACS Biomater. Sci. Eng.,* vol. 7, no. 12, pp. 5363-5396, 2021.
[http://dx.doi.org/10.1021/acsbiomaterials.1c00875] [PMID: 34747591]

[10] B. Dewan, S. Chaudhary, D. Singh, and M. yadav, "Investigating temperature reliability of RF performance metrics and linearity for double gate doping less TFET", *Eng. Res. Express,* vol. 6, no. 1, p. 015308, 2024.
[http://dx.doi.org/10.1088/2631-8695/ad23ca]

[11] Z. Noura, I. Shah, S. Aziz, A. Ahmed, D.W. Jung, L. Brahim, and R. ElMostafa, "Wearable Healthcare Monitoring Based on a Microfluidic Electrochemical Integrated Device for Sensing Glucose in Natural Sweat", *Sensors (Basel),* vol. 22, no. 22, p. 8971, 2022.
[http://dx.doi.org/10.3390/s22228971] [PMID: 36433566]

[12] M. Mohammad Mansour, Darweesh Saeed, and Soltan Ahmed, "Wearable devices for glucose monitoring: A review of state-of-the-art technologies and emerging trends", *Alexandria Engineering Journal,* vol. 89, pp. 224-243, 2024.

[13] S. Sreejith, L.M.I. Leo Joseph, and V.T. Sreedhar Kollem, "Biodegradable sensors: A comprehensive review", *Measurement,* vol. 219, pp. 0263-2241, 2023.

[14] C. Hwang, W.J. Lee, S.D. Kim, S. Park, and J.H. Kim, "Recent Advances in Biosensor Technologies for Point-of-Care Urinalysis", *Biosensors (Basel),* vol. 12, no. 11, p. 1020, 2022.
[http://dx.doi.org/10.3390/bios12111020] [PMID: 36421138]

[15] S.E. Naimi, B. Hajji, I. Humenyuk, J. Launay, and P. Temple-Boyer, "Temperature influence on pH-ISFET sensor operating in weak and moderate inversion regime: Model and circuitry,", *Sensors and Actuators B: Chemical,* vol. 202, pp. 1019-1027, 2014.

[16] D. Lee, and T. Cui, "A role of silica nanoparticles in layer-by-layer self-assembled carbon nanotube and In2O3 nanoparticle thin-film pH sensors: Tunable sensitivity and linearity", *Sensors and Actuators A: Physical,* vol. 188, pp. 203-211, 2012.

[17] S. Cao, P. Sun, G. Xiao, Q. Tang, X. Sun, H. Zhao, S. Zhao, H. Lu, and Z. Yue, "ISFET□based sensors for (bio) chemical applications: A review", *Electrochem. Sci. Adv.,* vol. 3, no. 4, p. e2100207, 2023.

[18] M. Sireesha, V. Jagadeesh Babu, A.S. Kranthi Kiran, and S. Ramakrishna, "A review on carbon nanotubes in biosensor devices and their applications in medicine", *Nanocomposites,* vol. 4, no. 2, pp. 36-57, 2018.
[http://dx.doi.org/10.1080/20550324.2018.1478765]

[19] D. Robert, Crapnell, Nina C. Dempsey-Hibbert, Marloes Peeters, Ascanio Tridente, Craig E. Banks, "Molecularly imprinted polymer based electrochemical biosensors: Overcoming the challenges of detecting vital biomarkers and speeding up diagnosis", *Talanta Open,* vol. 2, pp. 2666-8319, 2020.

[20] Y. Saylan, S. Akgönüllü, H. Yavuz, S. Ünal, and A. Denizli, "Molecularly Imprinted Polymer Based Sensors for Medical Applications", *Sensors (Basel),* vol. 19, no. 6, p. 1279, 2019.
[http://dx.doi.org/10.3390/s19061279] [PMID: 30871280]

[21] Keun-Yong Park, Sang-Bok Choi, Minho Lee, Byung-Ki Sohn, Sie-Young Choi, "ISFET glucose sensor system with fast recovery characteristics by employing electrolysis", *Sensors and Actuators B: Chemical,* vol. 83, no. 1, pp. 90-97, 2002.

[22] Eversense, "Eversense Continuous Glucose Monitoring System", *Product information page, accessed 2020,* IEEE Transactions on Biomedical Engineering, 2017.

[23] K'Watch Glucose, "K'Watch Glucose Smartwatch", In: *Product page.* vol. Vol. 3. Sensors and Actuators, 2021.

[24] B. Dewan, S. Chaudhary, D. Singh, and M. Yadav, "Label-free detection of breast cancer cell lines using dopingless heterojunction TFET considering non-ideal hybridization issue", *Mater. Sci. Eng. B,* vol. 302, p. 117192, 2024.
[http://dx.doi.org/10.1016/j.mseb.2024.117192]

[25] "MXenes in biosensing: "MXenes: Emerging 2D Materials for Biosensing Applications,"", *Nano Lett.,* vol. 22, no. 6, pp. 3210-3215, 2022.

[26] "Organic semiconductors: "Organic Semiconductors in Biosensing: Recent Progress and Future Directions,"", *ACS Nano,* vol. 15, no. 4, pp. 12895-12900, 2021.

[27] "Printed sensors: "Low-Cost Printed Electronics for Flexible Biosensing Devices,"", *Sens. Actuators B Chem.,* vol. 307, p. 127675, 2020.

[28] "AI in glucose monitoring: "AI Integration in Wearable Biosensors for Predictive Healthcare,"", *IEEE Trans. Biomed. Eng.,* vol. 68, no. 9, pp. 902-910, 2021.

[29] "AI-driven glucose monitoring: "Improving Sensor Accuracy with AI-Based Calibration Models,"", *J. Diabetes Sci. Technol.,* vol. 14, no. 3, pp. 678-685, 2020.

[30] "Non-invasive sweat monitoring: "Sweat-Based Glucose Monitoring Using ISFET Biosensors,", *Journal of Diabetes Research,* pp. 1-10, 2021.

[31] "Tear glucose monitoring: "Tear Fluid Glucose Monitoring: Advances in Sensor Design,"", *Biosens. Bioelectron.,* vol. 144, pp. 1-6, 2021.

[32] S. Jadoun, U. Riaz, and V. Budhiraja, "Biodegradable conducting polymeric materials for biomedical applications: a review", *Med. Devices Sens.,* vol. 4, no. 1, p. e10141, 2021.

CHAPTER 7

Design and Analysis of CMOS SRAM Cells for Read, Write, and SNM Optimization Using Even-Odd Transistor Configuration

Prajwal B. Pillewan[1,*], **Prabhat Singh**[2] and **Dharmendra Singh Yadav**[1]

[1] *National Institute of Technology Kurukshetra, Thanesar, Haryana, India*

[2] *Indian Institute of Information Technology, Senapati, Manipur, India*

Abstract: This chapter examines 6T, 8T, and 10T SRAM cell designs and their compatibility with 45nm, 90nm, and 180nm technological nodes. It primarily examines the power consumption, speed, and signal strength of SRAM cell designs. The study uses tools to see how each SRAM cell design works under different conditions. For read, write, and hold operations of the SRAM cell designs, it compares the signal to the noise margin. The findings aid in understanding the advantages and disadvantages of each design, using technologies such as 45nm, 90nm, and 180nm to guide future advancements in memory technology.

Keywords: SRAM cell, Power, Delay, Read delay, Write delay, Hold margin, SNM.

INTRODUCTION

SRAM is a type of memory that does not require constant refreshing to retain data. This implies it can swiftly access the data it stores. SRAM is crucial for the design of power-hungry computer chips [1, 2]. Microprocessors, game consoles, portable electronics, and other compact devices frequently contain SRAM. SRAM, a type of memory known as cache, is typically used in these devices to help the chip retain information it frequently accesses. Memory is a great option for a variety of purposes, as it offers features such as data access, low power consumption, and the ability to avoid constant refreshes [3, 4]. Today, static RAM is part of computer chips and is used to store data that the chip needs to access quickly. Static RAM becomes more difficult to maintain as technology shrinks. Static RAM becomes less stable, more susceptible to interference, and uses power while not in use, among other issues. Researchers have developed static RAM

[*] **Corresponding author Prajwal B. Pillewan:** National Institute of Technology Kurukshetra, Thanesar, Haryana, India; E-mail: pillewan.prajwal1212@gmail.com

Girdhar Gopal, Meena Panchore, Arun Kishor Johar & Tarun Varma (Eds.)

designs that differ from the standard six-transistor design to address these issues. The 6T, 7T, 8T, 9T, and 10T static RAM bitcell topologies will all be compared in this chapter. The study examines factors like stability, reading and writing time, power consumption, overall power consumption, and space usage. This makes it easier for us to understand what we give up when we choose one type of cell over another. SRAM is ideal for systems that require quick responses because of its exceptional speed in retrieving the information we need. When the power is cut off, SRAM memory forgets what it has learned. Even when not in use, SRAM consumes some power. When in use, it often consumes less power than DRAM. SRAM is an option due to its speed and low power consumption. It is also used in embedded applications where speed and energy economy are crucial, as well as networking hardware such as switches and routers.

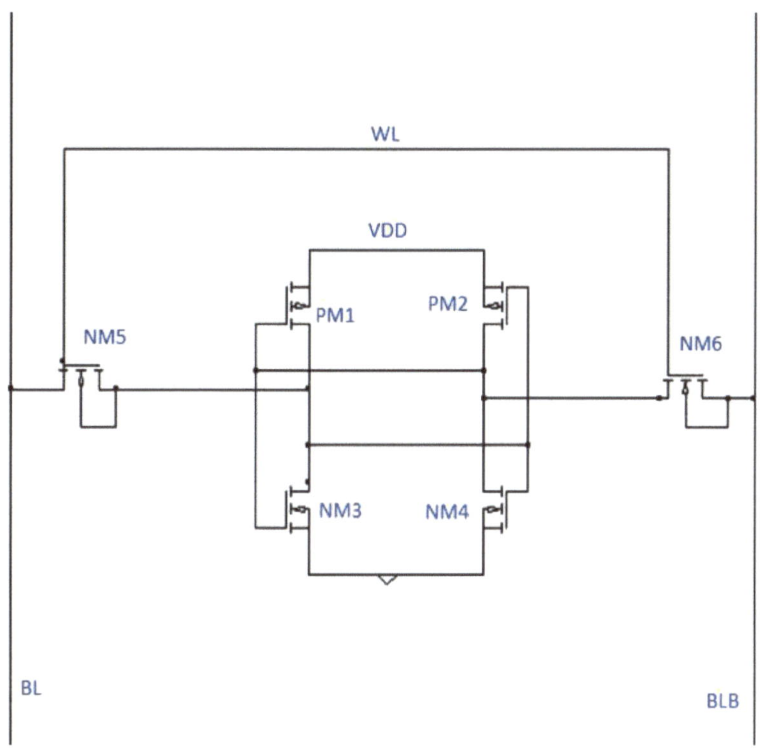

Fig. (1). Circuits diagram of a 6 T SRAM Cell.

The 6T SRAM cell is what people usually use to store a bit of data in random-access memory. This cell works because it maintains stability through internal feedback. In Fig. (**1**), you can see that the 6T SRAM cell has two CMOS inverters connected to retain information. The 6T SRAM cell can hold a logic 1 or a logic 0 for a long time, as long as it receives the power it needs. The 6T SRAM cell is very good at doing this. These inverters help store information and keep it safe when there are no external signals. The storage nodes are called Q and QB. They show the true and complementary outputs of the inverters.

There are two transistors, M5 and M6, that connect these nodes to the bit lines BL and BLB. The word line (WL) determines whether a cell can be accessed. It is, like a switch that controls the operation of the transistors M5 and M6. It determines whether the cell is open or closed based on the word line [5, 6]. When the word line is driven high (WL = 1), the access transistors turn on, enabling the connection between the internal nodes and the bit lines. This state allows data to be written into or read from the cell. When WL is low (WL = 0), the cell is disconnected from the bit lines, placing it in a hold condition where the cross-coupled inverters independently retain the stored data without external influence. The reason is that the data can be stored without constant refreshing. When you want to write some data, the bit lines BL and BLB are given voltage levels that match the data you want to write. If you turn on WL, these voltage levels are strong enough to override the cell's feedback loop, causing the values stored in Q and QB to be updated. The bit lines are like the path that the new data takes to get into the cell when you are writing something. This means the old stored value can be replaced with some information. If BL is driven low and BLB is high, Q becomes '0' and QB becomes '1'; reversing these signals results in the opposite state. Proper write operation requires that the access transistors be strong enough to override the existing latch state. Before any read or write operation, the bit lines are typically precharged to a known voltage level, usually V_{DD}. Precharge ensures proper sensing operation, prevents signal disturbance, and avoids conflicts between internal and external signals. When we read something from memory, the memory gets ready first. Then the word line is turned on. The bit line connected to the place with a zero starts to lose power slightly due to the pull-down network. The other bit line connected to the place with a 1 in it stays close to the power supply. The sense amplifier detects the power difference and determines whether the stored value is 1 or 0 [7, 8]. Accurate sensing of this differential voltage is essential to avoid read failures. A disturbance or voltage rise at node Q or QB during the read operation may force an unintended flip in the latch, resulting in erroneous data output.

When we write something, we need to change the value stored. To do this, we intentionally make one of the bit lines touch the ground. This makes a voltage difference. If we make WL high, it pulls node Q or QB low, turning off the pull-up transistor, and the other node can change state. For example, if we make BL low, this will affect PM1 and NM1. It will drive the node below the point at which it can work. This will turn off PM2. Turn on NM2. So, we get the logic level. This controlled process ensures that the cell is updated reliably. Overall, the

6T SRAM cell provides reliable data retention, fast read and write operations, and low standby power [9, 10]. However, maintaining stability during read and write operations becomes increasingly challenging as technology nodes scale down, due to reduced supply voltage and increased leakage and noise sensitivity. Ensuring robust performance requires careful optimization of transistor sizing, cell ratio, and access transistor strength.

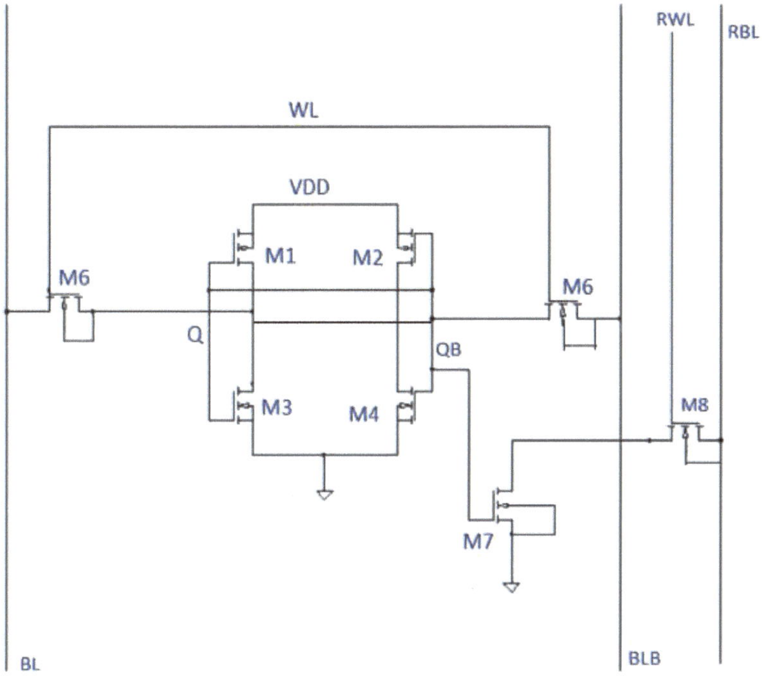

Fig. (2). Circuit diagram 8 T SRAM cells.

The 8T SRAM cell architecture (Fig. **2**) is one design we looked at. This 8T SRAM cell architecture uses two coupled inverters to form the storage element. The 8T SRAM cell architecture uses transistors M5 and M6 to help keep the data. When we want to read the data, the 8T SRAM cell architecture uses transistors M7 and M8. These transistors form a stage [11, 12]. This helps keep the stored data safe when we are reading it from the 8T SRAM cell architecture. The write word line (WL) controls write operations.

On the other hand, read operations are initiated by the read word line (RWL). This means we have control over the read and write functions. The 8T configuration is different from the 6T SRAM cell. It has two transistors, which makes it take up more space. This means that it uses space on a System,on-Chip platform [11 - 14]. The things around the memory cells, like the part that helps write, the part that helps sense, and the part that gets everything, are the same for both 6T and 8T memory cells. The extra parts in the 8T design improve its readability, even though it takes up more space. When we look at how the 8T cell works, we need to consider how long it takes to write and read, and how well it performs overall. One important thing we use to check whether the memory is working correctly is the Signal-to-Noise Margin (SNM), which is a measure of how reliable the SRAM is and is often shown as SNM [15]. SNM indicates the minimum noise voltage that can disturb the stored data and defines the cell's tolerance to electrical noise. In other words, it expresses how much noise the internal nodes can withstand while still reliably distinguishing between logic '1' and logic '0'. Higher SNM implies fewer read or write failures and improved immunity to variations. As noise levels increase in digital circuits, the allowable voltage range must remain wide enough to guarantee error-free operation, and sufficient SNM ensures this robustness [15, 16].

In Fig. (**3**), you can see the 10T SRAM cell. This is a type of memory cell commonly used in semiconductor memory designs. The name "10T" indicates the number of transistors in the 10T SRAM cell. There are 10T transistors in the 10T SRAM cell. The 10T SRAM cell consists of two cross-coupled inverters. These inverters are made up of PMOS and NMOS transistors. They help store information in the memory cell. These parts work together to create a loop that can remember either a '0' or a '1'. This depends on what was put in the 10T SRAM cell.

Additionally, the 10T SRAM cell includes two access transistors, typically connected to the bitline and bitline bar, facilitating read and write operations by enabling data transfer between the cell and the bitlines. In addition to the access transistors, the 10T SRAM cell incorporates two auxiliary transistors. These transistors are serially connected with the accessing transistors, which are typically managed by an independent wordline. The auxiliary transistors help keep the cross-coupled inverters separate from the bitlines during read operations. This helps prevent problems with the information that is already stored. When the computer writes information, it turns on the wordline connected to the access transistors. This allows the new information on the bitlines to get to the cross-coupled inverters. The cross-coupled inverters then take the information from the bitlines and update the information that is already stored in the cell with the new information. The auxiliary transistors and the cross-coupled inverters work together to make sure the information is handled correctly. Reading from the 10T SRAM cell involves activating the wordline connected to the access transistors, granting access to the cross-coupled inverters. The cell's state is then determined by sensing the voltage levels on the bitlines, reflecting the stored data. The additional auxiliary transistors in the 10T SRAM cell enhance stability and reduce the likelihood of read disturb compared to a 6T SRAM cell. The 10T SRAM cell isolates its cross-coupled inverters during read operations, effectively maintaining data integrity.

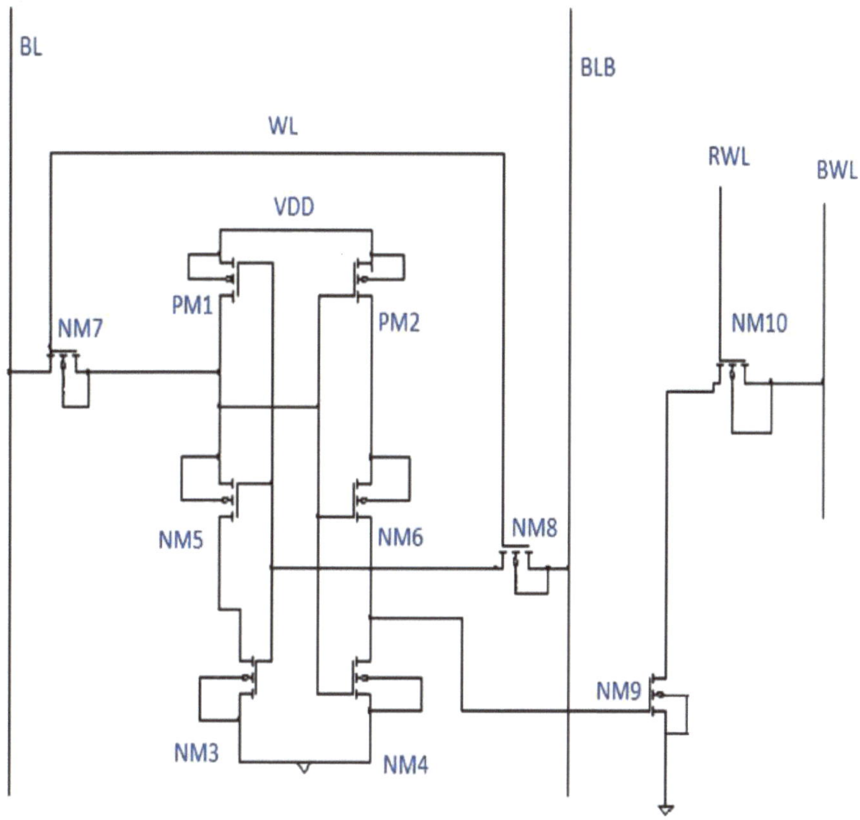

Fig. (3). Circuit diagram of 10 T SRAM Cell.

Fig. (4) illustrates the circuit architecture of the 7T SRAM cell. Similar to the traditional configuration, the core storage mechanism is formed by two cross-coupled CMOS inverters, each composed of transistors PM1, PM2, NM3, and NM4, which together form a bistable latch functioning as a NOT-gate pair. The memory nodes within the cell are accessed through two pass transistors, NM5 and NM6, which are controlled by the word line (WL). When the WL signal is turned on, the access transistors connect the storage nodes to the bit lines. This makes the cell ready for us to read from or write to. At the start of the read cycle, all bit lines are set to VDD, which means they are "1". After that, the bit lines are disconnected for a while, so they stay at the level they were set to. Once the address decoder picks the cell, we need the WL to be turned on, which connects the cell to the bit lines. The WL signal enables this connection. The cell is coupled to the bit lines via the WL signal. In this process, one bit line is intentionally pulled down to ground, creating a voltage difference that allows the stored value to be sensed. Data transfer occurs from the bit lines to the cross-coupled inverters. This lets the data transfer information from the cell [15]. When we perform a write operation, the WL that controls the access transistors is activated. This allows the bit lines to put values onto the internal storage nodes. The new values replace the previously stored data. The bit lines usually have many loads, and they switch on and off frequently. As a result, the bit lines tend to be slow. To minimize physical area, the MOSFET dimensions used in the cross-coupled inverter pair are typically kept small. As a result, a significant portion of power consumption in the 7T cell arises from the diffusion capacitance of the access transistors and from the long interconnect length of the bit lines. Updating the stored information is achieved by driving the bit lines hard enough to overwrite the cell's existing logic state.

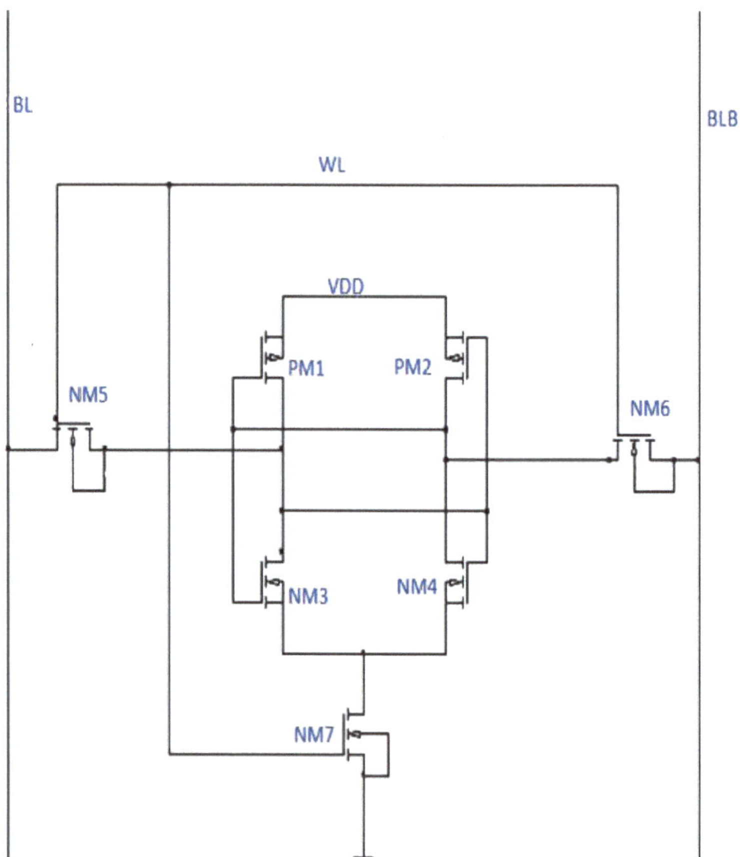

Fig. (4). Circuit diagram of7T SRAM Cell.

The old 6T, 7T, 8T, and 10T SRAM cells have some problems. These SRAM cells do not perform well when reading or writing data. They also use a lot of power when we write something. They use too much power overall. This is why people are working on a 9T SRAM cell. The new 9T SRAM structure is intended to improve on the 6T, 7T, 8T, and 10T SRAM cells. The proposed 9T SRAM cell architecture, illustrated in Fig. (**5**), incorporates nine transistors and six control lines, including Bit Line (BL), Bit Line-Bar (BLB), Write Word Line (WWL), Read Word Line (RWL), Zero Word Line (zWL), and Read Bit Line-Bar (RBLB). The read operation is enabled when the control signals RWL and RBLB are simultaneously active. The write operation is different; it uses the WWL signal along with the BL and BLB signals. The zWL signal is always off to prevent problems with the power supply or unexpected issues when the memory cell is working.

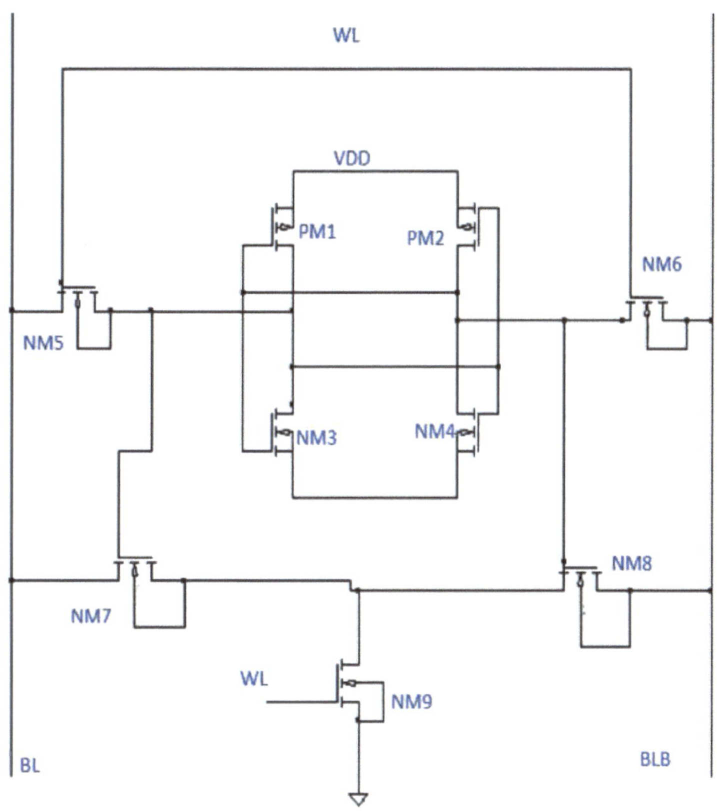

Fig. (5). Circuit diagram of 9T SRAM Cell.

When we perform a write operation, the word line connected to the access transistors is turned on. This creates a path for the data to flow between the bitlines and the internal cross-coupled inverters. This allows new data to replace previously stored data. The write operation uses the WWL signal and the bitlines to do this. In this structure, the PMOS transistors MP3 and MP4 are tied to the zWL control signal to ensure they remain continuously ON for stable feedback operation. At the same time, MP1 and MP2 are connected in series with them to maintain the data-holding capability. The internal storage nodes, denoted as H and L, function as the true and complementary output nodes. These nodes interface with the NMOS transistors MN3 and MN4, which serve as the read/write access transistors, allowing controlled transfer of information into or out of the cell.

The 9T cell has a cross-coupled inverter pair. It also has a read path. This is because it uses an access transistor. This helps to keep the read and write operations separate. The 9T cell does this to prevent read disturbs. Read disturbances are a problem in traditional 6T cells. They occur when bitlines and storage nodes interact during read operations. The 9T cell avoids this by keeping them separate. This makes the 9T cells better than the 6T cells. The 9T cell improves the isolation between the read and write operations. This is an advantage of the 9T cells over the traditional 6T cells. The read process is completed by activating the read word line (RWL), which enables sensing the voltage on the read bitline pair and thereby accurately determining the stored logic value. Through this enhanced architectural approach, the 9T SRAM cell achieves improved read stability, reduced write delay, and lower overall power consumption compared to earlier designs [17 - 19].

COMPARATIVE ANALYSIS

Read Delay Comparison

During the read operation, all three control signals—RBLB, RBL, and RWL—are driven to a HIGH state. Based on the stored data in the cell, either the BL or the RWL line begins to discharge from its initial precharged level. The resulting voltage variation, corresponding to the data stored in the cell, is subsequently sensed by the sense amplifier, which interprets the change to determine the correct output value.

Table **1** presents the read-delay performance of the PROP9T, CONV8T, and CONV6T SRAM cells measured at various supply voltages. To provide a clearer comparison of their behavior, the corresponding plot is shown in Fig. (**6**). The trend shows that the PROP9T architecture consistently reduces read delay compared to both CONV8T and CONV6T across all evaluated supply voltage levels [18, 19].

Table 1. Read delays at different supply voltages (at 180nm Technology Node).

Vdd(V)	6T (nSec)	7T (nSec)	8T (nSec)	9T (nSec)	10T (nSec)
0.2	0.535	0.528	0.603	0.553	0.529
1.0	0.514	0.517	0.532	0.532	0.461
1.5	0.527	0.522	0.532	0.530	0.432
1.8	0.527	0.530	0.539	0.535	0.421

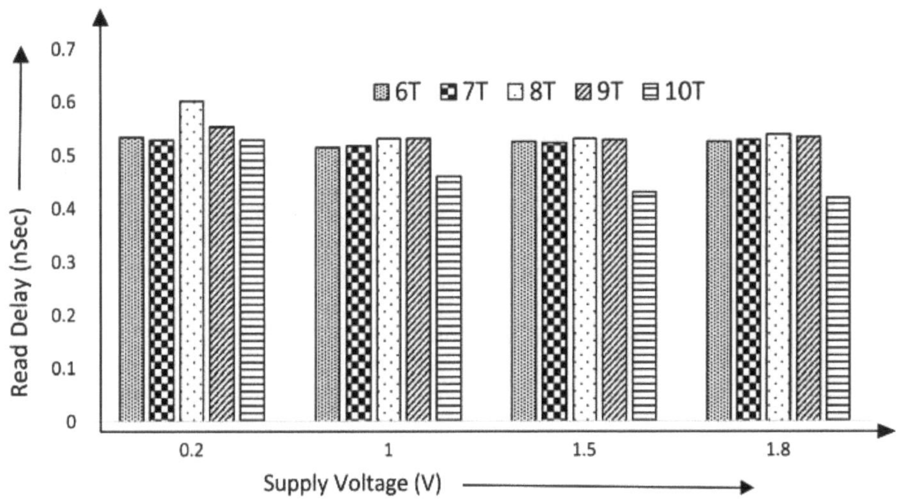

Fig. (6). Supply voltage vs read delay.

In Fig. (**6**), the relationship between read delay and supply voltage for the 6T, 7T, 8T, 9T, and 10T SRAM cells is depicted. The plotted results indicate that the read-delay characteristics for all five SRAM configurations are nearly identical, with no significant variation across the different supply voltage levels.

Write Delay Comparison

During the write operation, as discussed earlier, the Write Word Line (WWL) is driven to a HIGH state, and the data to be stored is applied to the Bit Line (BL). The duration measured from the moment WWL transitions to HIGH until the stored values at the internal storage nodes fully switch from their previous states is defined as the write delay. This delay represents the total time required to successfully overwrite the existing data in the SRAM cell.

Table **2** shows the time required to write data to 9T and 6T SRAM cells at different power levels. To make the difference easier to see, we created a graph shown in Fig. (7). It shows that it always takes longer to write data to the 9T SRAM cell than to the 6T SRAM cell, regardless of the power used. This means that the 9T SRAM cell is better at writing data. The 9T SRAM cell functions well because it can write data faster. This is good for systems that use SRAM because it allows them to operate more efficiently. The 9T SRAM cell is really good at writing data. This makes electronic systems work better. The observed improvement confirms the importance of investigating alternative SRAM structures and underscores the need for ongoing research to optimize memory cell design beyond traditional 6T configurations [20, 21].

Table 2. Write delays at different supply voltage (at 180nm Technology Node).

Vdd(V)	6 T (nSec)	7 T (nSec)	8 T (nSec)	9 T (nSec)	10 T (nSec)
0.2	0.532	0.489	0.495	0.465	0.495
1.0	0.318	0.285	0.302	0.222	0.293
1.5	0.151	0.164	0.206	0.121	0.161
1.8	0.126	0.126	0.194	0.104	0.127

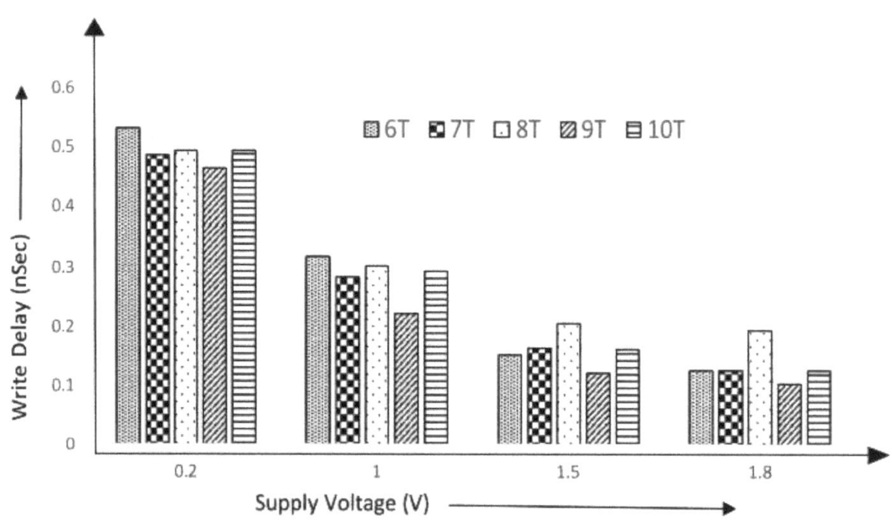

Fig. (7). V_{dd} (supply voltage) vs delay (write delay).

In Fig. (7) above, the write delay of SRAM cells—including 6T, 7T, 8T, 9T, and 10T—decreases as the supply voltage is reduced. This trend demonstrates a clear relationship between voltage scaling and improved write performance across all the considered SRAM configurations.

Delay Calculation

The time required for the data nodes in an SRAM cell to switch from one logic level to another—referred to as SRAM delay—is typically measured as the time difference between the 10% and 90% points of the voltage transition. Using this, a standardized measurement approach ensures uniformity, precision, and reliable comparison of performance characteristics among various SRAM architectures and applications [22].

In the SRAM architecture, the bit line stays charged when we read a '1'. This means the SRAM architecture can read fast. The time it takes to read is a lot less. When we compare the new 7T design to the old 6T SRAM cell, the SRAM architecture takes longer to perform the other tasks it needs to handle. The SRAM architecture is still really good at reading. This performance difference is reflected in Table **3**, which presents the delay values for various operations of both the standard 6T cell and the proposed 7T SRAM cell, as illustrated in Fig. (**8**). The marginal increase in delay can be attributed to the additional circuitry and structural complexity introduced in the 7T configuration to enhance its read performance and stability [23].

Table 3. Delay comparison of 7T and 6T (at 90nm Technology Node).

Operation	6T (p Sec)	7T (p Sec)
WRITE '0'	72.7	85.6
WRITE '1'	104.6	139.34
READ '0'	25.4	35.0
READ '1'	22.6	~0

The 6T SRAM cell helps us reach the cross-coupled inverters directly. This is really good because it means we can quickly read from the 6T SRAM cell. When we turn on the word line and measure the difference in voltage across the bit lines, the 6T SRAM cell can quickly give us what we need. The 7T SRAM cell differs from the 6T SRAM cell. The 7T SRAM cell has parts that can make it take a bit longer to read. This extra time is really small when we are reading from the 7T SRAM cell. The 7T SRAM cell is really good at reading. The 6T SRAM cell and the 7T SRAM cell do a job when they have to read something. If you look at the graph, you can see that it takes more time to write a zero and read a zero with the 6T SRAM cell than it does with the 7T SRAM cell. This is something to note about the 6T and 7T SRAM cells. Conversely, the delays for the write '1' and read '1' operations are higher in the 7T SRAM and comparatively lower in the 6T SRAM [24 - 28].

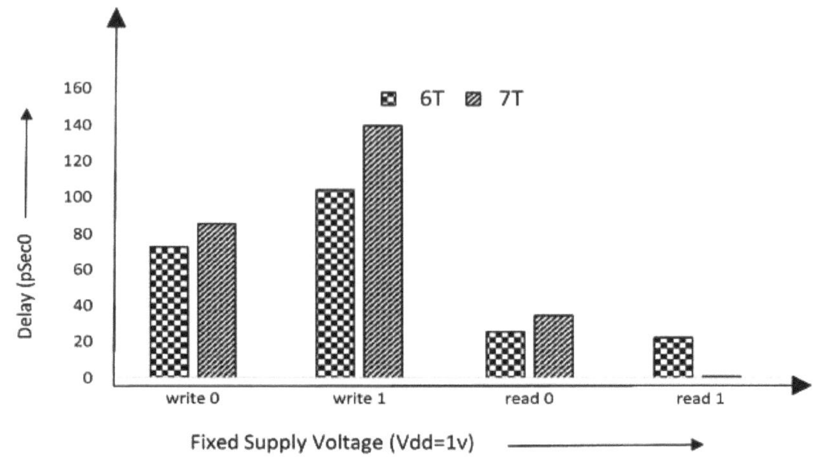

Fig. (8). Read and write delay for 6T and 7T for fixed voltage.

Static Noise Margins

The Static Noise Margin is particularly important for determining how robust an SRAM cell is during read or write operations. To determine the Read Static Noise Margin, we use the butterfly curve technique. This is where we find the square that fits inside the curve, which tells us how stable the SRAM cell is. We use the Static Noise Margin to evaluate this stability, and the Read Static Noise Margin is a part of that. The Write Noise Margin of the memory cell is determined by varying the bit-line voltage from the power supply voltage down to 0 V. This is done to find the voltage that can change the cell's state. The Write Noise Margin and Read Static Noise Margin values, for the 9T SRAM bit-cell made with 45 nm technology, are listed in Table **4**. Fig. (**9**) Shows how long it takes to read and write data in 6T and 7T SRAM cells made with this technology. The Write Noise Margin of SRAM cells is a factor to consider when designing memory cells.

Table 4. Read and Write Signal to Noise Margin 8T and 9T (at 45nm Technology Node).

Parameter	8T SRAM	9T SRAM
Technology	45nm	45nm
RSNM	77.29(mV)	204.54(mV)
WSNM	607.09(mV)	439.07(mV)

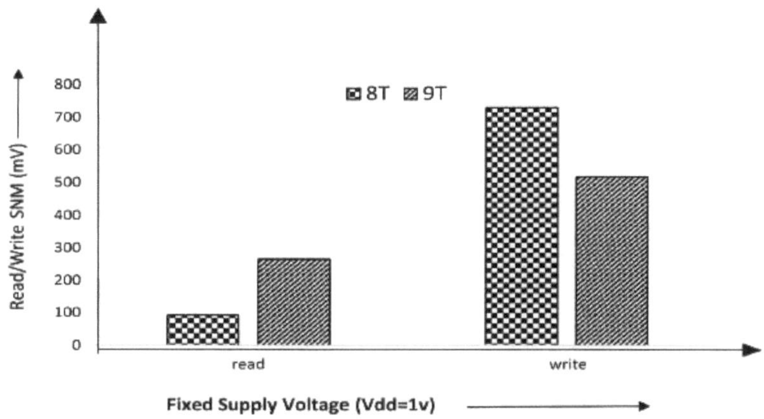

Fig. (9). Read and Write Delay for 6T and 7T for 45nm Technology node.

As transistor dimensions continue to scale down, SNM decreases for all SRAM configurations, reflecting growing sensitivity to noise and process variations. The 9T SE-DF architecture excels at reading data. It does this better than any design we looked at. On the other hand, the WRE-8T is not very good at reading data. It has the read quality of all the designs we checked. The same thing happens when we look at writing data. The 9T SE-DF cell is the best at writing data. The 9T-SF structure is the worst. We made some pictures to show how well the designs work when scaled down. You can see these pictures in Figs. (**8** and **9**). The pictures match what we found out from our simulations, which are in Table **5**. The 9T SE-DF architecture and the 9T SE-DF cell are really good at reading and writing data. These results highlight the importance of SNM as a key indicator of performance and reliability in the design of SRAM cells for advanced technology nodes [26 - 28].

The read Static Noise Margin (SNM) of an 8T SRAM cell is generally higher than that of a conventional 6T cell because the additional transistors enhance cell stability by isolating the internal storage nodes during read operations, as illustrated in Fig. (**8**). In a similar way, the write SNM of an 8T SRAM cell improves relative to the 6T configuration since the extra transistors provide better control and isolation while forcing new data into the cell. The read SNM of a 9T SRAM cell is usually better than that of the 9T SRAM cell's counterpart, the 8T cell. This is because the 9T SRAM cell is stable and can withstand noise disturbances. The write SNM of a 9T SRAM cell is also better than that of an 8T SRAM cell. This is because the 9T SRAM cell includes devices that help with write control and reduce interference during data movement. Table **5** compares the Read and Write Signal-to-Noise Margins (SNM) for 8T and 9T SRAM cells implemented using the 90 nm technology node. It lists the measured noise margins for both read and write operations for each cell type, enabling a clear evaluation of stability differences between the two configurations. The graphical trend presented in Fig. (**10**) shows the read and write SNM (in mV) evaluated at a constant supply voltage. Specifically, the write SNM of the 8T cell is observed to be greater than that of the 9T cell, while its read SNM remains lower than the corresponding value for the 9T SRAM cell [29 - 35].

Table 5. Read and write signal to noise margin 8T and 9T.

Parameter	8T	9T
Technology	90nm	90nm
RSNM	94(mV)	263.48(mV)
WSNM	729.87(mV)	516.63(mV)

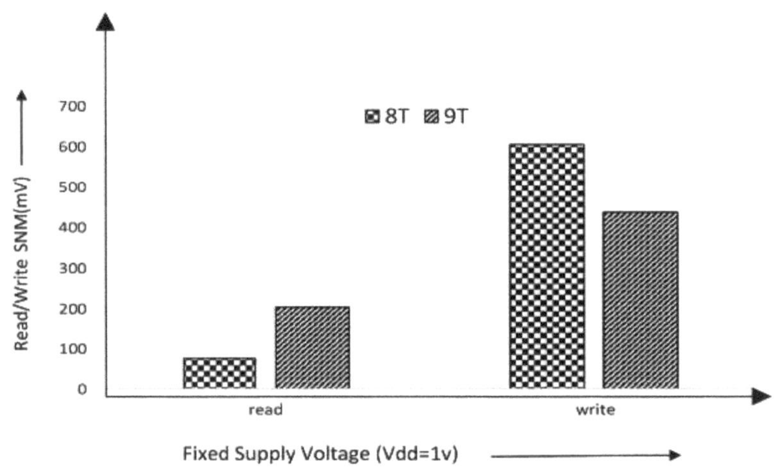

Fig. (10). Read and Write Delay for 8T and 9T for 90 nm Technology.

At the 95 nm technology node, the Static Noise Margin (SNM) is particularly important for SRAM design. The Static Noise Margin is a measure of how well a memory cell resists external noise or disturbances during read or write operations. The Static Noise Margin is the voltage difference that can disrupt the stable states of zero and one, which are necessary for the memory to work correctly, as shown in Fig. (**10**). It helps the memory cell tolerate noise during these operations. In this technology node, factors such as transistor sizing, circuit layout, and operating environment significantly influence the Read SNM of an 8T SRAM cell. The Read SNM of an 8T cell is usually lower than that of configurations like 9T or 10T cells. This is because the 8T cell is not as stable due to its size. The Write SNM of an 8T cell is comparable to that of cells with transistors. It is not as good as bigger SRAM designs. This is because the 8T cell is not very stable when we are writing to it. The 8T cell has this problem because of its size. Conversely, the Read SNM of a 9T SRAM cell in the same technology node is typically higher than that of smaller configurations, such as 6T and 8T. The 9T SRAM cell is better because it has transistors. These extra transistors help keep the cells separate and make the stored data more stable. This means the 9T SRAM cell can handle noise fluctuations. The Write SNM of a 9T SRAM cell is usually better than the Write SNM of a 9T SRAM cell with transistors. This is because the 9T SRAM cell can make write transitions even when there are problems. The 9T SRAM cell can do this. The trend illustrated in the graph confirms that both the Read and Write SNM values for the 9T SRAM cell are superior to those observed in the 8T SRAM cell, highlighting the advantages of increased device count and improved structural robustness.

CONCLUSION

When comparing 6T, 7T, 8T, 9T, and 10T SRAM cells, we need to consider stability, area, power, and complexity. 6T SRAM cells are simple. They have a good balance of area, speed, and power. That is why 6T SRAM cells are widely used. 6T SRAM cells can have some stability problems. 7T SRAM cells take up a little area and use more power than 6T SRAM cells. 7T SRAM cells are better when it comes to read stability. 8T SRAM cells are more stable than 6T and 7T SRAM cells. This is because 8T SRAM cells have read and write paths. This means 8T SRAM cells occupy more area than 6T and 7T SRAM cells. 9 T cells offer better noise margin and robustness, albeit at the cost of greater area and power. Although they use the greatest space and energy, 10T cells offer the best stability and noise immunity.

AI DISCLOSURE

During the preparation of the work, the author(s) used AI tools and the Grammarly tool to enhance the writing style and readability of the content. The authors clarify that AI was not involved in generating the substantive content and that they take full responsibility for the publication.

REFERENCES

[1] G. Saini, "A stable and power-efficient SRAM cell", *2015 International Conference on Computer, Communication and Control (IC4). IEEE,* 2015.

[2] E. Seevinck, F.J. List, and J. Lohstroh, "Static-noise margin analysis of MOS SRAM cells", *IEEE J. Solid-State Circuits,* vol. 22, no. 5, pp. 748-754, 1987. [http://dx.doi.org/10.1109/JSSC.1987.1052809]

[3] K. Anami, M. Yoshimoto, H. Shinohara, Y. Hirata, and T. Nakano, "Design consideration of a static memory cell", *IEEE J. Solid-State Circuits,* vol. 18, no. 4, pp. 414-418, 1983. [http://dx.doi.org/10.1109/JSSC.1983.1051965]

[4] M. Kutila, A. Paasio, and T. Lehtonen, "Comparison of 130 nm technology 6T and 8T SRAM cell designs for Near-Threshold operation", *2014 IEEE 57th International Midwest Symposium on Circuits and Systems (MWSCAS). IEEE,* 2014. [http://dx.doi.org/10.1109/MWSCAS.2014.6908567]

[5] D. Mittal, and V.K. Tomar, "Performance Evaluation of 6T, 7T, 8T, and 9T SRAM cell Topologies at 90 nm Technology Node", *2020 11th International Conference on Computing, Communication and Networking Technologies (ICCCNT). IEEE,* 2020. [http://dx.doi.org/10.1109/ICCCNT49239.2020.9225554]

[6] S. Shaik, and P. Jonnala, "Performance evaluation of different SRAM topologies using 180, 90, and 45 nm technology", *2013 International Conference on Renewable Energy and Sustainable Energy (ICRESE),* 2013. [http://dx.doi.org/10.1109/ICRESE.2013.6927819]

[7] R. Kolhal, and V. Agarwal, "A power and static noise margin analysis of different SRAM cells at 180nm technology", *2019 3rd International Conference on Electronics, Communication and Aerospace Technology (ICECA). IEEE,* 2019. [http://dx.doi.org/10.1109/ICECA.2019.8821868]

[8] A. Manna, and V.S. Kanchana Bhaaskaran, "Improved read noise margin characteristics for single bit line SRAM cell using adiabatically operated word line", *2017 International Conference on Nextgen Electronic Technologies: Silicon to Software (ICNETS2). IEEE,* 2017. [http://dx.doi.org/10.1109/ICNETS2.2017.8067965]

[9] Debasis Mukherjee, Hemanta Kr Mondal, and B. V. R. Reddy, "Static noise margin analysis of SRAM cell for high speed application", *International Journal of Computer Science Issues (IJCSI) 7.5,* p. 175, 2010.

[10] A. Goyal, A. Tomar, and A. Goyal, "Effect of W/L ratio on SRAM cell SNM for high-speed application", *International Research Journal of Engineering and Technology,* vol. 4, no. 4, pp. 2326-2332, 2017.

[11] S. Kumar, V.A. Tikkiwal, and H. Gupta, "Read SNM free SRAM cell design in deep submicron technology", *2013 International Conference on Signal Processing and Communication (ICSC),* 2013.

[12] N. Gierczynski, "A new combined methodology for write-margin extraction of advanced SRAM", *2007 IEEE international conference on microelectronic test structures,* 2007.

[13] N. Rahman, and B. Prasad Singh, "Static-noise-margin analysis of conventional 6T SRAM cell at 45nm technology", *Int. J. Comput. Appl.,* vol. 66, no. 20, pp. 19-23, 2013.

[14] P.K. Pandey, and D.S. Yadav, "Parametric Inspection of 6T SRAM Cells", *2021 5th International Conference on Trends in Electronics and Informatics (ICOEI). IEEE,* 2021. [http://dx.doi.org/10.1109/ICOEI51242.2021.9453036]

[15] S. Kumar, K. Saha, and H. Gupta, "Run time write detection in SRAM", *2015 International Conference on Signal Processing and Communication (ICSC). IEEE,* 2015. [http://dx.doi.org/10.1109/ICSPCom.2015.7150671]

[16] M. Kumar, and J.S. Ubhi, "Performance evaluation of 6T, 7T & 8T SRAM at 180 nm technology", *2017 8th international conference on computing, communication and networking technologies (ICCCNT). IEEE,* 2017. [http://dx.doi.org/10.1109/ICCCNT.2017.8204092]

[17] D. Suneja, N. Chaturvedi, and S. Gurunarayanan, "A comparative analysis of read/write assist techniques on performance & margin in 6T SRAM cell design", *2017 International Conference on Computer, Communications and Electronics (Comptelix),* 2017. [http://dx.doi.org/10.1109/COMPTELIX.2017.8004051]

[18] A. Gadhe, and U. Shirode, "Read stability and Write ability analysis of different SRAM cell structures", *Int. J. Eng. Res. Appl.,* vol. 3, no. 1, pp. 1073-1078, 2013. [IJERA].

[19] J. Ashwani, "Performance Analysis of Different SRAM Cells and Proposed 9T SRAM Cell", *2023 8th International Conference on Communication and Electronics Systems (ICCES). IEEE,* 2023. [http://dx.doi.org/10.1109/ICCES57224.2023.10192668]

[20] B. Madiwalar, and B.S. Kariyappa, "Single bit-line 7T SRAM cell for low power and high SNM", *2013 International Mutli-Conference on Automation, Computing, Communication, Control and Compressed Sensing (iMac4s),* 2013.

[21] H. Na, and T. Endoh, "A new compact SRAM cell by vertical MOSFET for low-power and stable operation", *2011 3rd IEEE International Memory Workshop (IMW). IEEE,* 2011. [http://dx.doi.org/10.1109/IMW.2011.5873204]

[22] B. Alorda, "Static-noise margin analysis during read operation of 6t sram cells", *DCIS Proceedings,* 2009.

[23] E. Seevinck, F.J. List, and J. Lohstroh, "Static-noise margin analysis of MOS SRAM cells", *IEEE J. Solid-State Circuits,* vol. 22, no. 5, pp. 748-754, 1987. [http://dx.doi.org/10.1109/JSSC.1987.1052809]

[24] K. Takeda, Y. Hagihara, Y. Aimoto, M. Nomura, Y. Nakazawa, T. Ishii, and H. Kobatake, "A read-static-noise-margin-free SRAM cell for low-VDD and high-speed applications", *IEEE J. Solid-State Circuits,* vol. 41, no. 1, pp. 113-121, 2006. [http://dx.doi.org/10.1109/JSSC.2005.859030]

[25] P.N. Vamsi Kiran, and Saxena Nikhil, "Design and analysis of different types SRAM cell topologies", *2015 2nd International Conference on Electronics and Communication Systems (ICECS). IEEE,* 2015.

[26] A. Agarwal, H. Li, and K. Roy, "DRG-cache: A data retention gated-ground cache for low power", *Proceedings of the 39th annual design automation conference,* 2002. [http://dx.doi.org/10.1145/513918.514037]

[27] S. Birla, "Analysis of the data stability and leakage power in the various SRAM cells topologies", *Int. J. Eng. Sci. Technol.,* vol. 2, no. 7, pp. 2936-2944, 2010.

[28] F. Hamzaoglu, K. Zhang, Y. Wang, H.J. Ahn, U. Bhattacharya, Z. Chen, Y-G. Ng, A. Pavlov, K. Smits, and M. Bohr, "A 3.8 GHz 153 Mb SRAM design with dynamic stability enhancement and leakage reduction in 45 nm high-k metal gate CMOS technology", *IEEE J. Solid-State Circuits,* vol. 44, no. 1, pp. 148-154, 2009. [http://dx.doi.org/10.1109/JSSC.2008.2007151]

[29] E. Grossar, M. Stucchi, K. Maex, and W. Dehaene, "Read stability and write-ability analysis of SRAM cells for nanometer technologies", *IEEE J. Solid-State Circuits,* vol. 41, no. 11, pp. 2577-2588, 2006. [http://dx.doi.org/10.1109/JSSC.2006.883344]

[30] C. M. R. Prabhu, T. G. Sargunam, and Singh Ajay Kumar, "High performance data aware (HPDA) SRAM cell for IoT applications", *ARPN Journal of Engineering and Applied Sciences 14.1,* pp. 91-94, 2019.

[31] Y. Takahashi, "Low-power adiabatic 9T static random access memory", *J. Eng.,* vol. 2014, no. 6, pp. 259-264, 2014.

[32] V.S. Bhaaskaran, "Energy recovery performance of quasi-adiabatic circuits using lower technology nodes", *India international conference on power electronics 2010 (IICPE2010),* 2011.

[33] M. Ansari, H. Afzali-Kusha, B. Ebrahimi, Z. Navabi, A. Afzali-Kusha, and M. Pedram, "A near-threshold 7T SRAM cell with high write and read margins and low write time for sub-20 nm FinFET technologies", *Integration (Amst.),* vol. 50, pp. 91-106, 2015. [http://dx.doi.org/10.1016/j.vlsi.2015.02.002]

[34] A. Datta, A. Goel, R.T. Cakici, H. Mahmoodi, D. Lekshmanan, and K. Roy, "Modeling and circuit synthesis for independently controlled double gate FinFET devices", *IEEE Trans. Comput. Aided Des. Integrated Circ. Syst.,* vol. 26, no. 11, pp. 1957-1966, 2007.

[http://dx.doi.org/10.1109/TCAD.2007.896320]

[35] B. Ebrahimi, S. Zeinolabedinzadeh, and A. Afzali-Kusha, "Low standby power and robust FinFET based SRAM design", *2008 IEEE Computer Society Annual Symposium on VLSI. IEEE,* 2008. [http://dx.doi.org/10.1109/ISVLSI.2008.8]

CHAPTER 8

Advancements in Power Optimization for CMOS Complementary Transistor Configurations

Prajwal B. Pillewan[1,*], **Prabhat Singh**[2] and **Dharmendra Singh Yadav**[1]

[1] *National Institute of Technology Kurukshetra, Thanesar, Haryana, India*

[2] *Indian Institute of Information Technology Senapati, Manipur, India*

Abstract: This chapter focusses on a comparison of SRAM cell design with six, eight, and ten cell designs and discusses them in the context of various technology nodes, such as 45nm, 90nm, and 180nm. The emphasis is put on essential performance aspects, such as power consumption, speed, and signal quality. Through the application of complex analyzing tools, the study examines the performance of each design under the various conditions, specifically the signal-to-noise margin when performing the read, write, and hold operations. The findings are useful in indicating the strengths and weaknesses of each design, giving an insight into the possible impact of different technologies on the future of memory technology.

Keywords: SRAM Cell, Power, Delay, Read Delay, Write Delay, Hold Margin, SNM.

INTRODUCTION

SRAM is recognized as a type of memory that is special, and that has several advantages, including the possibility to access each bit individually and the absence of a periodic refresh need. It is particularly significant to low-power design within VLSI systems, and particularly because of its significant role in the detailed design of high-efficiency chips with low power consumption [1 - 5]. This complexity attracts the interest of the challenges in balancing efficiency and performance in energy. SRAM is a major constituent of cache and memory applications that is commonly implemented in gaming devices, microprocessors, and portable electronics [6]. It is a choice of many applications in various technologies that operate on some of the attributes, such as the rapid transfer of data, minimal power supply, and the lack of periodic refreshments [7 - 12]. On-chip data storage and cache memory in the current integrated circuits consist of

* **Corresponding author Prajwal B. Pillewan:** National Institute of Technology Kurukshetra, Thanesar, Haryana, India; E-mail: pillewan.prajwal1212@gmail.com

Girdhar Gopal, Meena Panchore, Arun Kishor Johar & Tarun Varma (Eds.)

static RAM. SRAM is, however, more challenged with more technology nodes being reduced, including reduced stability, higher leakage currents, and being predisposed to various noise factors. To solve these issues, researchers have investigated alternative designs of SRAM cells beyond the traditional 6-transistor (6T) design. This chapter gives a thorough comparison of SRAM cells built with 6T, 7T, 8T, 9T, and 10T architectures. In this paper, the author seeks to explain the trade-offs associated with each design by referring to critical performance indicators such as stability, read delay/power, write delay/power, power consumption, and area efficiency [13 - 16]. The speed of SRAM access makes it suitable for situations where quick data access is required. However, like other types of RAM, SRAM is volatile, so when power is disconnected, the data stored in it will be lost. SRAM consumes power when unused due to transistor leakage, though it generally consumes less power when active than DRAM. Also, SRAM can be used in networking devices such as switches and routers and embedded systems, where its speed and low power consumption come in especially handy [17 - 25].

STRUCTURE AND RESULT DISCUSSION

An essential part of static random-access memory (SRAM) for data storage is a 6T SRAM cell. The different nodes, control signals, and parts connected to the SRAM cell are shown in Fig. (**1**). Two coupled CMOS inverters that function as NOT gates make up the cell. Transistors NM5 and M6 control access to the memory section, which is connected to the bit lines BL and BLB. These access transistors permit communication with the bit lines when the word line (WL) is set to one (enabled); on the other hand, the memory stays in a hold state when WL is set to zero (disabled). The BL and BLB lines, which act as input lines in this situation, are manipulated to input data into the memory during a write operation. On the other hand, these lines serve as output lines during a read operation. Furthermore, SRAM cells frequently undergo precharging, which guarantees that the bit lines are in a known state before any read or write operations [26 - 31]. This precharging procedure preserves signal integrity and helps avoid contention problems. Both read and write operations make use of the WL line. Voltage increases that could cause unexpected output states in the other inverter can occasionally be used to identify performance issues during read operations at the Q/QB inverter nodes. The memory segment must maintain certain values in these circumstances, such as Q being high (1) and QB being low (0), or vice versa. When the word line (WL) is enabled to one (1), the access transistors are activated, enabling the bit lines BL and BLB to operate as output lines during a read operation. The bit lines are precharged using the voltage Vdd as a result of the voltage differential. The storage node (Q or QB) with a value of 0 discharges, which is subsequently picked up by a sense amplifier to finish the read process. In a similar manner, values are assigned to Q and QB during a write operation. The bit lines BL and BLB, which function as input lines, can be accessed by turning on WL. In order to effectively manage these lines, one of the bit lines is grounded. This grounding produces a voltage difference that influences PM1. PM2 is off when the voltage is lower than the minimum threshold of M2, and generates the desired output. It is a sequence of events that enables the write action in the memory [32 - 38].

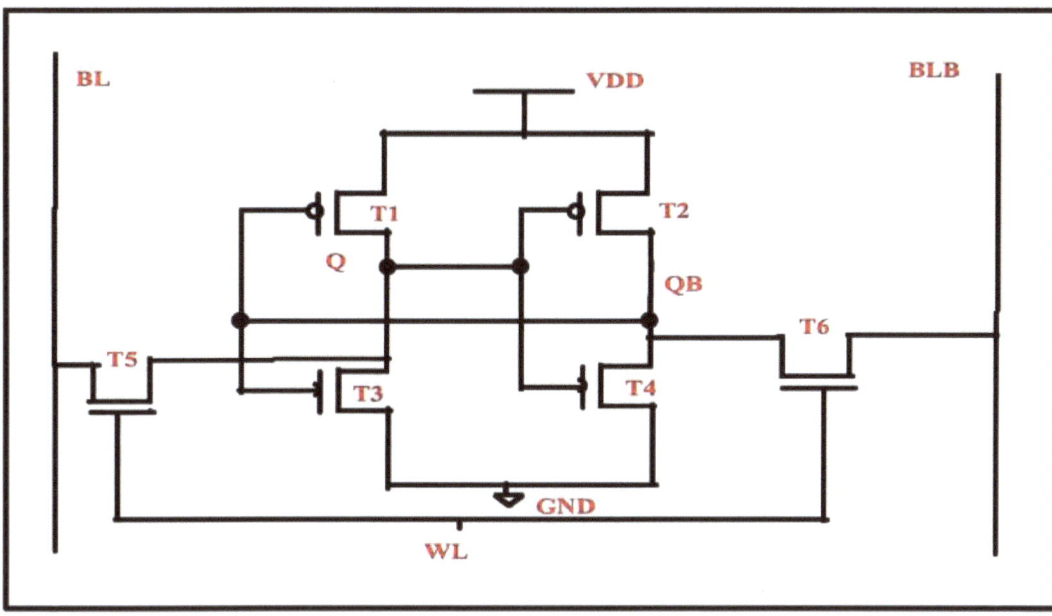

Fig. (1). Circuit diagram of a 6 T SRAM Cell.

A thorough comparative analysis is conducted using the 8T SRAM cell architecture, which is depicted in Fig. (**2**). According to this architecture, transistors M5 and M6 are linked to inverters that follow each other. The read buffer is applied to transistors M7 and M8. When read processes utilize read word lines (RWL), word lines (WWL) are utilized in write processes. The circuit footprint of the 8T SRAM cell is larger than the 6T SRAM cell because the former one incorporates two more transistors, which contribute significantly to the consumption of the System-on-Chip (SoC) space. To finalize the entire design, the two types of SRAM cells also have components such as a write driver, a sensing amplifier, and precharge circuitry. Parameters such as write delay and read delay are significant when it comes to evaluating the performance of SRAM.

Signal-to-Noise Margin (SNM) is an essential parameter that can be used to measure the stability of SRAM [39 - 41]. The SNM demonstrates the ability of integrated circuits (ICs) to withstand noise when transmitting logic 0 and logic 1, and the small amount of noise necessary to alter the state of a cell. The tolerable voltage variations are also increased when circuit noise is considered, and the data transmission is error-free but will maintain the required noise margin [12 - 16].

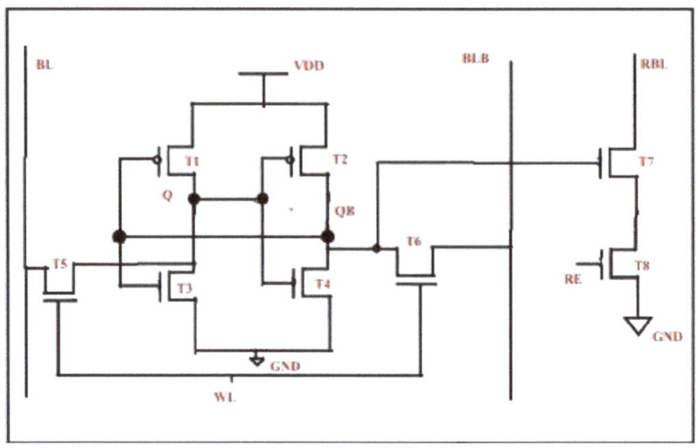

Fig. (2). Circuit diagram 8T SRAM Cells.

Fig. (**3**) shows the 10T SRAM cell. This is a memory cell design that people use a lot in semiconductor memory applications. The 10T SRAM cell has ten transistors. That is what the term 10T means. The 10T SRAM cell has two coupled inverters. Each inverter is made up of PMOS and NMOS transistors. These are the inverters that store information. They form a loop that assists the 10T SRAM cell to remember whether it is 0 or 1. The 10T SRAM cell stores this depending on what it was initially set to. Along with the two access transistors that are tied to the bit line and the bit line bar in the operation of read and write, there are two auxiliary transistors present in the 10T SRAM cell. These auxiliary transistors are hooked in series with the access transistors and are generally driven by an independent word line. The primary task of these things is to ensure that the cross-coupled inverters are not brought into contact with the bit lines during reading. This assists in preventing the occurrence of things going amiss in the event of reading something. In the case of writing something, the word line associated with the transistors' access is activated. This allows the transfer of the data off the bitlines into the cross-coupled inverters. The cross-connected inverters are then updated with the data on the bit lines. This entails that the old information in the cell is replaced with the information in the bit lines. In this process, the cross-coupled inverters are very crucial. The connected word line that is used to read the contents of the 10T SRAM cell that is connected to the access transistors is turned on. This allows access to the cross-coupled inverters. We can then determine what is stored by examining the levels of voltage on the bitlines. The transistors are found in the 10T SRAM cell, thus complicating the design. This may result in difficulties when we seek to plan the 10T SRAM cell and assemble it with circuit components. The 10T SRAM cell is more problematic due to this.

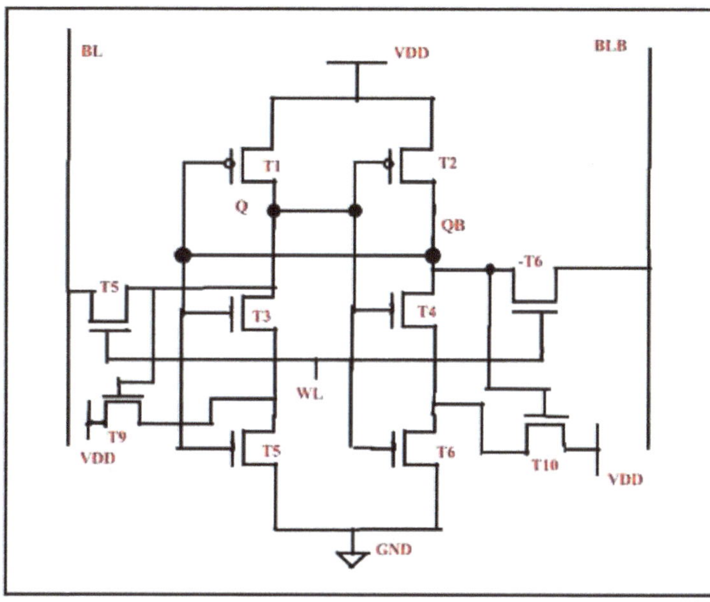

Fig. (3). Circuit diagram of 10 T SRAM Cell.

Performance Advantages: Regardless of the complexity, the 10T SRAM cell has important performance advantages, such as stability and noise immunity, and it can be used in high-performance applications. The inclusion of auxiliary transistors in the 10T SRAM cell enhances stability and reduces the likelihood of read disturb compared to a 6T SRAM cell. By effectively isolating the cross-coupled inverters during read operations, the 10T SRAM cell helps maintain data integrity [1 - 8].

Fig. (4) shows what the 7T SRAM cell looks like. The 7T SRAM cell is made up of two CMOS inverters. These inverters are formed by transistors PM1, PM2, NM3, and NM4. They work like a NOT gate.

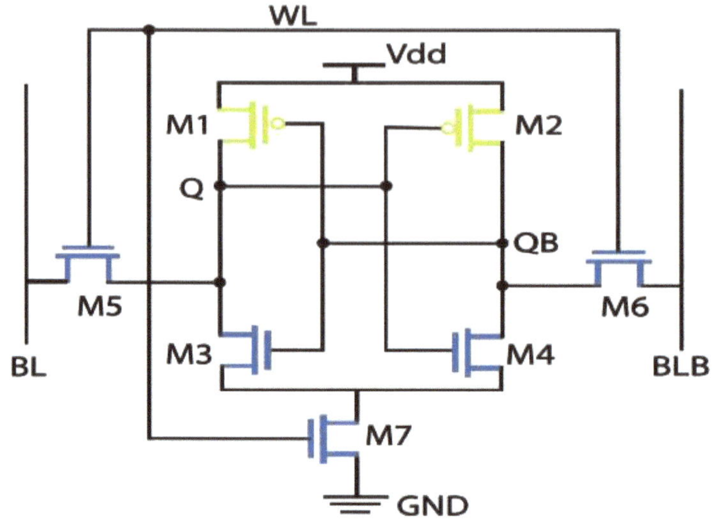

Fig. (4). Circuit diagram 7T of SRAM Cell.

When we want to read something from the 7T SRAM cell, the memory part of the 7T SRAM cell gets connected to two-word lines. These word lines are called WL. The connection happens through access transistors. The access transistors are NM5 and NM6. They get turned on when we want to read something from the 7T SRAM cell. The driver starts by setting all the bit lines to Vdd, which means they have a value of "1". Then, it stops connecting them, so the bit lines stay at this level. Next, the address decoder turns on one of the word lines, which connects the bit cell to the bit lines. It is really important that one of the two-bit lines that are connected to the cell is connected to the ground. This is made possible by the bit lines and the word lines, as well as the bit cell. The bit lines must be programmed to the value of Vdd, and thereafter, the word lines perform their part in connecting the bit cell to the bit lines. The design of the 7T SRAM cell is such that it is balanced in terms of its performance as well as the amount of space it occupies. This renders the 7T SRAM cell design truly good, in things where space is limited. The SRAM cell 7T design has two CMOS inverters. These inverters are fabricated out of transistors termed as PM1, PM2, NM3, and NM4. They work like a NOT gate. When we read data from the 7T SRAM cell, the memory part of the 7T SRAM cell is connected to two-word lines. This connection is made through access transistors called NM5 and NM6. These access transistors are turned on to help move the data. When you start to write something, the word line that is connected to the access transistors gets turned on. This lets the SRAM cell get the data from the bit lines. The new data then goes to the cross-coupled inverters. This means the old data, in the SRAM cell, is replaced with the data that comes from the bit lines. The bit lines use a lot of power, which can make the SRAM cell work more slowly. This can affect how fast the whole SRAM cell works. The SRAM cell is what gets affected by the slow bit lines. To make the cell smaller, the MOSFETs in the two inverters are made to be pretty small. The driver starts by putting all the bit lines at the level, as Vdd, which means they have a value of "1".Then it stops connecting them, so the bit lines just stay in this state. The address decoder picks one of the word lines. Turns it on, which connects the bit cell to the bit lines. The bit lines do not change quickly because they use a lot of power. To minimize the overall size of the cell, the MOSFETs within the two inverters are designed to be relatively small. The primary contributors to power consumption are the diffusion capacitance of the pass transistors and the length of the bit lines, which can lead to increased energy usage during data updates within the bit regions [16 - 20].

The new 9T SRAM cell is an improvement over the old 6T, 7T, and 8T SRAM cells. The 9T SRAM cell solves some problems we had with the old cells. For example, it helps with delays when we read or write data. It also uses power when we write data, and it does not waste as much power as the old cells.As you can see in Fig. (**5**), the 9T SRAM cell has nine transistors and six control lines. The 9T SRAM cell has features like the Bit Line, which we call the BL for short, the Read Word Line, which we call the RWL, the Write Word Line, which we call the WL, and the Zero Word Line, which we call the ZWL. The 9T SRAM cell is an improved design compared to the older 6T, 7T, and 8T SRAM cells. The Bit Line Bar, which is also known as the BLB, and the Read Bit Line Bar, which is also known as the RBLB, are important. When we do a read operation, the control lines for the RBLB and the RWL are turned on. On the other hand, when we perform a write operation, the control lines for the WL, the BL, and the BLB are used.The ZWL control line is always kept at zero so that nothing bad happens with the VDD. The 9T SRAM cell has three transistors that help with read and write operations. These access transistors are really important for the Bit Line Bar and the Read Bit Line Bar to work properly. These transistors are usually linked to the bit lines and word lines. This helps move data between the bit lines and the cell when we read or write something. In order to access data in the 9T SRAM cell, we switch on the word line, which is connected to the access transistors. This allows us to obtain the cross-coupled inverters. We calculate what is in the 9T SRAM cell by observing the voltage levels on the bit lines. The voltage levels give us information on what information the 9T SRAM cell stores. The 9T SRAM cell is a concept that helps traditional SRAM architectures perform better. It resolves certain issues, such as the inconveniences caused by using excess power to read/write data and being unstable. There are 9 transistors in the 9T SRAM cell. It is effective in ensuring that the data is secure. It also minimizes the possibility of an error occurring when you read data on a 9T SRAM cell. This issue is known as read disturb phenomena, and this can occur when reading data in the 9T SRAM cell. It is a cool feature of the 9T SRAM cell. It employs various control lines. These lines allow the 9T SRAM cell to be managed more easily when the cell is reading or writing. There is also another line named the Zero Word Line or ZWL, as it is commonly known, and it is very important. 0WL acts to stabilize the 9T SRAM cell. It achieves this by preventing the odd and even changes in voltage that are not anticipated to occur. The reason is that it implies that some components of the 9T SRAM cell, such as some of the transistors, will remain unchanged. This assists in increasing the reliability of the 9T SRAM cell. The 9T SRAM cell is a memory cell, and reliability is extremely significant in the case of the 9T SRAM cell. Other components of the architecture of the 9T SRAM cell are three access transistors, which are strategically connected to bit lines and word lines. This setup enables Ethernet transfer of data when reading and writing. When you begin reading something that you have memorized, you turn on the right word line. This allows you access to the inverters that are cross-coupled. The system then measures the voltage on the bit lines to determine the data that is stored. This will ensure that you receive the information. When you wish to write something, the connected word line to the transistors, which are the access transistors, is switched on. This allows information to be transferred out of the bit lines to the storage nodes. Thus, it is the updated data that is written in the memory cell. This implies that the current information is always held by the memory cell. When you write data to the memory cell, the information in the memory cell is updated. The bit lines and the word line collaborate in ensuring that the data is in the memory cell. The 9T SRAM cell design reduces the effects of write power, hence it is applicable in low-power applications. Moreover, the architecture of the 9T SRAM cell can be scaled and therefore will be flexible to the next technology node. Since the development of semiconductor technology has been going on, performance maintenance, power consumption, and area reduction are essential to achieve. The 9T design is indeed instrumental in achieving such objectives. It is an option for memory applications in many applications, such as high-performance computing and mobile gadgets. The 9T SRAM cell is an enhancement to the SRAM technology. It works better, is more stable, and more efficient in power consumption than older versions.The 9T design has some interesting attributes, making it a great choice for future development of memory solution options. The reason why the 9T SRAM cell is a choice is that it possesses these features and can accomplish things that other cells are unable to accomplish. The 9T SRAM cell supports multiple control lines, such as Write Word Line (WL), Read Word Line (RWL), Bit Line (BL), Bit Line Bar (BLB), and Zero Word Line (ZWL). This multi-line design gives perfect control over the cell functions to ensure that the right transistors are turned on in read and write procedures. An example is that, when an operation in the form of a read is taking place, RWL is enabled, allowing one to access the cross-coupled inverters and sense the voltage levels on the bit lines to read the data stored. On the other hand, the WL is switched on during a write operation so that data are directed between the bit lines and the storage nodes, effectively changing the cell's contents. Improved power efficiency of the 9T SRAM cell is one of the vital benefits of the design in the 9T SRAM cell. The design of the cell makes the transistors the right size, and the control lines prevent switching on and off when it is not required. This is quite useful when using devices powered by batteries, such as the 9T SRAM cell, since energy conservation is highly critical. The signal-to-noise margin is also present in the 9T SRAM cell, and hence it is more stable and is better able to handle noise when operating to ensure that the 9T SRAM cell is less prone to losing data. The 9T SRAM cell is used in many applications, including mobile devices, embedded systems, high-performance computing, and networking equipment. As far as devices are concerned, the 9T SRAM cell is indeed a wonderful item due to the fact that it consumes very little energy and is very stable. This has made it ideal for page caching in devices, which implies you will have access to your information in a short time, and your phone battery will last longer. The 9T SRAM cell, too, is wonderful when embedded systems are used, when you have to save space and conserve power. The 9T SRAM cell provides a good memory storage option in these systems. It can also be used in high-performance computing because its improved performance features, speed, and reliability are also needed when dealing with large volumes of data in a short time. Also, 9T SRAM cells may also be applied in networking stages, *e.g.*, routers and switches, where latency and accessing data speed are paramount to the network functioning at its best [30 - 39].

POWER CALCULATION COMPARISON

The mean electricity of a source is calculated by the product of the current discharged by the voltage produced. This informs us of the true use of electricity, the source. The electricity utilized by an SRAM cell consists of components. It obtains power at the source of electricity. It also obtains electricity through the line that is being written, which is known as the word line. Electricity is also used when the bit line and the bit line bar are charged and discharged. With a type of SRAM cell that is intended to be known as a 7T SRAM cell, there is also electricity utilized by the read line. This integrated method of power consumption analysis enables comprehensive analysis of the SRAM cell's power requirements, enabling an optimization strategy that reduces total power consumption while ensuring efficient operation [20, 21].

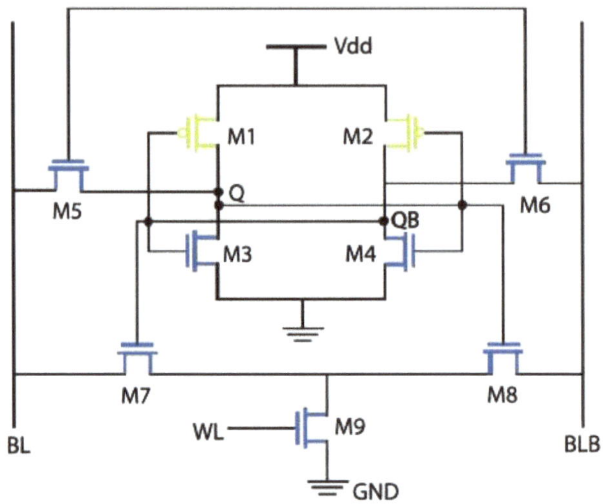

Fig. (5). Circuit diagram of 9T SRAM Cell.

Table 1. Percentage Power Improvement In 7T From 6T (90 nm Technology Node).

Operation	6T (uW)	7T (uW)	Improvement (%)
WRITE '0'	9.35	7.29	22.03
WRITE '1'	9.35	7.73	17.33
READ '0'	8.62	7.11	17.52
READ '1'	8.52	6.70	21.36

When we look at the ideas in the suggested design and compare a regular SRAM cell to a suggested 7T SRAM cell, we can see that the total power used includes many factors (as shown in Fig. **6** and Table **1**). The suggested design uses power in ways like the power used by the bit lines (Bl and Blb), the word lines (WL), the read lines (RL), and the transistors that are inside the 7T SRAM cell. The 7T SRAM cell has its own way of using power. I found it really interesting that the new 7T SRAM cell uses more power than the old 6T SRAM cell. When you read from the 6T SRAM cell, both bit lines get a charge from VDD. Then the charge values go away. If the stored data is '0', the bit line loses its charge. The 7T SRAM cell is different because it only uses power twice when you read from it. This is a difference from the 6T SRAM cell. The 7T SRAM cell is better because it does not waste much power. As a result, in contrast to the conventional 6T SRAM cell's use of power, the power used during read operations is cut in half. One-bit line releases when the procedure is finished, and the other bit line releases according to the information kept in the cell [40, 41].

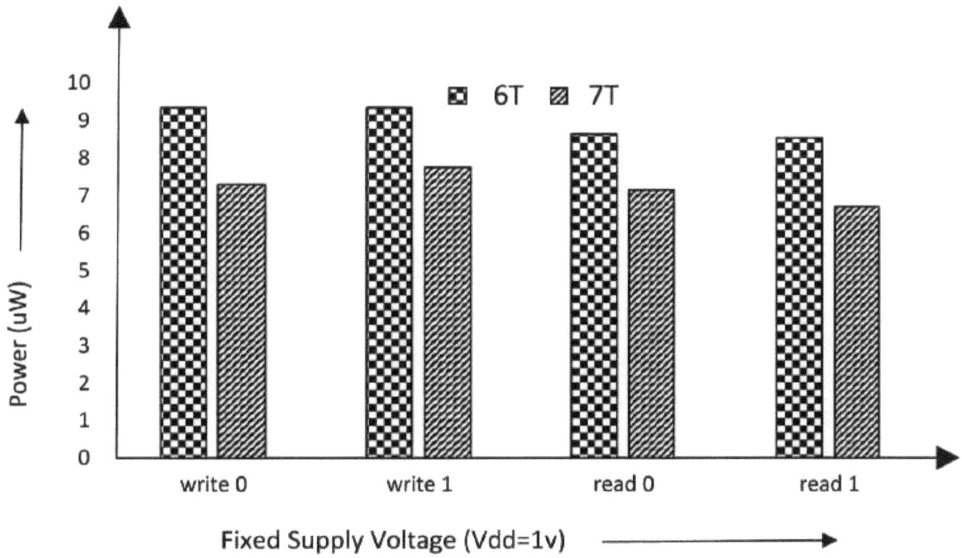

Fig (6). Read/Write power Vs supply voltage for 6T and 7T.

COMPARE POWER AND DELAY FOR 45NM TECHNOLOGY

The Cadence Virtuoso Tool was used to develop the proposed 10T SRAM cell. We used the Cadence SPECTRE simulator to generate waveforms. To make the read cycle faster and reduce leakage, an integrated circuit is used in the 10T SRAM cell. The output node of the 10T SRAM cell is connected to the disturbance circuits before the access control transistor.This way, the 10T SRAM cell can read data without relying on the bit lines or the access transistors.The Cadence Virtuoso Tool and the Cadence SPECTRE simulator help to make this happen.This design of the 10T SRAM cell significantly reduces leakage current and power consumption. The new setup makes things work faster, which means it uses power, especially when it has to read a lot of information. This really helps the cell use less energy overall. If you look at Table **1**, you can see how well the usual SRAM cells perform with 6T, 7T, 8T, 9T, and 10T capacities. The results of the study are compared with these results, as shown in Fig. (**7**) and Table **2**. The SRAM cells perform as expected, and the study's results show how they stack up against this performance. The results from the chapter notably exceed those reported in the referenced work, showcasing enhanced stability and reduced power loss in the proposed 10T SRAM cell design [27].

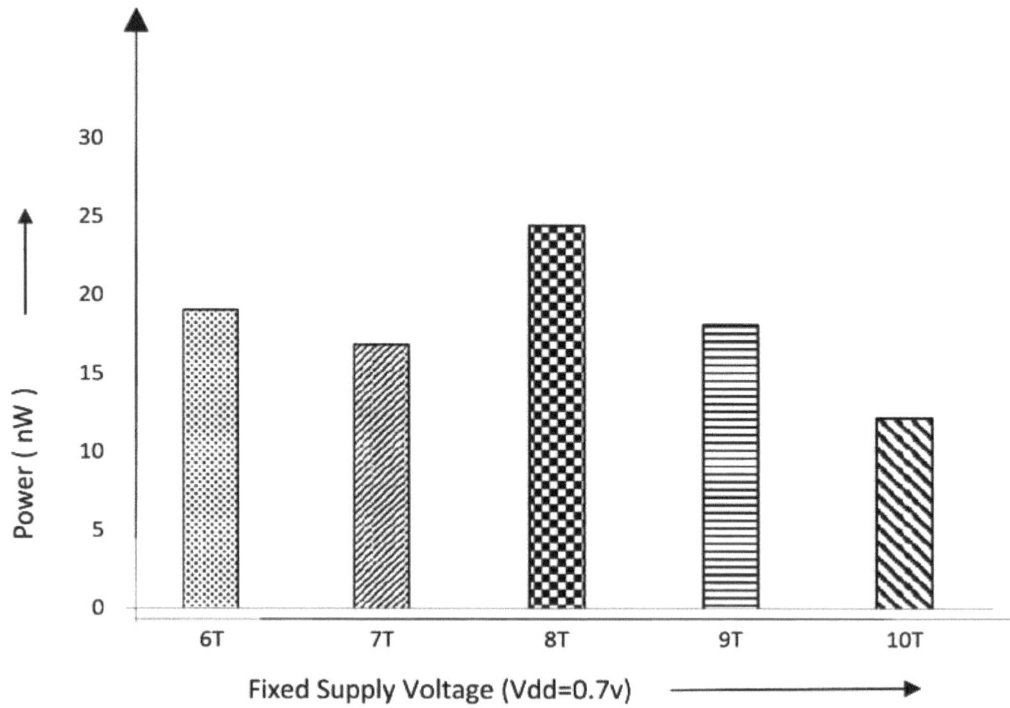

Fig (7). Compare the power of 6T, 7T, 8T, 9T, and 10T for 45nm technology.

Table 2. Result for difference SRAM cells (at 45 nm Technology Node).

S. N o.	Parameter	6 T	7 T	8 T	9 T	10 T
1	Supply Voltages	0.7 V	0.7 V	0.7 V	0.7 V	0.7 V
2	Power	19.1 nW	16.8 nW	24.4 nW	18.1 nW	12.1 nW
3	Read Delay	110 pSec	92.8 pSec	81.7 pSec	84.4 pSec	50 pSec

Fig. (7) shows that the power consumption is usually moderate. This is because there are a number of transistors. The power consumption shown in (Fig. 7) is mainly due to leakage currents and power loss that happens when we read and write data. The delay when we read data from a 6T SRAM cell is relatively low compared to more complex architectures. This is because we can directly access the cross-coupled inverters in the 6T SRAM cell. We can quickly retrieve the data from the 6T SRAM cell. The speed at which data is retrieved depends on factors such as the size of the transistors, the length of the connections, and differences in the manufacturing process of the 6T SRAM cell. In contrast, a 7T SRAM cell generally consumes slightly more power than a 6T cell due to an additional access transistor. However, the difference in power consumption may not be significant, particularly in low-power applications. The read delay of a 7T SRAM cell, mainly due to the additional access transistor, is marginally higher than that of a 6T cell. Nevertheless, this increase in read delay typically has a minimal impact on the overall system's performance.

So, an 8T SRAM cell needs more power than a 6T or 7T cell. This is because it has transistors. These extra transistors help make the 8T SRAM cell more stable and less prone to problems. They also use more power, as you can see in Fig. (**8**). The 9T SRAM cells are more complex than the 6T, 7T, and 8T cells. So they use more power. The reason is that they have transistors too. These transistors make the 9T SRAM cells more stable and capable of performing operations. Again, this means they use more power. The high transistor count and complexity of the 10T SRAM cell usually mean it consumes more power than other comparable architectures. Other transistors are used to achieve greater stability and functionality, which again consumes more power. In read delay, an 8T SRAM cell can have slightly higher delay than a 7T cell due to having more transistors. Power consumption and performance properties of different SRAM cell designs, namely the 6T, 7T, 8T, 9T, and 10T designs, are discussed in this chapter. The discussion brings to the fore recent developments in SRAM technology, including the introduction of the 7T SRAM cell, which shows a drastic drop in power consumption compared to the old 6T design. This has been reduced mainly because charging and discharging have been optimized for read and write operations. Although the 7T SRAM cell is effective at balancing power efficiency and performance, the more advanced 8T and 9T cells are better in terms of power consumption, as they have more transistors designed to boost stability and reduce read disturb. The 10T SRAM cell is the most power-consuming of the architectures under analysis, yet it offers the best stability and performance. The 6T SRAM cell has the lowest read delay, and the delay of the 7T cell is slightly higher, but this does not have a great impact on the general performance. The read delay also tends to increase with the number of transistors in the 8T, 9T, and 10T designs. In general, the results highlight that the design of SRAM cells must continue to be optimized to achieve an efficient compromise among power consumption, performance, and stability. The 7T SRAM cell may be a good bet for highly energy-conscious applications, and the 10T design can deliver good results for applications that require high stability. The next wave of research would be to further lower power consumption while maintaining the same performance, taking into account the increasing demand for energy-efficient memory solutions in current electronic devices.

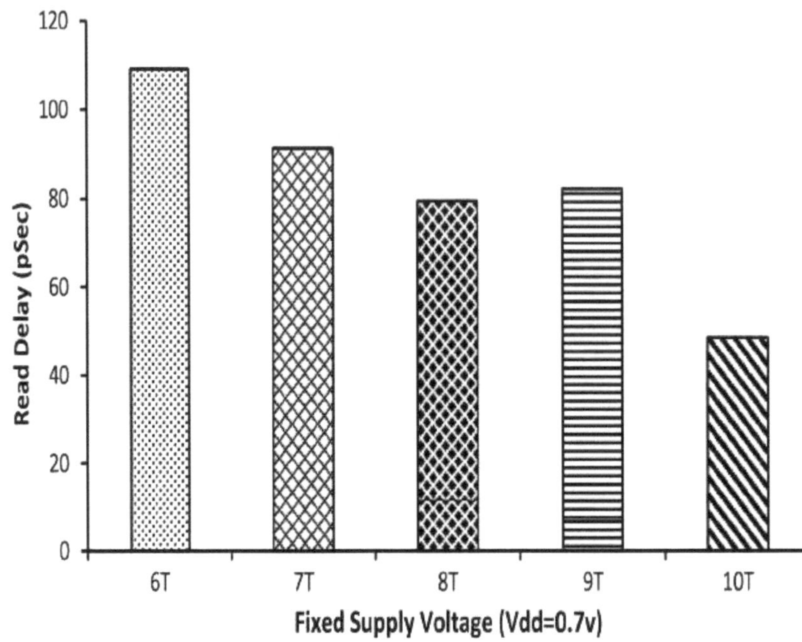

Fig. (8). Power Comparison of 6T, 7T, 8T, 9T, and 10T at 45nm technology node.

The 7T SRAM cell consumes less power than the previous 6T SRAM cell design. The reason is that a 7T SRAM cell can charge and discharge effectively whether it is reading or writing data. The SRAM cells are becoming better and better. This is a good example of the 7T SRAM cell. The 7T SRAM cell is better than the 6T design and is a good option because it consumes less power. The 8T and 9T versions are more complex and consume more power due to the additional transistors. These additional transistors are used to stabilize the 8T and the 9T SRAM cells. They reduce read disturb. The most superior is the 10T SRAM cell in terms of stability and performance. But the 10T SRAM cell utilises the power of all the SRAM cells that had been accessed. The power consumption is in the 10T SRAM cell. The 6T SRAM cell has the lowest read delay, and the 7T cell has a slightly higher delay, but this does not have a significant impact on overall performance. The read delay also tends to increase with the number of transistors in the 8T, 9T, and 10T designs.

In general, the results indicate that the SRAM cell design should be continuously optimized to achieve efficient power consumption, performance, and stability, whereas the 10T SRAM cell can be used when stability is crucial. Future studies should focus on minimising energy use without compromising performance, which will meet the increasing demand for energy-efficient memory solutions in contemporary electronic gadgets.

CONCLUSION

It is of great importance that SRAM cell (6T, 7T, 8T, 9T, and 10T) architectures consume power and perform well. The 7T SRAM cell consumes much more power than the 6T design. This is because it reads and writes data more efficiently. The 8T and 9T cells are more stable. They consume more power because they contain more transistors, but the 10T SRAM cell is the most stable. This is one issue to consider when discussing SRAM cell architectures, such as the 6T, 7T, 8T, 9T, and 10T architectures. The 6T cell has the best read delay, and the 7T cell shows a slight increase in access time. The 7T SRAM cell is quite good, when it comes to saving power. In making all the designs, we should make them use power. The 6T SRAM cell is the fastest in terms of read time. The 7T SRAM cell is slightly slower. The 8T, 9T, and 10T designs contain parts, which is why they are slower when we attempt to read something. The 7T SRAM cell is available because it conserves energy and is not extremely slow. On the whole, the results highlight the importance of continuously optimizing the SRAM cell design to achieve a reasonable balance between power, performance, and stability. The 7T SRAM cell is a potentially attractive choice when energy efficiency is important, whereas the 10T design can be applied in situations where stability is required. Further investigation is necessary in the future to reduce power consumption while still ensuring high performance, in response to the increasing adoption of energy-efficient memory in contemporary electronic gadgets.

AI DISCLOSURE

During the preparation of the work, the author(s) used AI tools and the Grammarly tool to enhance the writing style and readability of the content. The authors clarify that AI was not involved in generating the substantive content and that they take full responsibility for its publication.

REFERRENCES

[1] Yasuhiro Takahashi, Nazrul Anuar Nayan, Toshikazu Sekine, and Michio Yokoyama, "Low-power adiabatic 9T static random access memory", *J Eng,* IET, 2014.

[2] V.S. Kanchana Bhaaskaran, *Energy Recovery Performance of Quasi-Adiabatic Circuits using Lower Technology Nodes* IICPE, 2010.

[3] Mohammad Ansaria, Hassan Afzali-Kusha, Behzad Ebrahimi, Zainalabedin Navabi, Ali Afzali-Kush, and Massoud Pedram, "A near-threshold 7T SRAM cell with high write and read margins and low write time for sub-20 nm FinFET technologies", *Integr VLSI J,* 2015.

[4] Animesh Datta, Ashish Goel, Riza Tamer Cakici, Hamid Mahmoodi, Dheepa Lekshmanan, and Kaushik Roy, "Modeling and Circuit Synthesis for Independently Controlled Double Gate FinFET Devices", *IEEE Trans. On computer -aided design of Integrated Circuits and systems,* vol. 26, no. 11,

2007.
[http://dx.doi.org/10.1109/TCAD.2007.896320]

[5] B. Ebrahimi, S. Zeinolabedinzadeh, and A. Afzali-Kusha, "Low Standby Power and Robust FinFET Based SRAM Design", *IEEE Computer Society Annual Symposium on VLSI,* 2008.

[6] "Evert seevinck, senior member, IEEE, Frans j. List, and jan lohstroh, member, IEEE: Static-Noise Margin Analysis of MOS SRAM Cells", *IEEE J. Solid-State Circuits,* vol. SC-22, no. 5, 1987.

[7] B. Calhoun, and A. Chandrakasan, "A 256 kb subthreshold SRAM in 65 nm CMOS", *Proceedings of IEEE ISSCC,* pp. 480-481, 2006.

[8] Seung-Hwan Kim, and J.G. Fossum, "Design Optimization and Performance Projections of Double-Gate FinFETs With Gate–Source/Drain Underlap for SRAM Application", *IEEE Trans. Electron Dev.,* vol. 54, no. 8, pp. 1934-1942, 2007.
[http://dx.doi.org/10.1109/TED.2007.901070]

[9] R.V. Joshi, R.Q. Williams, E. Nowak, K. Kim, J. Beintner, T. Ludwig, and I. Aller,

[10] P.K. Pandey, and D.S. Yadav, "Parametric Inspection of 6T SRAM Cell", *2021 5th International Conference on Trends in Electronics and Informatics (ICOEI), Tirunelveli, India,* pp. 51-57, 2021.
[http://dx.doi.org/10.1109/ICOEI51242.2021.9453036]

[11] S. Chintaluri Arun Teja, Radha, H Victor du John, P Nagabushanam, "Analysis of Power in SRAM Cell with Various Pull up, Pull down and Pass Gate Transistors", *2022 IEEE 11th International Conference on Communication Systems and Network Technologies (CSNT),* pp. 141-145, 2022.

[12] V. Choudhary, and D.S. Yadav, "Analysis of Power, Delay and SNM of 6T & 8T SRAM Cells", *2021 5th International Conference on Electronics, Communication and Aerospace Technology (ICECA),* pp. 78-82, 2021.
[http://dx.doi.org/10.1109/ICECA52323.2021.9676022]

[13] A. Joy, M.A. Joseph, and J. Kuruvilla, "Implementation of a 4*4 10T SRAM Array using Domino Schmitt Trigger", *2022 International Conference on Industry 4.0 Technology (I4Tech), Pune, India,* 2022.
[http://dx.doi.org/10.1109/I4Tech55392.2022.9952907]

[14] S. G, M. k. S, K. N. Sathish, A. M, R. Indhumathi and E. E. Nithila, "Performance Evaluation of Auxiliary Threshold Low Power 6T SRAM Cell", *2022 6th International Conference on Trends in Electronics and Informatics (ICOEI), Tirunelveli, India,* pp. 140-144, 2022.
[http://dx.doi.org/10.1109/ICOEI53556.2022.9777142]

[15] V. Choudhary, and D.S. Yadav, "Analysis of Power, Delay and SNM of 6T & 8T SRAM Cells", *2021 5th International Conference on Electronics, Communication and Aerospace Technology (ICECA), Coimbatore, India,* pp. 78-82, 2021.
[http://dx.doi.org/10.1109/ICECA52323.2021.9676022]

[16] S. Sen, S. Saha, A. Biswas, and A. Dasgupta, "A Tanner Based Approach to analyse the comparisons between 6T and 8T SRAM", *2021 10th International Conference on Internet of Everything, Microwave Engineering, Communication and Networks (IEMECON), Jaipur, India,* pp. 01-05, 2021.
[http://dx.doi.org/10.1109/IEMECON53809.2021.9689129]

[17] C. Premalatha, K. Sarika, and P.M. Kannan, "A comparative analysis of 6T, 7T, 8T and 9T SRAM cells in 90nm technology", *2015 IEEE International Conference on Electrical, Computer and Communication Technologies (ICECCT),* 2015.
[http://dx.doi.org/10.1109/ICECCT.2015.7226147]

[18] Colin David Karat, K Soorya Krishna, "Design of SRAM array using 8T cell for low power sensor network", *2015 5th Nirma University International Conference on Engineering (NUiCONE),* pp. 1-6, 2015.

[19] C. Roy, and A. Islam, "Comparative analysis of various 9T SRAM cell at 22-nm technology node", *2015 IEEE 2nd International Conference on Recent Trends in Information Systems (ReTIS),* 2015.

[20] M.H. Tu, J.Y. Lin, M-C. Tsai, C-Y. Lu, Y-J. Lin, M-H. Wang, H-S. Huang, K-D. Lee, W-C. Shih, S-J. Jou, and C-T. Chuang, "A singleended disturbfree 9T subthreshold SRAM with cross-point dataaware write word-line structure, negative bit-line, and adaptive read operation timing tracing", *IEEE J. Solid-State Circuits,* vol. 47, no. 6, pp. 1469-1482, 2012. [http://dx.doi.org/10.1109/JSSC.2012.2187474]

[21] N. Hanchate, and N. Ranganathan, "LECTOR: a technique for leakage reduction in CMOS circuits", *IEEE Trans. Very Large Scale Integr. (VLSI) Syst.,* vol. 12, no. 2, pp. 196-205, 2004. [http://dx.doi.org/10.1109/TVLSI.2003.821547]

[22] S. Yadav, Y. Bansal, B. Joseph, and R.K. Kavitha, "Low-Power Dual-Vt 7T SRAM Bit-Cell With Reduced Area and Leakage", *2022 IEEE Delhi Section Conference (DELCON),* pp. 1-5, 2022.New Delhi, India [http://dx.doi.org/10.1109/DELCON54057.2022.9753131]

[23] G. Ravikishore, and N.M. Nandhitha, "6T-SRAM Design to Optimize Delay Using Finfet Technology", *2021 Third International Conference on Intelligent Communication Technologies and Virtual Mobile Networks (ICICV),* pp. 540-544, 2021.Tirunelveli, India [http://dx.doi.org/10.1109/ICICV50876.2021.9388559]

[24] A. Sachdeva, and V.K. Tomar, "Noise margin enhancement of Conventional 6T SRAM Cell by aspect Ratio Optimization", *2021 10th IEEE International Conference on Communication Systems and Network Technologies (CSNT), Bhopal, India,* pp. 264-268, 2021. [http://dx.doi.org/10.1109/CSNT51715.2021.9509552]

[25] G. Prasad, "Novel low power 10T SRAM cell on 90nm CMOS", *2016 2nd International Conference on Advances in Electrical, Electronics, Information, Communication and Bio-Informatics (AEEICB),* pp. 109-114, 2016. [http://dx.doi.org/10.1109/AEEICB.2016.7538408]

[26] D. Mittal, and V.K. Tomar, "Performance Evaluation of 6T, 7T, 8T, and 9T SRAM cell Topologies at 90 nm Technology Node", *2020 11th International Conference on Computing, Communication and Networking Technologies (ICCCNT),* pp. 1-4, 2020. [http://dx.doi.org/10.1109/ICCCNT49239.2020.9225554]

[27] V. Singh, S.K. Singh, and R. Kapoor, "Static Noise Margin Analysis of 6T SRAM", *2020 IEEE Int. Conf. Innov. Technol. Inocon,* pp. 6-9, 2020. [http://dx.doi.org/10.1109/INOCON50539.2020.9298431]

[28] V. B V and C. R, "A Comparative Study of SRAM Cells", *Ijireeice,* vol. 3, no. 11, pp. 92-94, 2015. [http://dx.doi.org/10.17148/IJIREEICE.2015.31119]

[29] Y. Berg, O. Mirmotahari, J.G. Lomsdalen, and S. Aunet, "High speed ultra low voltage CMOS inverter", *Proc. - IEEE Comput. Soc. Annu. Symp. VLSI Trends VLSI Technol. Des. ISVLSI,* pp. 122-127, 2008. [http://dx.doi.org/10.1109/ISVLSI.2008.23]

[30] D. Mittal, and V.K. Tomar, "Performance Evaluation of 6T, 7T, 8T, and 9T SRAM cell Topologies at 90 nm Technology Node", *2020 11th Int. Conf. Comput. Commun. Netw. Technol. ICCCNT,* pp. 4-8, 2020. [http://dx.doi.org/10.1109/ICCCNT49239.2020.9225554]

[31] R. Faraji, and H.R. Naji, "Adaptive technique for overcoming performance degradation due to aging on 6T SRAM Cells", *IEEE Trans. Device Mater. Reliab,* vol. 14, pp. 1031-1040, 2014. B. Dennington, "Low power design from technology challenge to great products," Proc. Int. Symp. Low Power Electron. Des., vol. 2006, p. 213, 2006 [http://dx.doi.org/10.1109/TDMR.2014.2360779]

[32] A. Kumar, "SRAM cell design with minimum number of transistor", In: *2014 Recent Adv. Eng. Comput. Sci.*, 2014, pp. 6-8. [http://dx.doi.org/10.1109/RAECS.2014.6799510]

[33] V. Rukkumani, M. Saravanakumar, and K. Srinivasan, "Design and analysis of SRAM cells for power reduction using low power techniques", *IEEE Reg. 10 Annu. Int. Conf. Proceedings/TENCON,* pp. 3058-3062, 2016.
[http://dx.doi.org/10.1109/TENCON.2016.7848609]

[34] K. Sharma, and M. Mehta, "Design and Performance Analysis of 6T SRAM on 130 nm", *Technology (Singap.),* pp. 363-368, 2019.

[35] A. Chandras, and V.S. Kanchana Bhaaskaran, "Sensing schemes of sense amplifier for singleended SRAM", *2017 Int. Conf. Nextgen Electron. Technol. Silicon to Software, ICNETS2,* pp. 379-384, 2017.
[http://dx.doi.org/10.1109/ICNETS2.2017.8067964]

[36] D. Anh-Tuan, K. Zhi-Hui, and Y. Kiat-Seng, "Hybrid-mode SRAM sense amplifiers: New approach on transistor sizing", *IEEE Trans. Circuits Syst. II Express Briefs,* vol. 55, no. 10, pp. 986-990, 2008.
[http://dx.doi.org/10.1109/TCSII.2008.2001965]

[37] A.S. Rajput, M. Pattanaik, and R. Tiwari, "Estimation of Static Noise Margin by Butterfly Method using Curve-Fitting Technique", *J. Act. Passiv. Electron. Devices,* vol. 13, no. 1, pp. 1-9, 2018.

[38] M. Yao, M. F. Cabanas-Holmen, and E. H. Cannon, *Direct Measurement Structure of SRAM SNM*pp. 273-276.

[39] C.D.C. Arandilla, A.B. Alvarez, and C.R.K. Roque, "Static noise Margin of 6T SRAM cell in 90-nm CMOS", *Proc. - 2011 UKSim 13th Int. Conf. Model. Simulation, UKSim,* pp. 534-539, 2011.
[http://dx.doi.org/10.1109/UKSIM.2011.108]

[40] S. Shaik, and P. Jonnala, "Performance evaluation of different SRAM topologies using 180, 90 and 45 nm technology", *2013 International Conference on Renewable Energy and Sustainable Energy (ICRESE),* 2013.
[http://dx.doi.org/10.1109/ICRESE.2013.6927819]

[41] R. Kolhal, and V. Agarwal, "A Power and Static Noise Margin Analysis of different SRAM cells at 180nm Technology", *2019 3rd International Conference on Electronics, Communication and Aerospace Technology (ICECA),* 2019.

CHAPTER 9

An Investigation of Junctionless Multigate Device and Its Application as 6T, 8T SRAM Cell at Sub-20 nm Technology Node

Devender Pal Singh[1,*], **Rashi Chaudhary**[2] and **Menka Yadav**[2]

[1] *Department of Electronics and Communication Engineering, Punjab Engineering College, Chandigarh, India*

[2] *Department of Electronics and Communication Engineering, Malaviya National Institute of Technology Jaipur, Jaipur, Rajasthan, India*

Abstract: In the electronic industry, the development of efficient FET devices with smaller feature sizes, lower power consumption, and improved performance has intensified competition. The key challenge remains in miniaturizing and creating energy-efficient devices. Multi-gate technologies like FinFETs have become a promising solution for further scaling, offering a compact design and superior current control. As technology continuous to progress, its importance increases, especially in narrowing the performance gap between processors and main memory. Therefore, SRAM has an important role in determining system performance, reliability, and power consumption. This chapter focuses on the design of 6T and 8T SRAM cells using a FinFET device. The key considerations in SRAM cell design include size, noise margin, and access time. Therefore, this chapter also explores SiGe-based SRAM cells, analysing their performance regarding noise margin and delay.

Keywords: FinFET-based SRAM, Junctionless finFET (JL-FinFET), Short-channel effects (SCEs), Low-power memory design.

INTRODUCTION

As the channel dimensions of conventional MOS transistors shrink to the deep nanoscale, below 20 nm, the continued advancement of Moore's law leads to undesirable short-channel effects (SCEs). These effects include higher leakage current, higher "Subthreshold Swing (SS)", greater "Drain Induced Barrier Lowering (DIBL)", and rising manufacturing cost [1 - 3]. The SS measures how quickly a device turns off below the threshold voltage (V_{th}) (V_{th} is the minimum gate voltage required to start conduction) [4, 5]. Ideally, SS should be around 60

* **Corresponding author Devender Pal Singh:** Department of Electronics and Communication Engineering, Punjab Engineering College, Chandigarh, India; E-mails: 2020rec9502@mnit.ac.in, devs.singh90@gmail.com

Girdhar Gopal, Meena Panchore, Arun Kishor Johar & Tarun Varma (Eds.)

mV/dec at room temperature, but this varies with the temperature. An "ideal" SS allows the >device to >switch OFF >rapidly, reducing >the leakage >current >and preserving the drive current. However, as transistor dimensions scale down, the device's performance deteriorates; SS flattens, V_{th} decreases, and the device becomes harder to turn OFF by reducing gate-source voltage (V_{GS}). DIBL is a phenomenon where drain potential further depletes the channel, worsens at nanoscale levels, resulting in inversion at lower V_{th}, increased leakage, and high static power dissipation. To continue scaling devices beyond physical limits, new device structures and materials are being explored. These advancements fall into two categories: (1) enhancing the CMOS platform by integrating additional technologies ("More Moore") and (2) introducing innovative computing paradigms or new device architectures ("Beyond CMOS") [6]. Beyond CMOS technologies may offer necessary breakthroughs in device architecture and techniques for the next era of computing [6]. Fig. (**1**) shows an overview of the scaling trends for beyond CMOS and More-Moore.

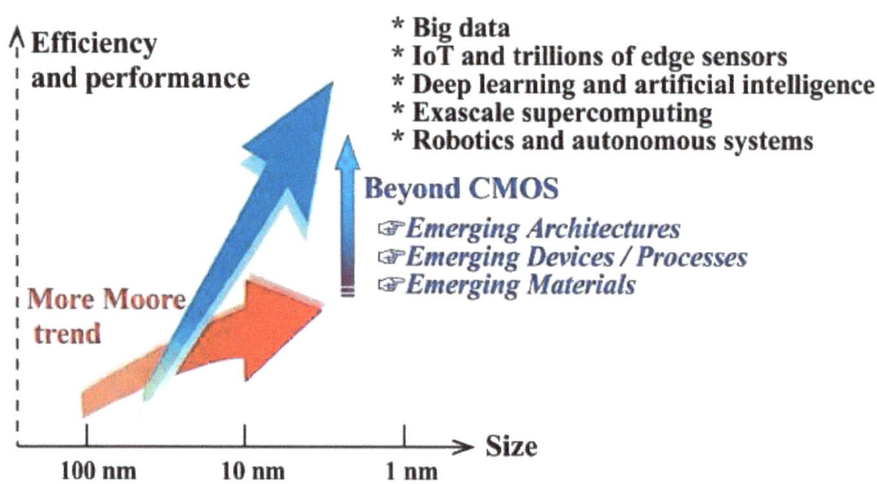

Fig. (1). Scaling trend and beyond CMOS scope [6].

This chapter provides an overview of the junctionless FinFET devices and compares them with the inversion mode device. This chapter also explores the application of FinFET devices as an application in designing SRAM cells. Furthermore, the analysis of 6T and 8T SRAM cells based on a FinFET device consisting of a stack channel (Si-SiGe-Si) is performed.

MULTI-GATE DEVICES

In highly scaled MOSFETs, the drain voltage can affect the channel's electrostatics, causing the gate to lose control over it. This limits the gate's ability to fully switch OFF the device in OFF mode, leading to an increase in OFF current (I_{OFF}) between the source and drain [7, 8]. One way to address this issue is by using thinner gate oxides and high-dielectric (high-k) materials, which increase the gate-source capacitance (C_{GS}). However, this approach has limitations due to the rise in the gate-induced drain leakage (GIDL) [7, 8]. Multi-gate FETs serve as an alternative to traditional FETs, providing larger CGS and better isolation of the drain potential from the channel. These devices outperform planar FETs in controlling short-channel effects (SCEs) like SS and DIBL, resulting in less degradation in V_{th} and lower I_{OFF} with device scaling. Among multi-gate structures, double-gate and tri-gate devices are preferred over conventional MOSFETs for their superior ability to reduce SCEs [8 - 10]. Further research has been conducted on multi-gate devices, including Tunnel FETs (TFETs), and FinFETs [11 - 14].

FinFETs have attracted significant attention in recent decades due to their superior resistance to SCEs compared to conventional (planar) MOSFETs [15]. Fig. (**2**) depicts the differentiation between a planar MOSFET with a horizontal channel and a FinFET with verticle channel. In the planar device, the key parameters are gate length (L_G) and device width (W), whereas in FinFETs, the critical parameters are L_G, fin height (H_{fin}), and fin thickness (t_{fin}) [16]. The FinFET structure gives rise to a unique characteristic known as "width quantization", which means the FinFET width depends on fin height and thickness. The effective fin width (W_{eff}) is calculated as: $W_{eff}=2H_{fin}+t_{fin}$ [17, 18].

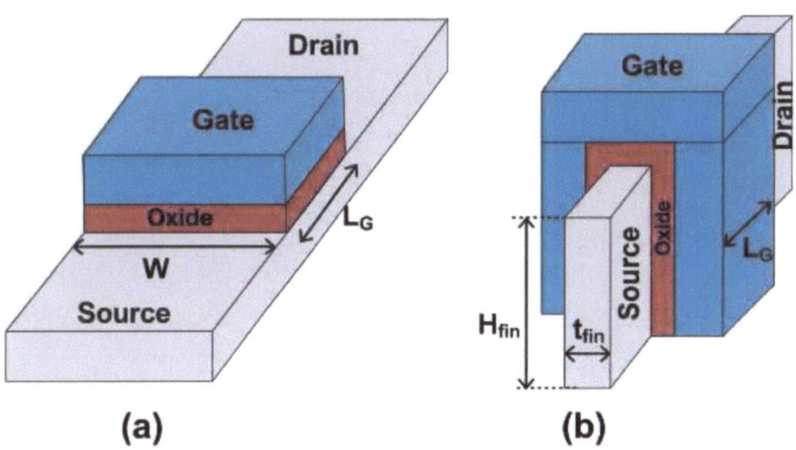

Fig. (2). (a) Planar MOSFET (b) FinFET device.

FinFETs may be built on both bulk and SOI substrates, as depicted in Fig. (**3**). SOI FinFETs have a "buried oxide layer (BOX)" made of silicon dioxide (SiO_2) between the fin and substrate. The oxide layer serves as an isolation barrier, boosting device performance by lowering parasitic capacitance and increasing isolation [18].

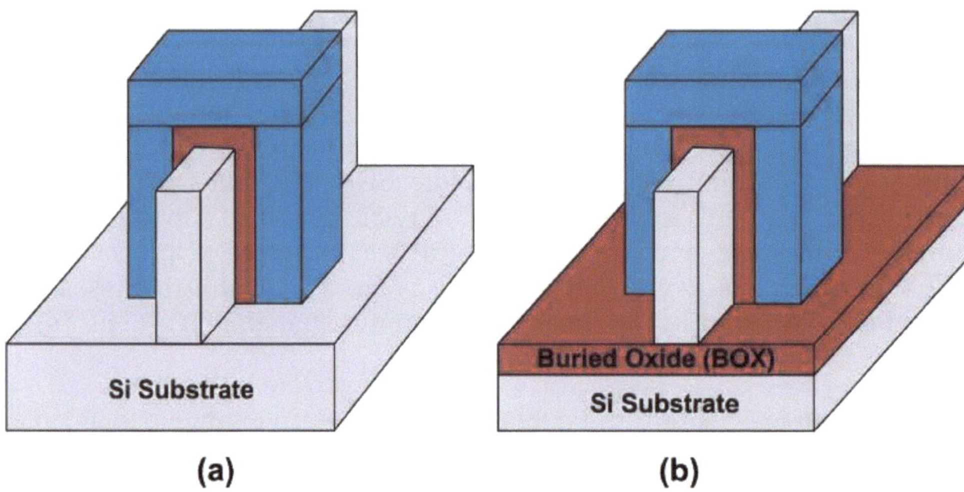

Fig. (3). (a) Bulk FinFET (b) Silicon-on-insulator (SOI) FinFET.

JUNCTIONLES FET

As devices decrease in size below 20 nm, apprehensions regarding random dopant fluctuations and work-function variations intensify. To address the manufacturing challenges of traditional transistors, a new architecture known as Junctionless (JL) transistors was introduced. In a JL FET, doping is uniform throughout the source, channel, and drain regions, eliminating the need for junctions. In inversion-mode (IM) devices, drain current flows along the channel surface, while in junctionless devices, current flows through the semiconductor's bulk. This design helps overcome key issues like junction leakage and breakdown that are common in IM devices, thereby improving overall reliability. Fig. (**4**) contrasts junction-based and junctionless devices, showing that JL transistors maintain consistent doping concentrations across the source, channel, and drain, whereas IM devices exhibit lower channel doping than the source and drain.

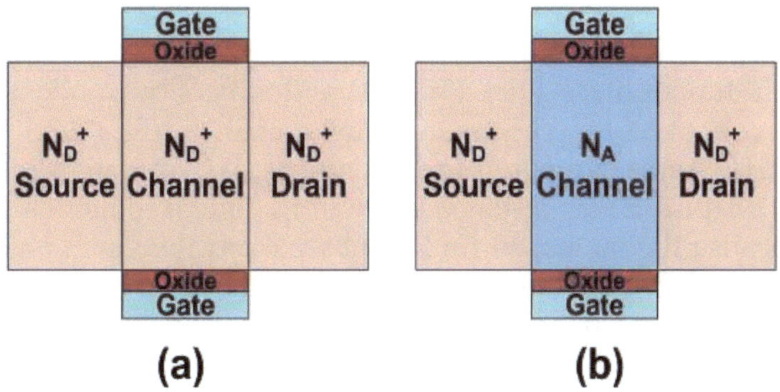

Fig. (4). (a) Junctionless (JL) device with uniform concentration in source, channel, and drain (b) Inversion mode (IM) device with channel concentration less than source and drain.

JL Device Working

Junctionless (JL) devices operate in three distinct regions: full depletion, semi-depletion, and accumulation. Fig. (**5**) illustrates the general operation of a JL FET [19, 20]. In Fig. (**5a**), the JL device is shown with "uniform" doping concentration across the source, channel, and drain. At zero gate-source voltage (V_{GS}), the channel is fully depleted of charge carriers, keeping the device in the OFF state, as shown in Fig. (**5b**). When the V_{GS} surpasses the "threshold voltage (V_{th})", the device enters the semi-depletion region. In this state, the reduced depletion in the centre of the semiconductor allows carrier movement through the bulk, turning the device ON, as illustrated in Fig. (**5c**). Finally, as V_{GS} surpasses the flat-band voltage (V_{FB}), carriers accumulate at the surface, enabling conduction freely across the channel, as depicted in Fig. (**5d**).

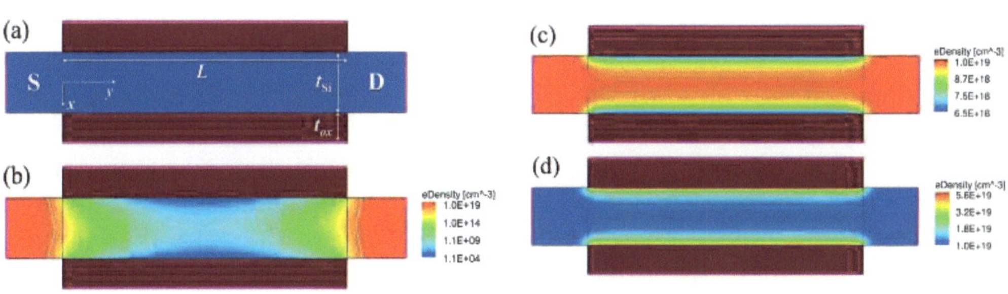

Fig. (5). (a) Schematic depiction of JL FET with uniform doping concentration throughout source, channel, and drain (b) Electron density distribution in full depletion ($V_{GS} < V_{th}$) (c) Electron density depletion in partial-depletion ($V_{th} < V_{GS} < V_{FB}$) (d) Electron density distribution in accumulation ($V_{GS} > V_{FB}$).

Comparison of IM and JL FinFET

Fig. (**6a**) illustrates the variation in I_{OFF} and the I_{ON}/I_{OFF} ratio at different fin heights for both Junctionless (JL) FinFET and inversion mode (IM) FinFET devices, with V_{DS} set at 0.6 V. The results show that JL-FinFETs exhibit lower I_{OFF} and a higher I_{ON}/I_{OFF} ratio compared to IM devices. As fin height increases, both JL-FinFET and IM-FinFET experience a rise in I_{OFF} and a reduction in the I_{ON}/I_{OFF} ratio. This is because the increased fin height weakens the gate's control over the channel, leading to higher leakage currents. Taller fins also provide more surface area for leakage paths, which contributes to increased leakage current. Notably, the lowest I_{OFF} is seen at a fin height of 12 nm, after which it rises with fin height. These results imply that JL-FinFETs outperform IM-FinFETs, especially at lower fin heights.

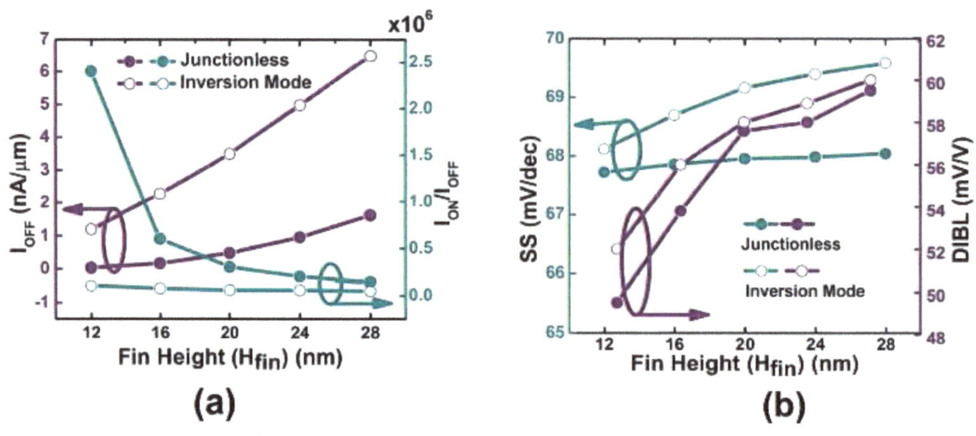

Fig. (6). JL-FinFET and IM-FinFET comparison at different fin heights (a) I_{OFF} and I_{ON}/I_{OFF} (b) SS and DIBL variation.

Fig. (**6b**) demonstrates how fin height affects SS and DIBL in both devices. JL-FinFETs exhibit lower SS and DIBL than IM-FinFETs; however, these metrics improve significantly with fin height. Taller fins allow the gate to regulate the channel more effectively, leading to higher capacitance. This increased capacitance might impede charging and discharging of the gate, thus impacting SS. Additionally, taller fins limit electrostatic control of the gate over the channel, allowing the drain voltage to have more impact, resulting in enhanced DIBL.

STACKED CHANNEL-BASED JUNCTIONLESS FINFET DEVICE

This work explores two types of channel structures for a junctionless FinFET: silicon (Si) and a S i-SiGe-Si configuration, as illustrated in Fig. (**7**). Fig. (**7a**) portrays the device structure with a stack of Si-SiGe-Si, where the channel is composed of a stack of Si-$Si_{0.75}Ge_{0.25}$-Si (here 0.75 and 0.25 indicate Si and Ge mole fractions, respectively). In this, the SiGe layer is positioned between two Si layers, whereas Fig. (**7b**) provides a 2D depiction of the Si Channel JL FinFET. The parameters, including gate length, fin thickness, and material, are aligned with IRDS 2022 "5 nm" process node [6], as detailed in Table **1**. The doping concentration in all regions is taken as 10^{19} cm^{-3}. As per the width quantization rule, the effective width (W_{eff}) of the device is calculated using the formula $W_{eff} = 2H_{fin} + t_{fin}$ [16]. The devices are designed using VisualTCAD, and simulations are performed using Genuis [21]. Our work [22] includes a detailed investigation of the Si-Ge-Si channel device by considering digital and analog parameters.

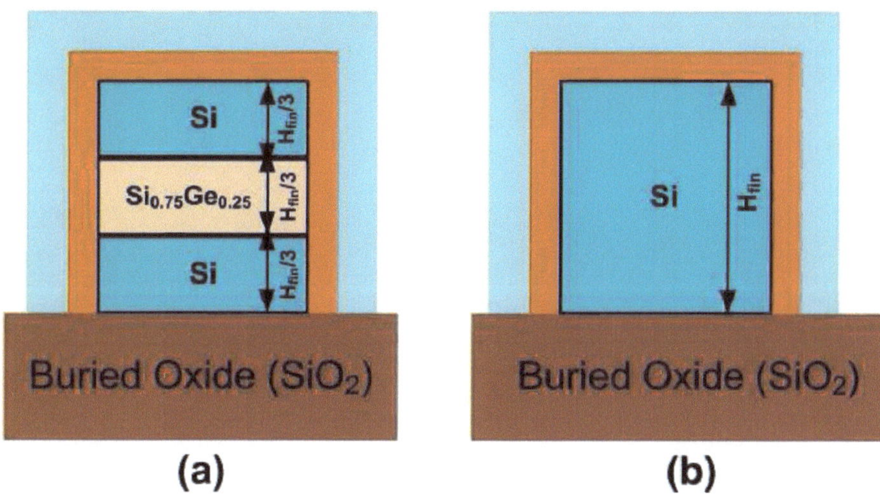

Fig. (7). Cross-sectional depiction of JL FinFET (a) Si-Ge-Si channel (b) Si channel.

Table 1. Parameters used for the device.

Parameter	Si-SiGe-Si channel	Si channel
Gate length (L_G) (nm)	20	20
Fin thickness (t_{fin}) (nm)	7	7
Fin height (H_{fin}) (nm)	36	39
Height of each layer (nm)	12	-
Doping (cm^{-3})	10^{19}	10^{19}

Fig. (**8a**) depicts the energy band diagrams of the Si channel and stacked-channel devices, showing that the stacked-channel device has a lower energy band gap than the Si channel device, due to the insertion of SiGe material. The insertion of SiGe material introduces strain, which improves the mobility of charge carriers. The modified band-gap and density-of-states are written below (here, x is Ge mole fraction) [23]:

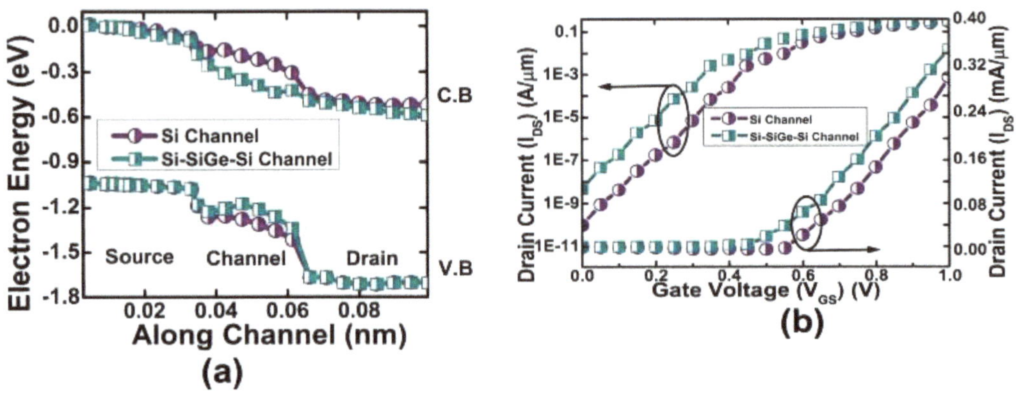

Fig. (8). (a) Energy band diagram (b) $I_{DS} - V_{GS}$ characteristics for Si and Si-SiGe-Si channel JL FinFET at V_{DS}=0.7 V.

$$\Delta E_g(SiGe) = 0.467x$$

$$N_V(SiGe) = \left(0.6x + 0.04(1-x)\right) * 10^{19} cm^{-3} \quad (1)$$

FINFET-BASED SRAM DESIGN

FinFETs are extensively utilized in advanced circuits for a range of applications, such as memory and various digital circuits [24]. A primary objective of memory research is to minimize the latency between the processor and primary memory, underscoring the importance of 'Static Random Access Memory (SRAM)' as technology advances. Today, SRAMs play a critical role in a system's overall power consumption and performance, as they occupy an important part of an integrated circuit. When designing SRAM cells, factors such as size, reliability, leakage current, and access time must be carefully considered. One of the challenges in achieving a high yield in SRAM production is the increasing variation in transistor performance that accompanies dimensional scaling [25, 26]. Additionally, since SRAM arrays spend substantial time in hold mode, standby power consumption is another significant concern.

Types of SRAM Cells

SRAM cells are a type of semiconductor memory used in logic circuits to store data in two stable states (0 and 1). The term "random access" indicates that each cell in an SRAM array can be read or written to in any order, without being affected by the sequence of previous accesses. An SRAM cell retains stored data as long as power is supplied, distinguishing it from dynamic RAM (DRAM), which requires periodic refreshing, and from non-volatile memory, such as flash memory, which does not need power to preserve data. In this chapter, we have primarily focused on the 6T and 8T memory cells designed using FinFET technology.

6T SRAM cell and working

A 6T memory cell is composed of six transistors arranged as cross-coupled inverters: T_1 and T_2 form the first inverter, while T_3 and T_4 make up the second inverter, as illustrated in Fig. (**9**). Each inverter contains one PMOS and one NMOS transistor. In Fig. (**9**), Q and Q_{bar} represent the output nodes of both inverters. Additionally, there are access transistors (NMOS) labelled N1 and N2 that facilitate the read and write functions of the bit-cell. The memory cell has two bit-lines: "Bit-Line (BL) and Bit-line bar (BL_{bar})", which are essential for accessing the stored data in the bit-cell. The "word-line (WL)" controls the operation of the access transistors N_1 and N_2 [27].

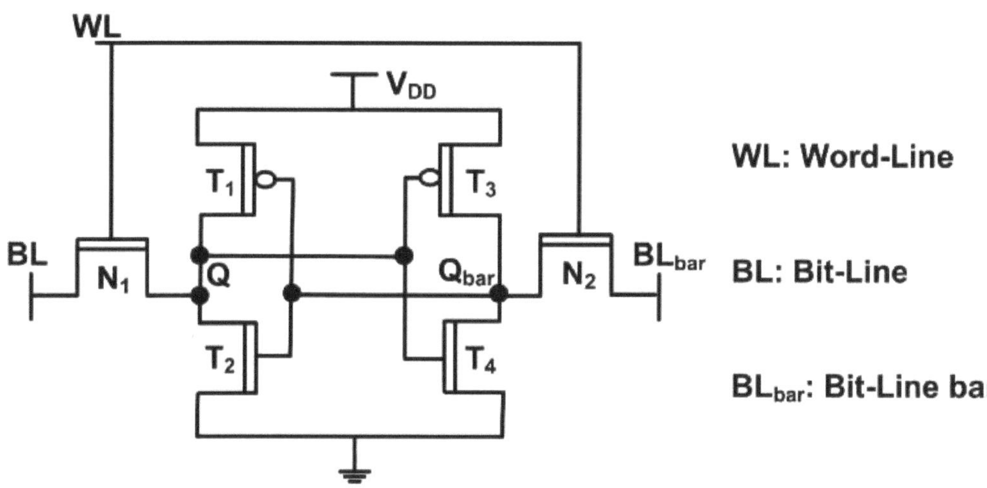

Fig. (9). 6T SRAM cell with cross-coupled inverters.

Hold Operation

In hold or standby mode, the "Word Line (WL)" is not activated (WL = 0), so the access transistors N1 and N2 are switched off. Consequently, this prevents the memory cell from being accessed or performing any read/write operations. The two cross-coupled inverters will keep feeding back into each other as long as they stay connected to the power source. This ensures that data is retained in the latch, keeping it stable without any external access or interference.

Read Operation

A general representation of the read '0' operation is portrayed in Fig. (**10a**). Initially, the capacitances C_1 and C_2 are precharged to V_{DD} using transistors P_1 and P_2, assuming that logic '0' is stored at node Q and logic '1' at Q_{bar}. With this configuration, transistors T_1 and T_4 are OFF, while T_2 and T_3 are ON. The access transistors N_1 and N_2 are activated by applying a logic '1' to the "Word Line (WL)".

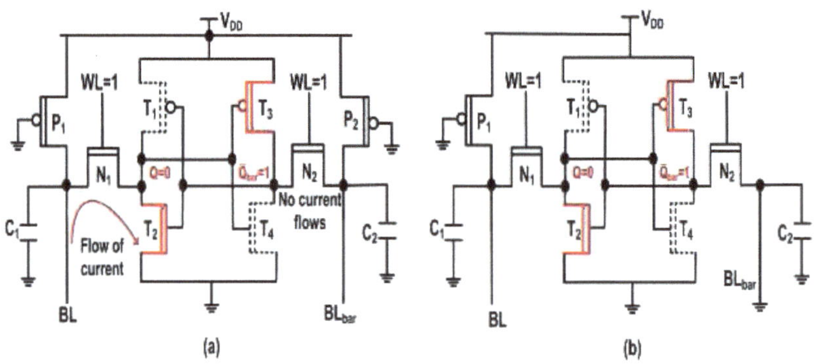

Fig. (10). 6T memory cell (a) Read '0' (b) Write '1' operation.

At this point, the source and drain of N_2 are at the same potential, so no current flows through that node. However, at N_1, the source and drain are at different potentials, which allows current to flow through N_1 along the path: $C_1 \rightarrow N_1 \rightarrow T_2 \rightarrow$ ground. Consequently, the voltage on the Bit Line (BL) begins to drop, leading to the discharge of C_1. This change is measured by the sense amplifier, indicating a read '0' operation.

Similarly, the read '1' operation can be performed under the assumption that Q stores logic '1'. In this case, the discharging of "Bit Line Bar (BL_{bar})" indicates a read '1' operation.

Write Operation

Fig. (**10b**) illustrates the generalized write function of the 6T memory cell. To implement the write '1' operation, we start with logic '0' stored at Q. In this configuration, transistors T_1 and T_4 are OFF, while T_2 and T_3 are ON. To begin the write operation, the transistors N_1 and N_2 are activated by applying a logic '1' to the "Word Line (WL)".

In this context, writing '1' means that the state of Q must change from '0' to '1', and the state of Q_{bar} must change from '1' to '0'. This is carried out by turning ON T_1 and T_4. For the write '1' operation, "Bit Line (BL)" and "Bit Line Bar (BL_{bar})" are connected to V_{DD} and "ground", respectively. This configuration results in the change of voltage levels at Q and Q_{bar}, as charge transfers occur: from BL to Q through N_1, and from Q_{bar} to BL_{bar} through N_2. As a result, Q transitions from '0' to '1', and Q_{bar} changes from '1' to '0', effectively completing the write '1' operation.

(b). 8T SRAM cell and Working

Fig. (**11**) shows the architecture of a standard 8T memory cell, which combines a 6T cell with two additional transistors, N_3 and N_4. The 8T bit-cell features distinct ports for read and write functions, identified as the "Read Word Line (RWL) and Write Word Line (WWL)", respectively.

For the read operation, the "Read Bit Line (RBL)" is precharged to V_{DD}, allowing data retrieval. In contrast, for the write operation, the "Write Bit Line (WBL) and Write Bit Line Bar (WBL_{bar})" are connected to V_{DD} and ground (or vice versa) to facilitate write '1' or write '0' operations. This design enhances the functionality and performance of the SRAM cell by providing dedicated read and write paths.

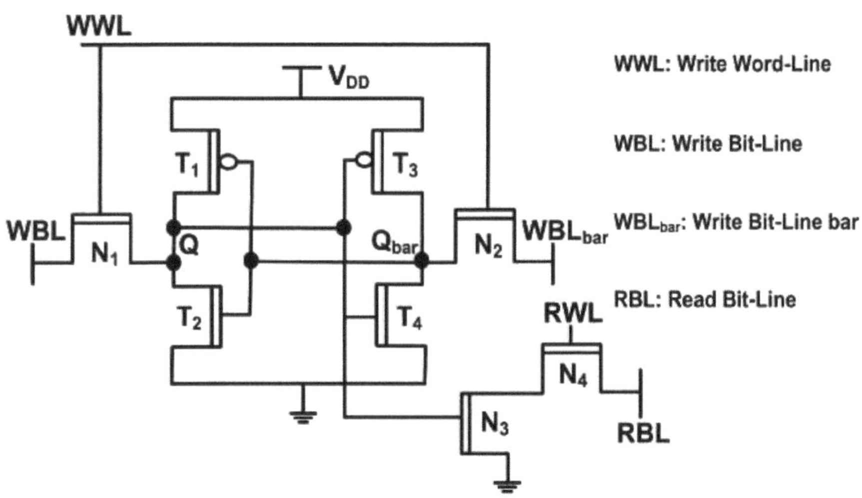

Fig. (11). 8T bit-cell consisting of a 6T memory cell and two additional transistors N_3, N_4.

Hold Operation

In hold or standby mode, both the "Write Word Line (WWL) and Read Word Line (RWL)" are applied as 0, resulting in the transistors N_1, N_2, and N_4 being turned OFF. This prevents the cell from being accessed and ensures that no read or write functions are performed. The two cross-coupled inverters will continue to maintain feedback to each other as long as they remain connected to the power supply. This configuration allows the stored data to be retained in the latch without any external interference or access.

Read Operation

A general representation of the read operation is illustrated in Fig. (**12a**). Initially, the "Read Bit Line (RBL)" is precharged to V_{DD} by keeping the "Read Word Line (RWL)" at low logic (RWL = 0) and connecting the "Write Word Line (WWL)" to 0 to turn OFF the transistors N_1 and N_2, ensuring they do not interfere with the read operation.

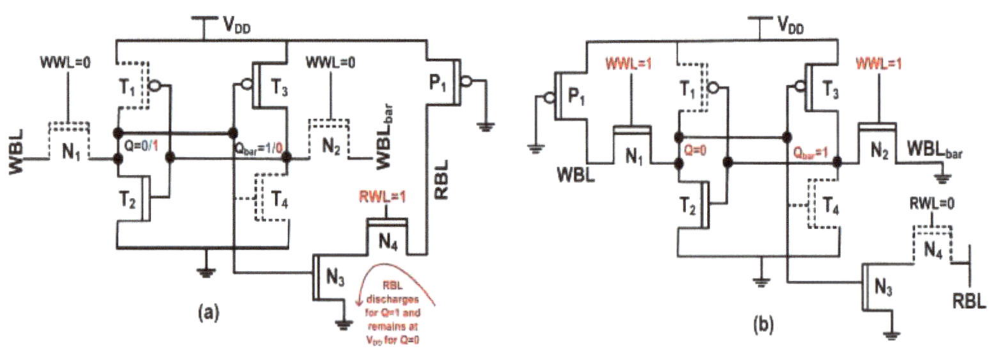

Fig. (12). (a) Read and (b) Write function of an 8T bit-cell.

When RWL is asserted to a high logic level, transistor N_4 is turned ON, allowing RBL to change based on the data stored in node Q. If Q holds a '0', transistor N_3 remains OFF, and RBL stays at its precharged state (V_{DD}) since there is no path to ground. This condition indicates a read '0' operation. Conversely, if Q stores logic '1', N_3 becomes conductive, and RBL discharges to ground through the path: RBL $\rightarrow N_4 \rightarrow N_3 \rightarrow$ ground. This results in a read '1' operation.

Write Operation

Fig. (**12b**) illustrates the generalized write operation of the 8T bit-cell. In this scenario, the "Read Word Line (RWL)" is connected to ground to prevent any read operations. For the write '1' operation, we assume that logic '0' is stored at Q and logic '1' at Q_{bar}. In this configuration, transistors T_1 and T_4 are OFF, while T_2 and T_3 are ON.

To initiate the write function, the devices N_1 and N_2 are activated by applying a logic '1' to the "Write Word Line (WWL)". Writing '1' means that the state of Q must change from '0' to '1' while Q_{bar} changes from '1' to '0'. This is accomplished by turning ON transistors T_1 and T_4.

The "Write Bit Line (WBL)" is attached to $V_{DD,}$ and the "Write Bit Line Bar (WBL_{bar})" is connected to ground to perform a write '1' operation. This configuration allows for voltage changes at Q and Q_{bar}, as charge transfers occur: from WBL to Q through N_1, and from Q_{bar} to WBL_{bar} through N_2. Consequently, Q transitions from '0' to '1', and Q_{bar} changes from '1' to '0', successfully completing the write '1' operation.

SRAM Performance Parameters

SRAM cells are evaluated using various performance parameters, with device sizing a fundamental design criterion for achieving adequate performance. In SRAM topology, two critical parameters for enhancing stability are the "**cell ratio (C.R.)** and **pull-up ratio (P.R.)**", defined as [28]:

1. **Cell Ratio (C.R.)**: It is the ratio of the pull-down transistors width to the access transistors width. For read stability, C.R > 1 is required.

2. **Pull-Up Ratio (P.R.)**: It refers to the ratio of pull-up transistor width to the access transistor width. For write stability, P. R<1 is required.

Other related SRAM performance parameters are discussed below:

Hold Static Noise Margin (HSNM)

"Static Noise Margin (SNM)" is the utmost DC noise voltage that a memory cell can sustain without affecting the stored data. SNM is important since it directly affects the reliability of a bit-cell. A larger SNM suggests greater stability and a reduced probability of error during read, write, and hold functions. In contrast, a low SNM implies that a bit-cell is more vulnerable to noise, increasing the chance of data corruption.

"Hold Static Noise Margin (HSNM)" specially evaluates the bit-cell's stability in hold or standby mode. As depicted in Fig. (**13a**), the HSNM is defined as the length of the side of the greatest square that can be inscribed within a "butterfly curve". This curve is obtained by combining the voltage transfer characteristics (VTCs) of the inverters of the bit-cell [27].

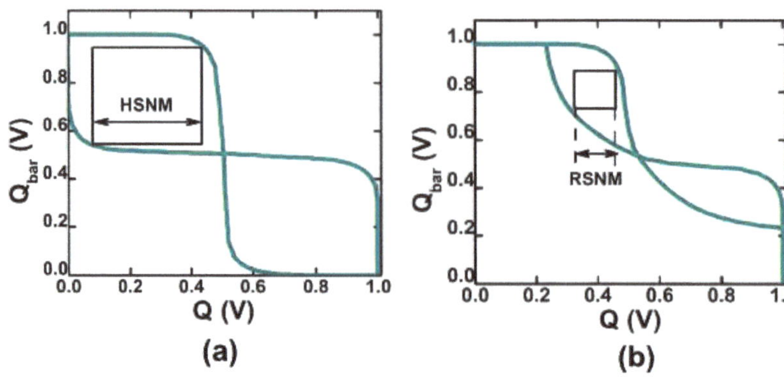

Fig. (13). General representation of (a) HSNM (b) RSNM for an SRAM cell.

Read Static Noise Margin (RSNM)

The read strength of a bit-cell is characterized by the “Read Static Noise Margin (RSNM)”, which is the length of the maximum square inscribed within a “butterfly curve.” The measurement of RSNM follows a similar approach to that of “Hold Static Noise Margin (HSNM)”, with the key difference being that the word line is connected to V_{DD} during the measurement to trace the butterfly curve.

During the read operation, the bit line either discharges or remains at its precharged state. This behaviour allows for the generation of the inverter's Voltage Transfer Characteristics (VTC), from which the RSNM can be determined, as depicted in Fig. (**13b**). The RSNM is crucial for assessing the SRAM cell's stability during read operations, providing insights into its susceptibility to data corruption under varying noise conditions.

Write Static Noise Margin (WSNM)

“Write Static Noise Margin (WSNM)” pertains to the ability to write information into a bit-cell. It quantifies the maximum noise voltage on the bit lines during a successful write operation. The data intended for writing is placed on the bit lines, while WSNM is the difference between the VDD voltage and the word line voltage at which the nodes Q and Q_{bar} transition states.

A lower WSNM value indicates greater difficulty in writing information into the cell, making it more susceptible to errors. Therefore, optimizing WSNM is essential for ensuring reliable data storage and retrieval in SRAM cells.

Read and Write Access Time

The read delay is the interval between the application of the “word-line (WL)” and the moment when data becomes available at the output node of the bit-cell. Specifically, it measures the time difference between WL activation and the time when the bit-line discharges to 50% of its initial value, which is sufficient for a sense amplifier to measure.

Conversely, write access time is the time required to write data into a bit cell after initiating the write function. It is measured from the moment the WL is activated until the charging node Q or Q_{bar} reaches 90% of V_{DD}. Both access times are critical for determining the overall performance and efficiency of SRAM cells in memory applications.

Subthreshold Leakage Power

Subthreshold leakage power is an essential metric in memory design, as it contributes significantly to the bit-cell's total power dissipation and can affect the circuit’s efficiency. This leakage occurs in the subthreshold region, where the device begins to operate weakly, leading to power consumption even when no activities are performed (such as reading or writing). Increased subthreshold leakage can lead to greater fluctuations in stored data over time, risking the memory cell’s stability.

FinFET SRAM cell: Literature

Table **2** shows a general review of various SRAM cells designed using a FinFET device.

Table 2. A review of FinFET SRAM cell.

SRAM cell type	Specifications	Reference
8T SRAM cell consisting of stacked p and n-FinFET for inverter.	Works on low voltage than the 6T SRAM cell due to additional p and n-FinFETs.	[29]
8T SRAM cell with tied-gate FinFETs.	Works on 44.3% reduced power supply voltage as compared to a 6T cell.	[30]
6T SRAM cell using Junctionless Accumulation Mode FinFETs.	Dual-k spacers improve read and write noise margin by 9.7%, 3.8%, respectively, and reduces 25% leakage compared to the 6T cell.	[31]
6T and 8T using shorted-gate FinFET.	Energy delay product reduces using power-gating or at near-threshold operation.	[32]
8T SRAM cell with virtual ground in read path.	Minimizes leakage through access transistors as it utilizes a single bit-line for the write function.	[33]

(Table 2) cont.....

SRAM cell type	Specifications	Reference
6T SRAM using asymmetric FinFET.	Improves noise margin due to built-in feedback mechanism.	[34]
6T cell using negative capacitance FinFET.	Negative capacitance improves stability metrics, read, write, and hold noise margin.	[35]
8T SRAM cell using shorted-gate FinFET.	Read-decoupling technique enhances read stability, and write-assist pull-down path cut scheme to improve write ability.	[36]

ANALYSIS OF SI-SIGE-SI CHANNEL-BASED SRAM CELL

The standard 8T bit-cell, as depicted in Fig. (**11**), features separate "read word line (RWL) and write word line (WWL)" connections. Initially, the "read bit-line (RBL)" is precharged to V_{DD} start the read function. To perform a write operation on the output nodes Q or Q_{bar}, the "write bit line (WBL) and write bit line bar (WBL_{bar})" are attached to V_{DD} or ground accordingly.

The RSNM curves for the 6T and 8T SRAM cells at V_{DD} = 1 V are illustrated in Fig. (**14a**), and it is evident that the 8T bit-cell possesses a greater noise margin than the 6T SRAM cell.

In Fig. (**14b**, **c** and **d**) a comparison is made among the 6T and 8T bit-cells in terms of "RSNM, HSNM, and WSNM", respectively. The lowest RSNM is obtained for the 6T memory cell due to the phenomenon known as "read disturb." During read operations, the access devices can adversely impact the "bit lines and storage node", leading to potential disturbances.

Furthermore, the results reveal that RSNM, HSNM, and WSNM values increase with larger supply voltages (V_{DD}) and decline when V_{DD} is reduced. A higher V_{DD} allows for greater voltage fluctuations within the circuit, thereby affecting the range of input voltages that may be recognised as acceptable logic levels. Additionally, when V_{DD} increases, the voltage margins for both logic '1' and logic '0' input levels widen, resulting in greater noise margin. In contrast, lowering V_{DD} reduces the voltage differential between the SRAM cell's stable states, making it more susceptible to noise.

Delay at reduced V_{DD} is also an important consideration in current semiconductor design. The electric field across the transistor channel likewise becomes less when the supply voltage drops. Read and write operations have longer delays as a result of the restriction in charge carrier mobility. Thus, in order to guarantee the stability of SRAM circuits, it is essential to examine these delays at a lower supply voltage.

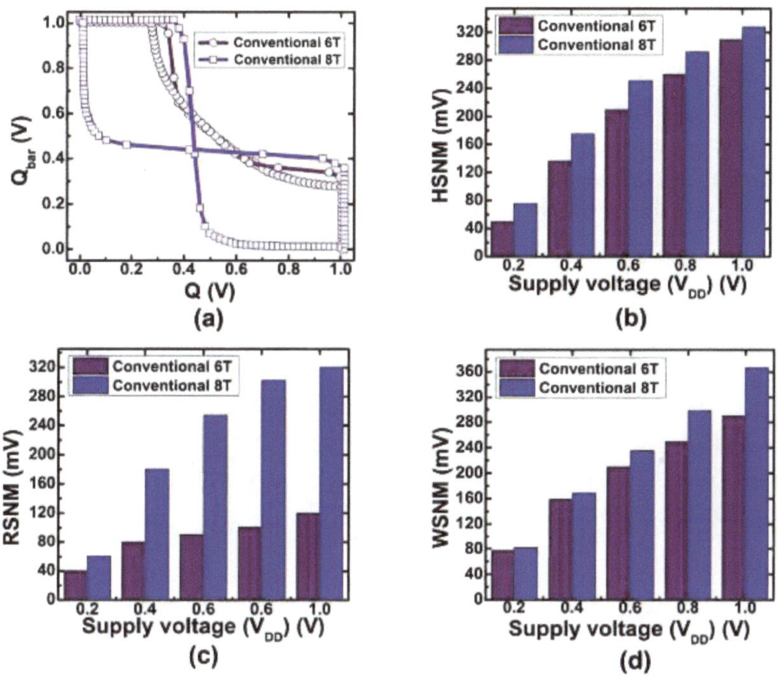

Fig. (14). (a) RSNM comparison of Si-SiGe-Si channel FinFET-based conventional 6T and 8T SRAM cell, (b) variation of HSNM, (c) variation of RSNM, (d) variation of WSNM with V_{DD}.

Fig. (15) depicts the read '1' and write '1' delays for different memory cells at a supply voltage of 0.2 V. In comparison to the 6T bit-cell, the standard 8T SRAM cell has a shorter read access time due to its decoupled read port and two additional devices for read operation. This performance improvement is due to the additional transistors in the 8T bit-cell, which improve data accessibility during read operations, thereby lowering the total read time.

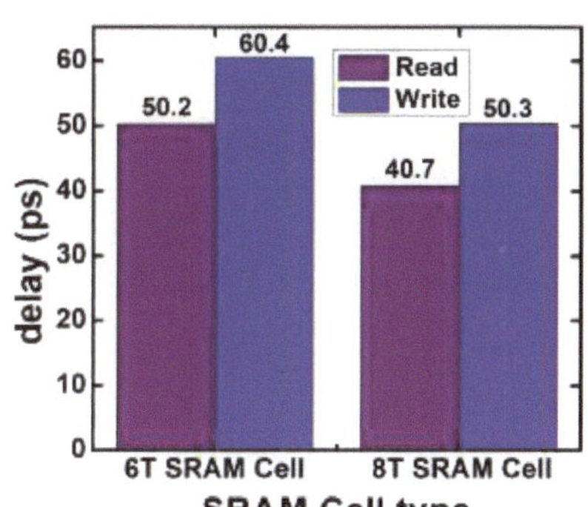

Fig. (15). Read and write delay comparison of Si-SiGe-Si channel-based FinFET 6T and 8T SRAM cell.

CONCLUSION

The chapter explores the analysis of junctionless multigate devices and their comparison with inversion-mode transistors. The chapter also investigates a Si-SiGe-Si channel-based junctionless FinFET that improves drain current by inserting SiGe material. As an application of the device, conventional 6T and 8T SRAM cells are also explored, in which the analysis of read, write, and static noise margin is compared for both the SRAM cells. The results show that the conventional 8T bit-cell improves the noise margin over the standard 6T cell.

AI DISCLOSURE

During the preparation of work, author(s) used AI tools and Grammarly tool to enhance the writing style and readability of the content. The authors clarify that AI was not involved in generating the substantive content and that they take full responsibility for the publication.

REFERENCES

[1] G.V. Angelov, D.N. Nikolov, and M.H. Hristov, "Technology and modeling of nonclassical transistor devices", *J. Electr. Comput. Eng.,* vol. 2019, pp. 1-18, 2019. [http://dx.doi.org/10.1155/2019/4792461]

[2] G.D. Hutcheson, *"Moore's law, lithography, and how optics drive the semiconductor industry," in Extreme Ultraviolet (EUV) Lithography IX* vol. Vol. 10583. SPIE, 2018, p. 1058303.

[3] G.E. Moore, "Cramming more components onto integrated circuits", *Proc. IEEE,* vol. 86, no. 1, pp. 82-85, 1998. [http://dx.doi.org/10.1109/JPROC.1998.658762]

[4] D.J. Frank, R.H. Dennard, E. Nowak, P.M. Solomon, Y. Taur, and H-S.P. Wong, "Device scaling limits of Si MOSFETs and their application dependencies", *Proc. IEEE,* vol. 89, no. 3, pp. 259-288, 2001. [http://dx.doi.org/10.1109/5.915374]

[5] Y-B. Kim, "Challenges for nanoscale mosfets and emerging nanoelectronics", *transactions on electrical and electronic materials,* vol. 11, no. 3, pp. 93-105, 2010. [http://dx.doi.org/10.4313/TEEM.2010.11.3.093]

[6] I. R. Committee, *International roadmap for devices and systems, 2022 edition*, 2022.

[7] C. Hu, "Gate oxide scaling limits and projection", In: *International Electron Devices Meeting* Technical Digest. IEEE, 1996, pp. 319-322.

[8] Yee-Chia Yeo, Tsu-Jae King, and Chenming Hu, "MOSFET gate leakage modeling and selection guide for alternative gate dielectrics based on leakage considerations", *IEEE Trans. Electron Dev.,* vol. 50, no. 4, pp. 1027-1035, 2003. [http://dx.doi.org/10.1109/TED.2003.812504]

[9] P. Solomon, K. Guarini, Y. Zhang, K. Chan, E. Jones, G. Cohen, A. Krasnoperova, M. Ronay, O. Dokumaci, and H. Hovel, *Two gates are better than one [double-gate mosfet process]* vol. 19. IEEE circuits and devices magazine, 2003, no. 1, pp. 48-62. [http://dx.doi.org/10.1109/MCD.2003.1175108]

[10] H-S. Wong, D.J. Frank, and P.M. Solomon, "Device design considerations for doublegate, ground-plane, and single-gated ultra-thin soi mosfet's at the 25 nm channel length generation", *International*

Electron Devices Meeting, IEEE, pp. 407-410, 1998.

[11] P. Vimala, T.S.A. Samuel, D. Nirmal, and A.K. Panda, "Performance enhancement of triple material double gate TFET with heterojunction and heterodielectric", *Solid State Electronics Letters,* vol. 1, no. 2, pp. 64-72, 2019. [http://dx.doi.org/10.1016/j.ssel.2019.10.001]

[12] A. Chauhan, G. Saini, and P.K. Yerur, "Improving the performance of dual-k spacer underlap Double Gate TFET", *Superlattices Microstruct.,* vol. 124, pp. 79-91, 2018. [http://dx.doi.org/10.1016/j.spmi.2018.10.006]

[13] M. Yadav, A. Bulusu, and S. Dasgupta, "Two dimensional analytical modeling for asymmetric 3T and 4T double gate tunnel FET in sub-threshold region: Potential and electric field", *Microelectronics J.,* vol. 44, no. 12, pp. 1251-1259, 2013. [http://dx.doi.org/10.1016/j.mejo.2013.08.011]

[14] B. Kumar, M. Sharma, and R. Chaujar, "Junctionless-accumulation-mode stacked gate GAA FinFET with dual-k spacer for reliable RFIC design", *Microelectronics J.,* vol. 139, p. 105910, 2023. [http://dx.doi.org/10.1016/j.mejo.2023.105910]

[15] J. Chang, M. Guillorn, P. Solomon, C-H. Lin, S.U. Engelmann, A. Pyzyna, J. Ott, and W. Haensch, "Scaling of soi finfets down to fin width of 4 nm for the 10nm technology node", *in 2011 Symposium on VLSI Technology-Digest of Technical Papers. IEEE,* pp. 12-13, 2011.

[16] D. Hisamoto, T. Kaga, Y. Kawamoto, and E. Takeda, "A fully depleted lean-channel transistor (delta)- a novel vertical ultra thin soi mosfet", *International Technical Digest on Electron Devices Meeting,* IEEE, pp. 833-836, 1989. [http://dx.doi.org/10.1109/IEDM.1989.74182]

[17] J-P. Colinge, *FinFETs and other multi-gate transistors* vol. 73. Springer, 2008. [http://dx.doi.org/10.1007/978-0-387-71752-4]

[18] D. Bhattacharya, and N.K. Jha, *Finfets: From devices to architectures* vol. 2014. Advances in Electronics, 2014.

[19] Z. Chen, Y. Xiao, M. Tang, Y. Xiong, J. Huang, J. Li, X. Gu, and Y. Zhou, "Surface potential-based drain current model for long-channel junctionless double-gate mosfets", *IEEE Trans. Electron Dev.,* vol. 59, no. 12, pp. 3292-3298, 2012. [http://dx.doi.org/10.1109/TED.2012.2221164]

[20] A. Baidya, S. Baishya, and T.R. Lenka, "Impact of thin high-k dielectrics and gate metals on RF characteristics of 3D double gate junctionless transistor", *Mater. Sci. Semicond. Process.,* vol. 71, pp. 413-420, 2017. [http://dx.doi.org/10.1016/j.mssp.2017.08.031]

[21] Cogenda Genius Parallel 3D TCAD Simulation Tool User'S Guide Cogenda Genius 2.2: User's Guide.

[22] D. Singh, S. Chaudhary, B. Dewan, and M. Yadav, "Performance optimization of tri-gate junctionless FinFET using channel stack engineering for digital and analog/RF design", *Journal of semiconductors,* vol. 44, 2023. [http://dx.doi.org/10.1088/1674-4926/44/11/114103]

[23] M. Rahimian, A.A. Orouji, and A. Aminbeidokhti, "A novel deep submicron SiGe-on-insulator (SGOI) MOSFET with modified channel band energy for electrical performance improvement", *Curr. Appl. Phys.,* vol. 13, no. 4, pp. 779-784, 2013. [http://dx.doi.org/10.1016/j.cap.2012.12.005]

[24] M.U. Mohammed, A. Nizam, L. Ali, and M.H. Chowdhury, "FinFET based SRAMs in Sub-10nm domain", *Microelectronics J.,* vol. 114, p. 105116, 2021. [http://dx.doi.org/10.1016/j.mejo.2021.105116]

[25] N. Agrawal, Huichu Liu, R. Arghavani, V. Narayanan, and S. Datta, "Impact of variation in nanoscale silicon and non-silicon finfets and tunnel fets on device and sram performance", *IEEE Trans. Electron*

Dev., vol. 62, no. 6, pp. 1691-1697, 2015.
[http://dx.doi.org/10.1109/TED.2015.2406333]

[26] X. Zhang, J. Li, M. Grubbs, M. Deal, B. Magyari-K¨ope, B.M. Clemens, and Y. Nishi, "Physical model of the impact of metal grain work function variability on emerging dual metal gate mosfets and its implication for sram reliability", *2009 IEEE International Electron Devices Meeting (IEDM),* IEEE, pp. 1-4, 2009.

[27] J. Singh, S.P. Mohanty, and D.K. Pradhan, *Robust SRAM designs and analysis.* Springer Science & Business Media, 2012.

[28] R.B. Almeida, C.M. Marques, P.F. Butzen, F.R.G. Silva, R.A.L. Reis, and C. Meinhardt, "Analysis of 6 T SRAM cell in sub-45 nm CMOS and FinFET technologies", *Microelectron. Reliab.,* vol. 88-90, pp. 196-202, 2018.
[http://dx.doi.org/10.1016/j.microrel.2018.07.134]

[29] G. Pasandi, and S.M. Fakhraie, "An 8t low-voltage and low-leakage half-selection disturb-free sram using bulk-cmos and finfets", *IEEE Trans. Electron Dev.,* vol. 61, no. 7, pp. 2357-2363, 2014.
[http://dx.doi.org/10.1109/TED.2014.2321295]

[30] S.M. Salahuddin, and M. Chan, "Eight-finfet fully differential sram cell with enhanced read and write voltage margins", *IEEE Trans. Electron Dev.,* vol. 62, no. 6, pp. 2014-2021, 2015.
[http://dx.doi.org/10.1109/TED.2015.2424376]

[31] G. Saini, and S. Choudhary, "Improving the performance of SRAMs using asymmetric junctionless accumulation mode (JAM) FinFETs", *Microelectronics J.,* vol. 58, pp. 1-8, 2016.
[http://dx.doi.org/10.1016/j.mejo.2016.10.004]

[32] M.A. Turi, and J.G. Delgado-Frias, "Effective low leakage 6t and 8t finfet srams: using cells with reverse-biased finfets, near-threshold operation, and power gating", *IEEE Trans. Circuits Syst. II Express Briefs,* vol. 67, no. 4, pp. 765-769, 2020.
[http://dx.doi.org/10.1109/TCSII.2019.2922921]

[33] B.N.K. Reddy, C. Ramalingaswamy, R. Nagulapalli, and D. Ramesh, "A novel 8T SRAM with improved cell density", *Analog Integr. Circuits Signal Process.,* vol. 98, no. 2, pp. 357-366, 2019.
[http://dx.doi.org/10.1007/s10470-018-1309-z]

[34] R. Niaraki Asli, and S. Taghipour, "Reliable and high performance asymmetric FinFET SRAM cell using back-gate control", *Microelectron. Reliab.,* vol. 104, p. 113545, 2020.
[http://dx.doi.org/10.1016/j.microrel.2019.113545]

[35] S. Yadav, P.N. Kondekar, and B. Awadhiya, "Performance estimation of non-hysteretic negative capacitance FinFET based SRAM", *Microelectronics J.,* vol. 137, p. 105796, 2023.
[http://dx.doi.org/10.1016/j.mejo.2023.105796]

[36] E. Mani, P. Nimmagadda, S.J. Basha, M.A. El-Meligy, and H.A. Mahmoud, "A FinFET-based low-power, stable 8T SRAM cell with high yield", *AEU Int. J. Electron. Commun.,* vol. 175, p. 155102, 2024.
[http://dx.doi.org/10.1016/j.aeue.2023.155102]

CHAPTER 10

A Comparative Study of Split-Gate *vs.* Non-Split Gate Doping-Less Heterojunction Tunnel FET: Performance and Design Insights

Basudha Dewan[1,*]

[1] *Department of Electronics & Communication Engineering, Manipal University Jaipur, Jaipur, Rajasthan, India*

Abstract: In this chapter, an analysis of a split-gate, doping-free heterojunction tunnel field-effect transistor (SG-DG-HJ-TFET) compared with a heterojunction tunnel field-effect transistor with a non-split-gate (NSG) architecture will be presented. The Doping-Free Tunnel Field-Effect Transistor (DF-TFET) is a promising next-generation design, offering robustness against random dopant fluctuations and eliminating the need for high-temperature processing and expensive annealing. The performance of both configurations is analyzed and compared in terms of DC performance. Parameters such as drain current, electric field distribution, surface potential, sub-threshold swing (SS), current, *etc.*, are evaluated during this evaluation. The temperature dependency of both the TFET architectures is compared and reported in this work. The Silvaco ATLAS technology is used for two-dimensional design simulation to implement the suggested design. The fabrication flow of the proposed SG-DG-HJ-TFET is also described in detail.

Keywords: Band-to-band tunneling (BTBT), DopingLess, Potential, Short channel effects (SCEs), TFET.

INTRODUCTION

Reducing device power consumption requires a significant drop in the supply voltage (V_{DD}). To retain the same ON current (I_{ON}), the decrease in the threshold voltage (V_{TH}) is also necessary so that there is sufficient overdrive voltage ($V_{DD} - V_{TH}$) maintained. But when we reduce V_{DD} and V_{TH}, the OFF current (I_{OFF}) increases noticeably.

This surge in I_{OFF} is inevitable due to the persistent Sub-threshold Swing (SS) of Metal Oxide Semiconductor Field Effect Transistors (MOSFETs) with a theore-

* **Corresponding author Basudha Dewan:** Department of Electronics & Communication Engineering, Manipal University Jaipur, Jaipur, Rajasthan, India; E-mail: basudha.ece@gmail.com

Girdhar Gopal, Meena Panchore, Arun Kishor Johar & Tarun Varma (Eds.)

tical minimal threshold of 60 millivolts/decade at ambient temperatures. Because of such thermionic limitations on the MOSFETs' Sub-threshold Swing (SS), contemporary electronic devices that rely on MOSFET technology have trouble operating at reduced power supplies [1, 2]. A solution to overcome this problem is by suggesting devices with SS scaling <60 mV/decade. Tunnel Field-Effect Transistors (TFETs) are emerging as a promising alternative to conventional CMOS logic devices due to their unique characteristics [3 - 5]. Unlike MOSFETs, TFETs use BTBT as the primary carrier-injection mechanism rather than thermionic emission. This distinction allows TFETs to offer a variable and scalable sub-threshold Swing (SS) that is lower than 60 mV/decade [6 - 8]. Nevertheless, limitations of TFETs, including low I_{ON} (drain current) and ambipolarity, continue to hinder their potential to replace conventional MOSFETs. TFETs exhibit lower I_{ON} values than MOSFETs, which poses a significant obstacle to their overall performance and suitability as a replacement technology.

The ON-state performance of TFETs has been improved by a number of methods that emphasize both structural and material-based techniques [9 - 14]. These solutions aim to enhance the overall efficacy and functionality of TFET devices. An alternative approach, known as the charge-plasma-based structure, has been documented to enhance the effectiveness of TFET devices [15 - 17]. This structure avoids conventional doping of the source and drain regions. As an alternative, the charge concentration is brought about by introducing metal into contact with the semiconductor contacts. However, specific conditions related to the work function and film thickness must be met for this method to be effective. We present a new design in this work, termed the split-gate doping-less heterojunction TFET (SG-DL-HJ HTFET). A split-gate configuration significantly improves the ON-state performance of the dual-gate Tunnel Field-Effect Transistor.

The operational principle of TFET is explained in Section II. In Section III, the device's design and simulator details are presented. The findings and discussion are presented in Section IV. Section V reports on and discusses the results of the sensitivity analysis. Section V concludes this chapter.

New Approaches for Upcoming Technology Generations

Changes in channel material and topologies are essential for tackling the growing trajectory of static energy use for semiconducting devices at the forthcoming technological node. The following innovations are considered the most viable options to continue scaling technology down to dimensions in the nanometer range:

1. The next step in the scalability plan of action, based on 2015 ITRS estimates, is to replace the stretched silicon channels in MOSFETs with materials that have excellent mobility. This transition poses several challenges that need to be addressed. These challenges include enhancing the interface across large-K materials and Ge-III-V materials, reducing band-to-band tunnelling within narrow bandgap materials, and achieving exceptionally large-scale integration utilising a silicon medium's scalable process flow. These are just a few illustrations of the obstacles that must be overcome in this endeavor [18, 19].

1. Nanowires (NWs) offer a promising avenue for replacing traditional planar MOSFET channels, potentially enabling a further reduction in technology node size. This is achievable because NWs can be manufactured with diameters in the nanometer range [20]. NWs are particularly well-suited for gate- all-around structures, which have the potential to mitigate short-channel effects. Their vertical growth orientation aligns with the future trend of vertically integrating devices on semiconductor chips. To ensure the growth of defect-free nanowires, improvements in circuit yields and consistency are required. Additionally, supposing the transfer of NWs to other platforms becomes necessary, precise positioning becomes crucial. To avoid surface irregularities and defects, it is essential to develop effective surface treatment and passivation techniques.
2. Carbon Nanotubes (CNTs) are considered a promising choice for technology nodes below 10 nanometers due to their extremely thin bodies, approximately 1 nanometer in diameter. Compared to silicon, CNT-FETs can provide enhanced electron and hole mobility in the channel, especially at room temperature. However, challenges exist in terms of achieving VLSI integration with CNTs, including the need to purify and sort CNTs to ensure a relatively uniform diameter distribution [21].
3. Graphene Field-Effect Transistors (GFETs) hold significant potential for RF design due to their superior carrier mobility compared to CNTs. They enable the expansion of faster transistors with smaller dimensions. Graphene, which represents a solitary atomic layer, is essentially a zero-bandgap semiconductor. Consequently, the primary focus of research in this area is on developing effective methods to introduce a bandgap. This is crucial for achieving a substantial current gain (I_{ON}/I_{OFF}) suitable for radio-frequency and analogue applications [22].
4. Tunnel Field-Effect Transistors (TFETs) are highly regarded as among the most advanced technological advancements for applications requiring unelevated power and modest effectiveness as a result of their distinct electrically properties [23].

The operation mechanism of TFET is a distinctive carrier injection process enabling a sub-threshold slope of lower than 60mV/decade that is not achievable in the case of thermionic-based devices. This provides low-voltage efficiency and relatively high operating current. Devices that use TFETs embedded on energy-efficient chips will rely heavily on advances in III-V material manufacturing technologies. Group IIIV compound semiconductor devices have been shown to possess better electrical properties than their silicon counterparts [24].

TUNNEL FET TECHNOLOGY: STATE OF THE ART

A breakthrough in 1958 contributed to the efforts of L. Esaki, a Japanese researcher who was employed at Sony Corporation, and presented the device, which operated on the mechanism of the Band-to-Band Tunneling (BTBT) [25]. The tunnel diode differs fundamentally from other known transistors in its operation, which is based on quantum-mechanical tunneling.

Tunnel diodes: When we have a similar diode with two terminals but the junction between the two dissimilar metals is coated in a conducting medium so that we have a tunnel in the diode, we have the strange state of affairs of having a charge carrier move through the potential barrier just as though it had the energy to surmount it in the classical sense, and still it does not. This was the queering of carriers capable of tunneling across the barrier, in response to the space-charge depletion region in p-n junctions.

This situation was the case throughout the initial period of the tunnel diode, and most of it was due to a lack of understanding of the concept of quantum tunneling. Through the years, voluminous studies have made our understanding of the tunneling procedures tremendously richer. Consequently, tunneling has recently emerged as a potential method for achieving next-generation switches, and in particular, it seems that there is an advantage to low-voltage operation, which in most cases is less than 0.5 V.

Eager to investigate the internal field emission in semiconductor diodes, L. Esaki observed something revolutionary when he carried out his experiment on a heavily doped germanium (Ge) junction (semiconductor diodes). Under the condition of a forward bias at low temperatures, a non-monotonic behavior occurred in the current-voltage characteristic. This was explained by the fact that the tunneling process in germanium was elastic, and this leads to the negative differential resistance (NDR) effect. Significantly, it was revealed that electron energy did not get lost during the tunneling process. Fig. (**1**) illustrates a comparison of the input-characteristic behaviour of MOSFETs and Tunnel FETs.

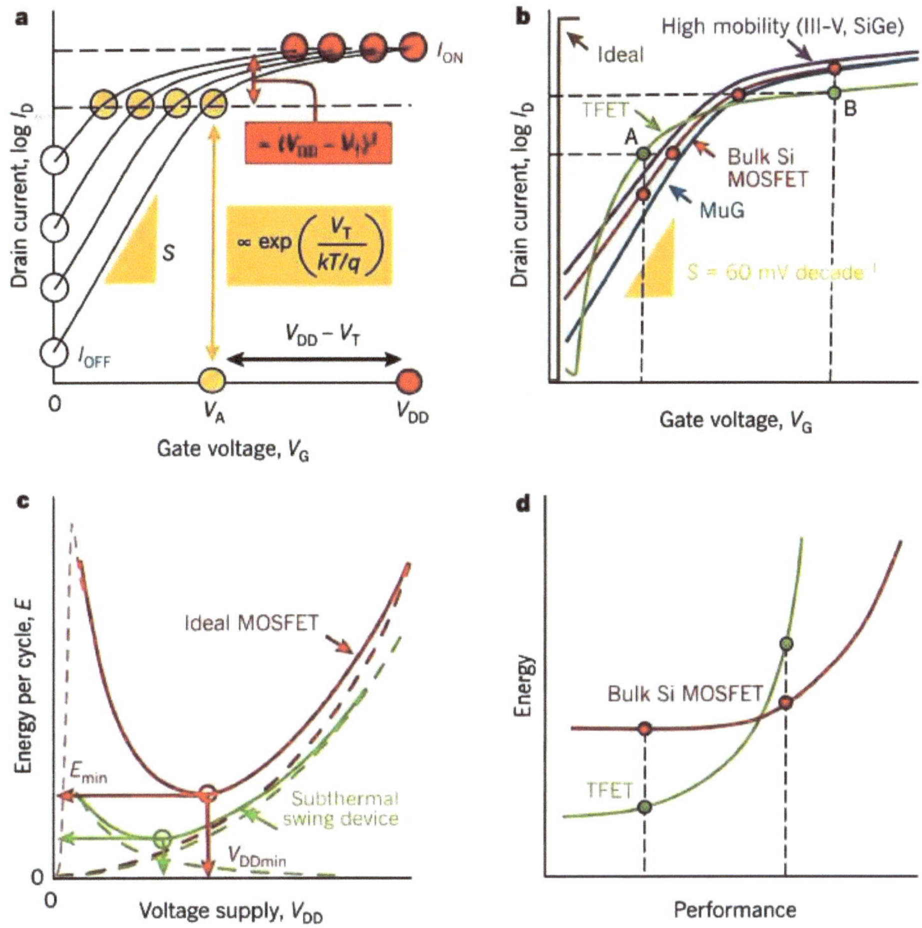

Fig. (1). Characteristics pertaining to input behavior for MOSFETs and Tunnel FETs [26].

The prototype was the tunneling effect in the semiconductor, which was found to be most influential, and it earned L. Esaki a Nobel Prize in Physics in 1973 as its pioneer. His invention was far-reaching, and other breakthroughs included radio transmission and reception, signal amplification, computing systems, and the conversion of DC to AC. These periods were among the earliest applications of the advantages of negative differential resistance (NDR) observed in the current-voltage characteristics of tunnel diodes [27].

Band-to-Band Tunneling (BTBT) Current

Classical physics assumes that any particle cannot overcome a potential barrier unless it has more energy than the barrier energy, usually by thermionic emission.

Nevertheless, the quantum mechanical examination of particles, in the form of wave functions (Psi), does not end suddenly at a potential barrier of fixed height U_0. The likelihood is not zero that the particles may be able to penetrate the barriers and pass through them; this process is called quantum tunneling [28]. There are two major forms of tunneling current in tunneling devices: Zener tunneling,where electrons change their position between the valence and conduction bands, and Esaki tunneling, where electrons tunnel between the conduction and valence bands. Fig. (**2**) presents the potential barrier profiles for (a) Zener tunneling and (b) Esaki tunneling.

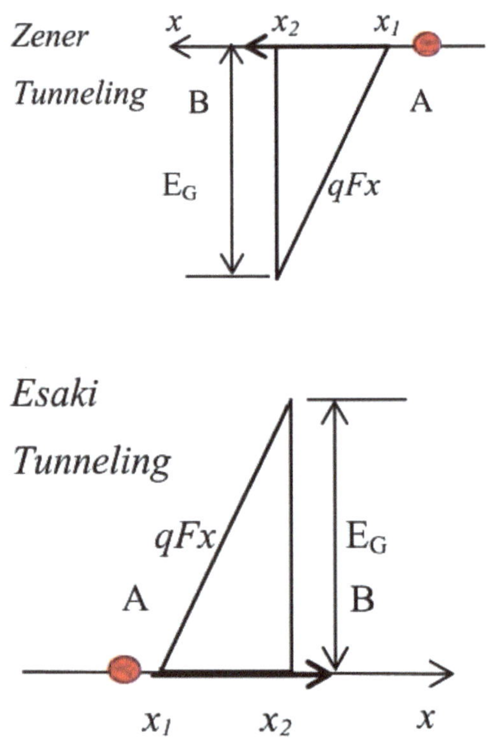

Fig. (2). Potential barrier for (a) Zener Tunneling, (b) Esaki Tunneling.

In tunneling devices, the current is proportional to the tunneling transmission probability. Band structure theory is also used to estimate the current due to tunneling, and in this case, we apply the Wentzel-Kramers- Brillouin (WKB) approximation, which is where we can obtain the formula of the transmission coefficient, which is calculable, see Equation (2) [29].

$$T_{BTBT} \cong exp\left[-\frac{4\sqrt{2m^*}E_g^{1.5}}{3\hbar qF}\right]$$

METHODOLOGIES OF DEVICE DESIGN AND SIMULATION

According to the guidelines provided by the International Roadmap for Devices and Systems (IRDS) [30], the two-dimensional structure of the SG-HJ-TFET is illustrated in Fig. (**3**). The device under consideration has a split-gate design that is comparable to a structure with two gates split into four gates.

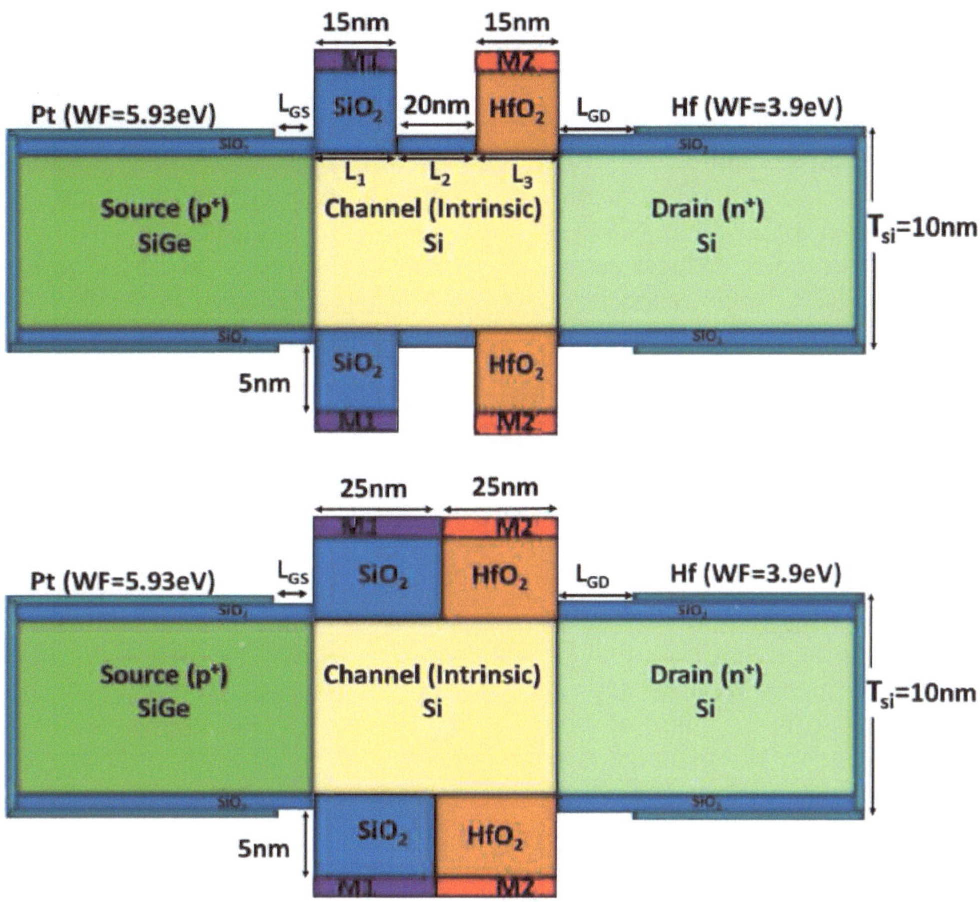

Fig. (3). 2-D schematic of (a) SG-DL-HJ-TFET, (b) NSG-DL-HJ-TFET.

Table 1. Physical Properties of the Device

Symbol	Parameter	Value
N_c $L_1/L_2/L_3$	Channel Doping (P^-) Gate Length	$10^{15}cm^{-3}$ 15nm/
		20nm/
		15nm
L_s/L_d Tsi ϕ_{M1}/ϕ_{M2} T_{HfO2} L_{gs}/L_{gd}	Source/ Drain Length Channel Thickness Gate Work Function HfO_2 Thickness Source/Drain Spacer	100 nm 10 nm 3.9/4.5 eV 4 nm 3 nm/15 nm

This modified split-gate TFET structure guarantees increased ON and subthreshold slope, making it ideal for biosensing applications. Three regions make up the entire channel length (LCH) = 50nm: region a has length L1 = 15 nm, region b has length L2 = 20 nm, and region c has length L3 = 15 nm. The HfO_2 oxide layer's thickness (Tox) is 6 nm, the source/drain dimension ($L_{S/D}$) is 100 nm, and the body's thickness (T_{Si}) is 10 nm. The drain spacing (L_{GD}) and source spacing (L_{GS}) have respective sizes of 3 nm and 15 nm. For the P^+ source and N^+ drain regionsimultaneously, platinum electrodes (work function of 5.93eV and hafnium 3.9eV, respectively) are used [31, 32]. Table **1** provides an extensive description of the device parameters. To achieve better outcomes with respect to I_{ON}, the suggested configuration with a molar proportion of Ge (x)=0.5 uses reduced bandgap $Si_{1-x}Ge_x$ for the source component [33 - 35]. Considering decreased ambipolarity, two different gate metals used are Tin (M_2=4.3eV) near the drain end and Aluminum (M_1=3.8eV) at the source end. The 10 nm thickness of the silicon substrate allows one to disregard quantum impacts. The manufacturing process of the proposed device begins with an appropriate selection of a silicon substrate that should satisfy the specifications. The wafer is then prepared using a new method to outline the active device area. Then photolithography method and photomasking are used to accurately pattern this area. To impregnate the n-type and the p-type impurities of the source and drain region, a so-called Spin-On Dopant (SOD) methodology is used. In this method, a uniform film of doping solution is coated on the substrate, and a high-velocity thermal annealing process is used to allow diffusion of the dopants. It is considered a simple and cost-effective method with little crystal damage in the case of an equal application [36].

During the following process, the wafers are cleaned with the help of the RCA method, which removes all types of surface-scalers. After this, the source, channel, and drain are covered with an oxide layer deposited by Atomic Layer

Deposition (ALD). The selective removal of the silicon oxide and hafnium oxide layers is performed usinga mixture of dry Reactive Ion Etching (RIE) and plasma-assisted methods. The junctions are then covered with metal depositions to form metal connections via electron beam evaporation, thereby achieving the required metal accuracy. Lastly, it undergoes a post-metallization process that aims towards improving mechanical adhesion between the metal layers and the silicon substrate.

Based on the SILVACO ATLAS TCAD simulator [37], the sensitivity of the proposed device is measured. TFETs are extremely sensitive to energy bandgap when it comes to their tunneling probability, which is why bandgap narrowing is included in the simulation. The Fermi-Dirac distribution model comes into application when the source and the drain of a semiconductor are highly doped, such that the semiconductor exhibits the behavior of being degenerate. The effect of recombination is properly simulated using the Shockley-Read-Hall (SRH) model. Also, the non-local Band-to-Band Tunneling (BTBT) model is switched on so that it can be used to study the dynamics of the charge carriers. The tunneling parameters, which are unique to silicon, are used for calibration: A = 3.29×10^{15} cm^3/s and B = 2.38×10^6 V/cm.

As shown in Fig. (**4**), the outcomes of the simulation have been calibrated to [38] in order to verify and standardize the mathematical models and methodology employed.

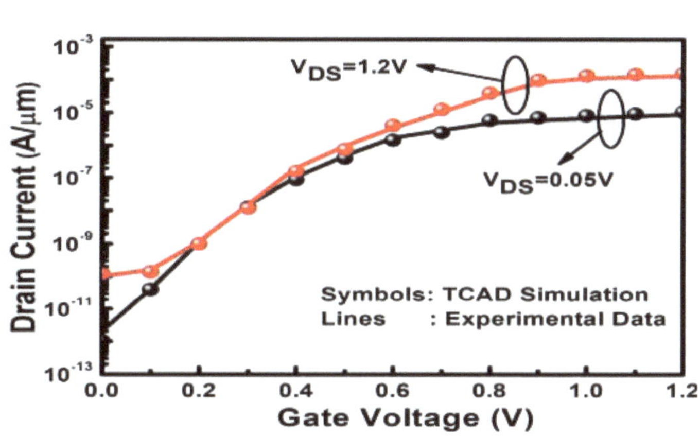

Fig. (4). TCAD simulation models have been calibrated with experimental data as presented in [25].

RESULTS AND DISCUSSIONS

The transfer characteristics for SG-DL-HJ-TFET and NSG-DL-HJ-TFET are plotted in Fig. (**5a**). The OFF current I_{OFF} is almost the same for both the TFET configurations, but there is a huge variation in the ON state current I_{ON}. SG-D-

-HJ-TFET has a peak I_{ON} of 2.11577 × 10^{-5}A/μm as compared to 2.15292 × 10^{-8}A/μm for the NSG-DL-HJ-TFET configuration. This shows that the use of the Split Gate (SG) configuration leads to an improvement of almost 10^3 times in I_{ON}, which makes this configuration quite fascinating. The output characteristics for both configurations are plotted in Fig. (**5b**). Similar trends are observed in the case of the transfer curve.

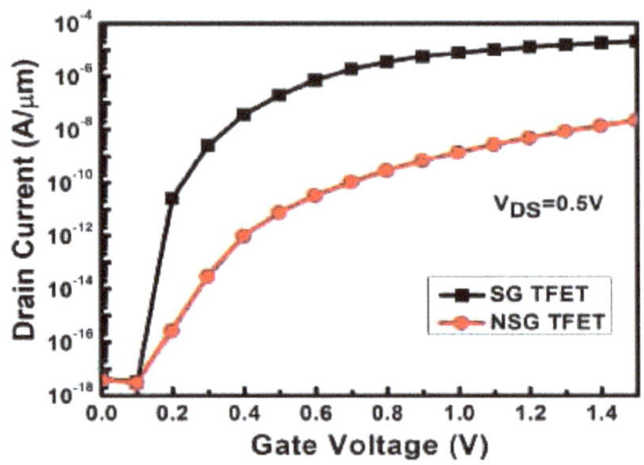

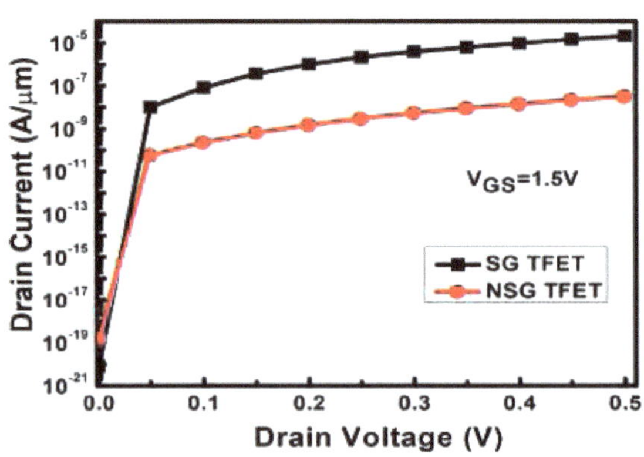

Fig. (5). (a) Transfer (b) Output characteristics for SG-DL-HJ-TFET and NSG- DL-HJ-TFET.

In the OFF state (V_{GS}=0V, V_{DS}=0.5V), a significant tunnel barrier prevents current from flowing. Fig. (**6a**) shows the energy band curve of both SG-DL-HJ-TFET and NSG-DL-HJ-TFET in ON-states V_{GS} = 1.5V and V_{DS} = 0.5V. W_1 is the tunnelling barrier for SG-DL-HJ-TFET, while W_2 is the tunnelling barrier for NSG-DL-HJ-TFET, as seen in the inset figure.

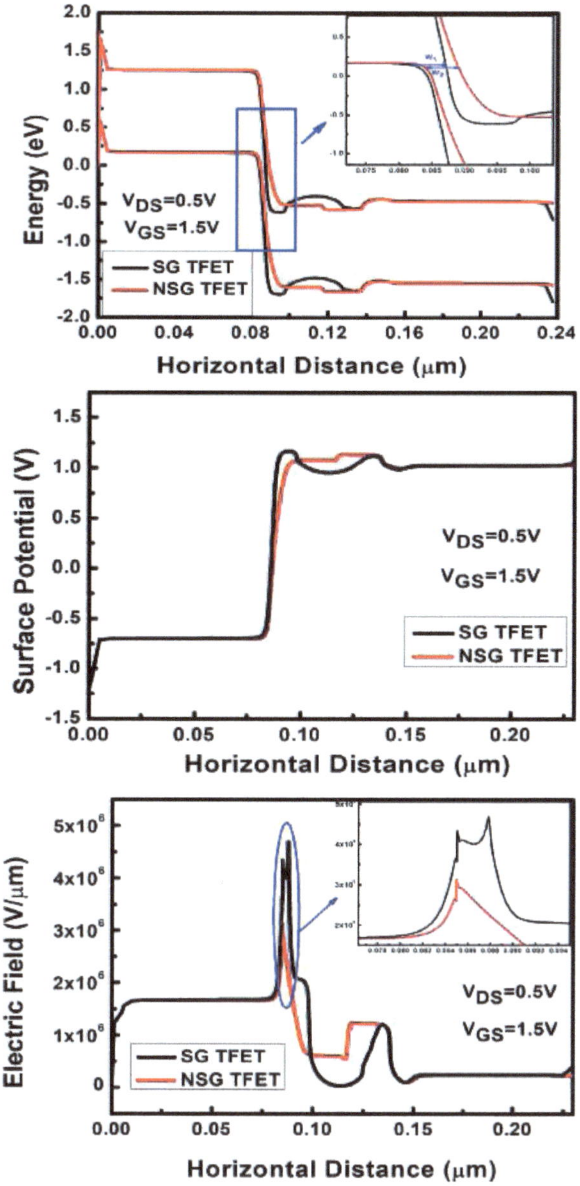

Fig.(6). (a) EBD (b) Surface Potential (c) Electric Field for Split Gate and Non-Split Gate Configurations.

At the source-channel junction, the surface potential is higher in SG-TFET thanin NSG-TFET, as clearly shown in Fig. (**6b**). In the case of SG TFET, the surface potential at the tunnelling junction is 1.16703 eV, which goes down to 1.02309 eV for NSG-TFET. Afterwards, as we move towards the drain terminal, the surface potential of SG-TFET drops below that of NSG-TFET. As we reach the drain channel junction, the surface potential of both configurations becomes equal. Similarly, the electric field at the tunnelling junction is higher for SG-TFET when compared to NSG TFET, as shown in Fig. (**6c**). For SG-TFET, the electric field at the tunnelling junction is 2.05×10^{6} V/μm, which is 9.06×10^{5} V/μm for NSG-TFET.

A major problem in several contemporary semiconductor applications, including sensing, thermal imaging, and the Internet of Things, is device stability throughout a broad range of temperatures. As a result, we examine the SG-D--HJ-TFET's efficiency over a wide range of temperatures in this study. The initial study indicates that the TFET is only weakly responsive to temperature [39, 40]. Bandgap energy is the main reason why TFET is sensitive to temperature. E_g is the component in the expression of the tunneling current, I_D in Equation (1) [41]:

$$I_D = exp\left[\frac{-4\sqrt{2m^*}\,E_g^{1.5}}{3e\hbar(E_g^* + \Delta\phi)}\sqrt{\frac{\epsilon_{Si}t_{ox}t_{Si}}{\epsilon_{ox}}}\right] \quad \textbf{(1)}$$

where $\Delta\phi$ represents the energy spectrum over which tunneling can take place, t_{ox} and ϵ_{ox} represent the depth of the oxide and permittivity, accordingly, m is the actual mass associated with the carrier, c is the carrier's charge, and $\hbar$ is the reduced Planck's constant. The permittivity and width of the silicon substrate are denoted by ϵ_{Si} and t_{Si}, respectively.

Additionally, temperature is specified by Eqn. (2) and affects the energy band gap of semiconducting substances [42].

$$E_g(T) = E_g(0) - \frac{\alpha T^2}{\beta + T} \quad \textbf{(2)}$$

where E_g (0) is the value of the energy band gap at zero kelvin, T is the temperature in absolute terms in kelvin, and α and β denote the properties of the substance. As temperatures increase, the E_g contracts and the I_D increases [43]. Thus, temperatures definitely affect the OFF-state (I_{OFF}), but at elevated V_{gs}, temperatures have a relatively minimal impact on the I_{ON}. As regards time behavior, a distinction should be made between Shockley-Read-Hall (SRH) recombination, which prevails at lower output powers, and other mechanisms that become increasingly important. V_{GS} has a high values during the OFF-state [44], and band-to-band tunneling (BTBT), which is less sensitive to changes in temperature during the ON-state.

This study examines how the suggested SG-DL-HJ-TFET DC performance specifications, such as I_{ON}, I_{OFF}, I_{ON}/I_{OFF} *ratio, SS, and VTH, are affected by a temperature* shift of 200 K to 400 K in a 50 K interval. Fig. (7) illustrates why changes in temperature affect the SG-DL-HJ-TFET's transfer characteristics. It may be demonstrated that I_{ON} is not significantly altered than I_{OFF}. Instead of being influenced by temperature, driving current during the ON-state is primarily BTBT- influenced. However, since the percentage of minority carriers grows substantially with temperature, the OFF-state exhibits notable tremors [45]. At T=200K, I_{OFF} is reported to be of the order 10^{-20} *A/μm,* which increases to almost 10^{-15} *A/μm* at T=400K.

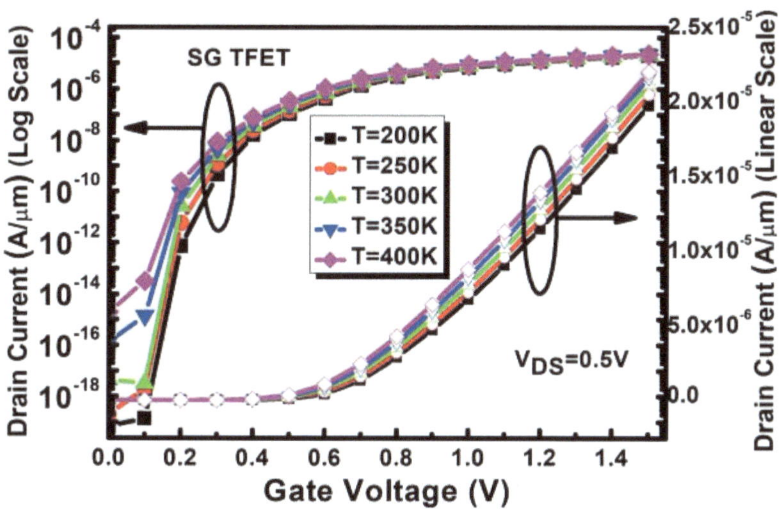

Fig. (7). Transfer characteristics of SG-DL-HJ-TFET for different temperatures.

Any device must have a significant I_{ON}/I_{OFF} ratio in order to achieve an improved switching frequency. To do it, I_{ON} and I_{OFF} should be elevated and small, correspondingly. When the temperature rises, there is little change in I_{ON}, and I_{OFF} increases substantially, leading to a decrease in the current ratio, as shown in Fig. (**8a**). Consequently, the current ratio is greater at lower temperatures than it is at the higher side. In order to attain a fast speed that yields to a large fan-out and less power use, steepest SS and smaller V_{TH} are preferred. Device effectiveness is hampered by SS, which rises with temperature. On the other hand, as V_{TH} decreases in the sub-threshold zone due to higher drain current, as depicted in Fig. (**8b**). Table **2** specifics of the various DC FoMs for SG-DL-HJ-TFET at various operating temperatures.

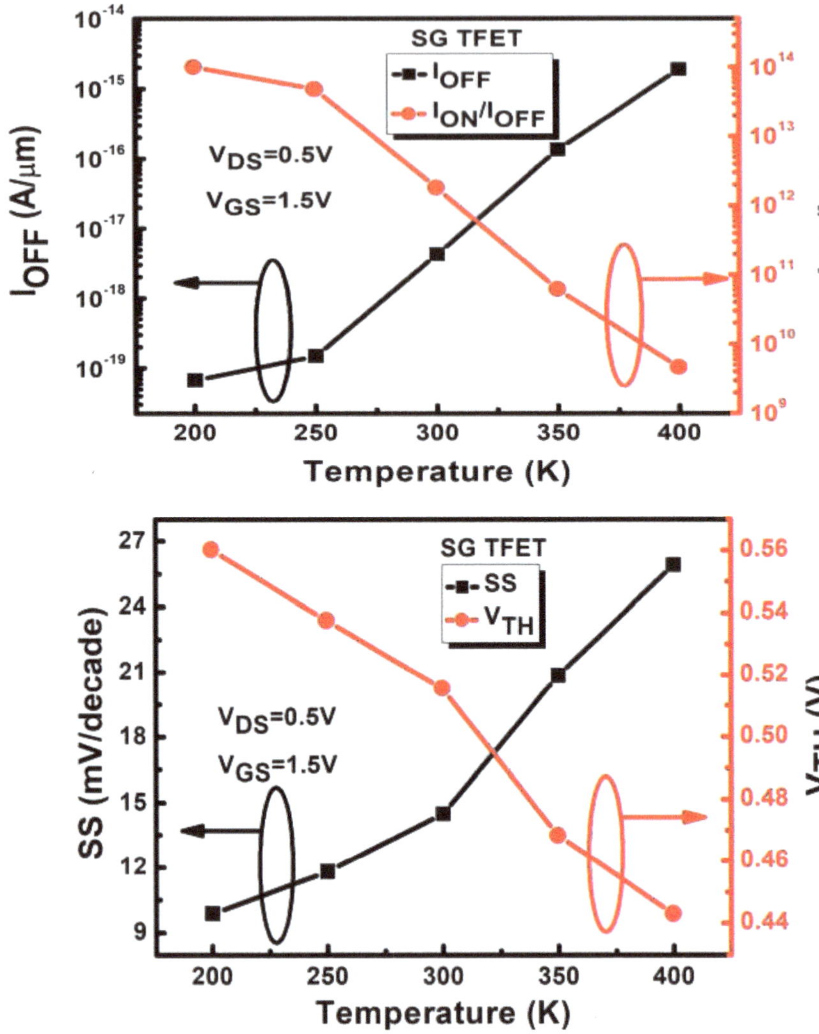

Fig. (8). Influences of temperature alterations on (a) I_{OFF} and I_{ON}/I_{OFF} (b) SS and V_{TH}.

Table 2. Temperature-dependent variability in DC FoMs for SG-TFET.

Temperature (K)	I_{OFF} (A/μm)	I_{ON}/I_{OFF}	SS (mV/dec)	V_{TH} (V)
200	6.741×10^{-20}	1.003×10^{14}	9.882	0.5602
250	1.494×10^{-19}	4.844×10^{13}	11.802	0.5376
300	4.252×10^{-18}	1.820×10^{12}	14.493	0.5157
350	1.336×10^{-16}	6.174×10^{10}	20.849	0.4683
400	1.894×10^{-15}	4.622×10^{9}	25.915	0.4430

The performance of NSG-DL-HJ-TFET is also analyzed at different operating temperatures so that a comparison can be drawn between NSG-DL-HJ-TFET and SG- DL-HJ-TFET. Fig. (**9**) depicts the transfer characteristics of the NSG-DL-J-TFET for temperatures ranging from 200K to 400K with a gap of 50K. Given that the drive current during the ON-state is primarily reliant on BTBT rather than temperature, a negligible change in I_{ON} is observed. But I_{OFF} varies from $2.006 \times 10^{-19} A/\mu m$ to $1.972 \times 10^{-15} A/\mu m$ as temperature goes from 200K to 400K. With increasing temperature, I_{OFF} increases, resulting in a reduction in the I_{ON}/I_{OFF} ratio, as seen in Fig. (**10a**). SS increases from 35mV/dec to 65mV/dec, and V_{TH} drops from 0.979V to 0.967V as temperature increases from 200K to 400K, clearly demonstrated in Fig. (**10b**). Table **3** presents the complete DC FoMs of the NSG-DL-HJ-TFET at various operating temperatures.

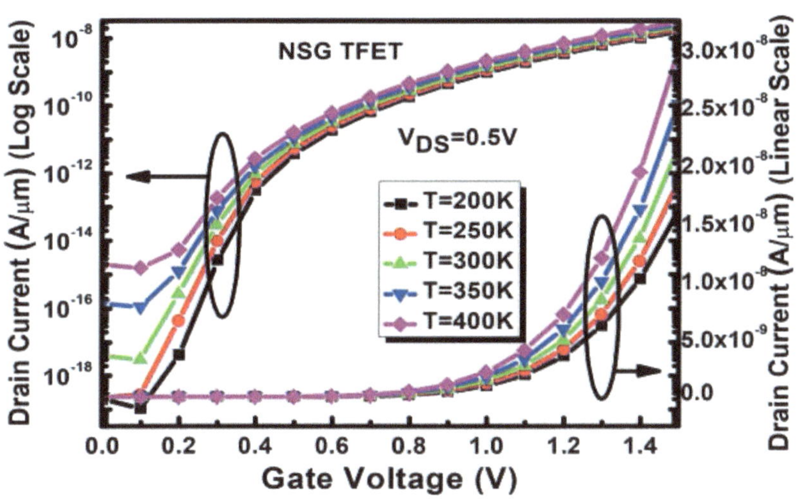

Fig. (9). The transfer curve of NSG-DL-HJ-TFET at various temperatures.

Table 3. Temperature-dependent variability in DC FoMs for NSG-TFET.

Temperature (K)	I_{OFF} ($A/\mu m$)	I_{ON}/I_{OFF}	SS (mV/dec)	V_{TH} (V)
200	2.442×10^{-19}	1.818×10^{12}	35.694	0.9799
250	2.911×10^{-18}	1.641×10^{11}	42.528	0.9777
300	3.572×10^{-17}	1.448×10^{10}	49.069	0.9758
350	2.625×10^{-16}	2.144×10^{9}	55.912	0.9715
400	1.444×10^{-15}	4.256×10^{8}	64.793	0.9671

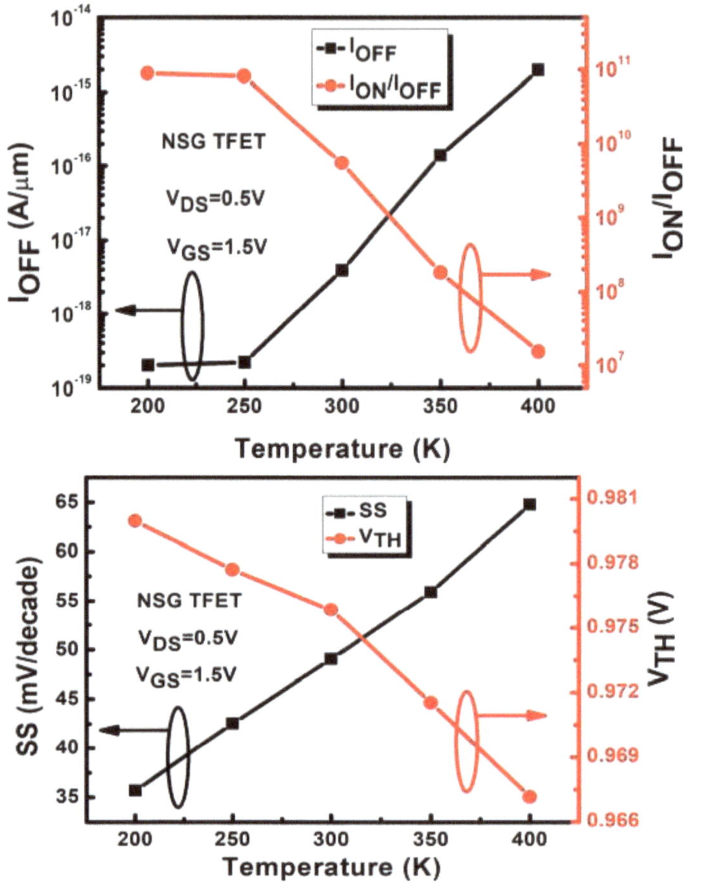

Fig. (10). Sub-threshold swing (SS) and threshold voltage changes VTH are studied with (a) positive, and (b) negative biomolecular charge densities.

CONCLUSION

This chapter introduces a novel design, the Split Gate Dopingless Heterojunction TFET (SG-DL-HJ-TFET), to enhance TFET ON-state effectiveness. The SG-DL-HJ-TFET structure demonstrates a significant improvement in I_{ON} compared to Non- Split Gate configurations. Simulation results and performance characteristics of SG-DL-HJ-TFET, including its stability across a wide temperature range, are discussed. Parameters used to measure the performance of the devices include drain current, electric field intensity, surface potential, sub-threshold swing (SS), I_{ON}/I_{OFF} Ratio, and threshold voltage (V_{TH}) for both the configurations- split-gate and non-split gate.

AI DISCLOSURE

During the preparation of the work, the author(s) used AI tools & Grammarly tool to enhance the writing style and readability of the content. The authors clarify that AI was not involved in generating the substantive content and that they take full responsibility for the publication.

REFERENCES

[1] D.J. Frank, R.H. Dennard, E. Nowak, P.M. Solomon, Y. Taur, and H-S.P. Wong, "Device scaling limits of Si MOSFETs and their application dependencies", *Proc. IEEE,* vol. 89, no. 3, pp. 259-288, 2001.
[http://dx.doi.org/10.1109/5.915374]

[2] S.A. Loan, S. Qureshi, and S.S.K. Iyer, "A novel partial-ground-plane-based mosfet on selective buried oxide: 2-d simulation study", *IEEE Trans. Electron Dev.,* vol. 57, no. 3, pp. 671-680, 2010.
[http://dx.doi.org/10.1109/TED.2009.2039545]

[3] A.M. Ionescu, and H. Riel, "Tunnel field-effect transistors as energy-efficient electronic switches", *Nature,* vol. 479, pp. 329-337, 20112011.
[http://dx.doi.org/10.1038/nature10679]

[4] Byung-Gook Park, B-G. Park, J.D. Lee, and T-J.K. Liu, "Tunneling field-effect transistors (tfets) with subthreshold swing (ss) less than 60 mv/dec", *IEEE Electron Device Lett.,* vol. 28, no. 8, pp. 743-745, 2007.
[http://dx.doi.org/10.1109/LED.2007.901273]

[5] B. Dewan, S. Chaudhary, D. Singh, and M. Yadav, "A physics-based model of dielectric modulated TFET biosensor considering non-ideal hybridization issue", *Micro and Nanostructures,* vol. 183, p. 207668, 2023.
[http://dx.doi.org/10.1016/j.micrna.2023.207668]

[6] B. Dewan, S. Chaudhary, and M. Yadav, "Electrically doped SiGe-heterojunction TFET based biosensor considering non-ideal hybridization issues: a simulation study", *Appl. Phys., A Mater. Sci. Process.,* vol. 127, no. 10, p. 791, 2021.
[http://dx.doi.org/10.1007/s00339-021-04933-8] [PMID: 34602753]

[7] G. Gopal, M. Kumawat, and T. Varma, *"Recent advancement in tfet-based biosensor addressing technique and outcome: A review," Biosensors: Developments* Challenges and Perspectives, 2024, pp. 119-145.

[8] S. Tiwari, and R. Saha, "Methods to reduce ambipolar current of various tfet structures: a review", *Silicon,* vol. 14, no. 12, pp. 6507-6515, 2022.
[http://dx.doi.org/10.1007/s12633-021-01458-w]

[9] M. Haris, S. A. Loan, Mainnuddin, and A. M. Alamoud, "Laterally Asymmetric Channel-Based Tunnel Fieldeffect Transistors: Design And Investigation", *Int. J. Electron.,* vol. 108, no. 2, pp. 284-304, 2021.
[http://dx.doi.org/10.1063/1.3367880]

[10] T. Krishnamohan, D. Kim, S. Raghunathan, and K. Saraswat, "Double-gate strained-ge heterostructure tunneling fet (tfet) with record high drive currents and 60mv/dec subthreshold slope", *2008 IEEE International Electron Devices Meeting,* IEEE, pp. 1-3, 2008.
[http://dx.doi.org/10.1109/IEDM.2008.4796839]

[11] Ashita, S.A. Loan, and M. Rafat, "A high-performance inverted-c tunnel junction fet with source–channel overlap pockets", *IEEE Trans. Electron Dev.,* vol. 65, no. 2, pp. 763-768, 2018.
[http://dx.doi.org/10.1109/TED.2017.2783764]

[12] B. Dewan, S. Chaudhary, and M. Yadav, "Split-Gate Dielectrically Modulated TFET For Ultrasensitive And Label-Free Breast Cancer Cell Detection", *Micro Nanostruct.,* p. 208351, 2025. [http://dx.doi.org/10.1016/j.micrna.2025.208351]

[13] M. Haris, S.A. Loan, Mainnuddin, and A.M. Alamoud, "Laterally asymmetric channel-based tunnel fieldeffect transistors: design and investigation", *Int. J. Electron.,* vol. 108, no. 2, pp. 284-304, 2021. [http://dx.doi.org/10.1080/00207217.2020.1793400]

[14] H. Kumar, and R. Basu, "Effect of active layer scaling on the performance of ge 1–x sn x phototransistors", *IEEE Trans. Electron Dev.,* vol. 66, no. 9, pp. 3867-3873, 2019. [http://dx.doi.org/10.1109/TED.2019.2925892]

[15] N. Kumar, and A. Raman, "Performance assessment of the charge-plasma-based cylindrical gaa vertical nanowire tfet with impact of interface trap charges", *IEEE Trans. Electron Dev.,* vol. 66, no. 10, pp. 4453-4460, 2019. [http://dx.doi.org/10.1109/TED.2019.2935342]

[16] A.K. Gupta, A. Raman, and N. Kumar, "Design and investigation of a novel charge plasma-based core-shell ring-tfet: analog and linearity analysis", *IEEE Trans. Electron Dev.,* vol. 66, no. 8, pp. 3506-3512, 2019. [http://dx.doi.org/10.1109/TED.2019.2924809]

[17] B. Dewan, S. Chaudhary, and M. Yadav, *A charge plasma based label free biomolecule detector using sige-heterojunction double gate tunnel fet* Silicon, 2021, pp. 1-10.

[18] A. Nainani, S. Raghunathan, D. Witte, M. Kobayashi, T. Irisawa, T. Krishnamohan, K. Saraswat, B.R. Bennett, M.G. Ancona, and J.B. Boos, "Engineering of strained iii–v heterostructures for high hole mobility", Baltimore, MD, USA, 2009, pp. 1-4 , [http://dx.doi.org/10.1109/IEDM.2009.5424267]

[19] S. Luryi, J. Xu, and A. Zaslavsky, *Future Trends in Microelectronics: From Nanophotonics to Sensors to Energy* John Wiley & Sons, 2010. [http://dx.doi.org/10.1002/9780470649343]

[20] Y.B. Kim, "Integrated circuit design based on carbon nanotube field effect transistor", *Transactions on Electrical and Electronic Materials,* vol. 12, no. 5, pp. 175-188, 2011. [http://dx.doi.org/10.4313/TEEM.2011.12.5.175]

[21] D. Akinwande, N. Petrone, and J. Hone, "Two-dimensional flexible nanoelectronics", *Nat. Commun.,* vol. 5, no. 1, p. 5678, 2014. [http://dx.doi.org/10.1038/ncomms6678] [PMID: 25517105]

[22] A.M. Ionescu, and H. Riel, *Tunnel field-effect transistors as energy-efficient electronic switches* vol. 479. nature, 2011, no. 7373, pp. 329-337. [http://dx.doi.org/10.1038/nature10679]

[23] U.E. Avci, D.H. Morris, and I.A. Young, "Tunnel field-effect transistors: Prospects and challenges", *IEEE J. Electron Devices Soc.,* vol. 3, no. 3, pp. 88-95, 2015. [http://dx.doi.org/10.1109/JEDS.2015.2390591]

[24] Shanzhi Chen, Hui Xu, Dake Liu, Bo Hu, and Hucheng Wang, "A vision of iot: Applications, challenges, and opportunities with china perspective", *IEEE Internet Things J.,* vol. 1, no. 4, pp. 349-359, 2014. [http://dx.doi.org/10.1109/JIOT.2014.2337336]

[25] B. Dewan, S. Chaudhary, and D. pal Singh, "Design And Simulation Of Ferroelectric-Enhanced Dual-Gate TFET Biosensor With SiGe Source And Nanocavity Modulation For Biomolecular Detection", *Micro Nanostruct,* p. 208372, 2025.

[26] D. Cavalheiro, "Ultra-Low Power Circuits Based On Tunnel Fets For Energy Harvesting Applications",

[27] S. Sze, and K. Ng, "Chapter 3 Metal-Semiconductors Contacts", In: *Phys. Semicond. Devices* John Wiley & Sons: Hoboken NJ, USA, 2007, pp. 134-196.

[28] E.O. Kane, "Theory of Tunneling", *J. Appl. Phys.,* vol. 32, no. 1, pp. 83-91, 1961. [http://dx.doi.org/10.1063/1.1735965]

[29] P. Bergveld, "The development and application of FET-based biosensors", *Biosensors,* vol. 2, no. 1, pp. 15-33, 1986. [http://dx.doi.org/10.1016/0265-928X(86)85010-6] [PMID: 3790175]

[30] M. Moore, *International Roadmap For Devices And Systems (IRDS™) Edition*, 2020.

[31] A. Bhattacharyya, M. Chanda, and D. De, "GaAs0.5Sb0.5/ In0.53Ga0.47As heterojunction dopingless charge plasma-based tunnel FET for analog/digital performance improvement", *Superlattices Microstruct.,* vol. 142, p. 106522, 2020. [http://dx.doi.org/10.1016/j.spmi.2020.106522]

[32] M.J. Kumar, and S. Janardhanan, "Doping-less tunnel field effect transistor: Design and investigation", *IEEE Trans. Electron Dev.,* vol. 60, no. 10, pp. 3285-3290, 2013. [http://dx.doi.org/10.1109/TED.2013.2276888]

[33] N. Thoti, and Y. Li, "Influence Of Fringing-Field On Dc/Ac Characteristics Of Si Ge Based Multi-Channel Tunnel Fets", *IEEE Access,* vol. 8, pp. 658-208, 2020.

[34] B. Dewan, S. Chaudhary, and M. Yadav, "A label-free dielectric-modulated biosensor using SiGe-heterojunction dual cavity dual metal electrically doped TFET", *Engineering Research Express,* vol. 5, no. 3, p. 035002, 2023. [http://dx.doi.org/10.1088/2631-8695/acdec7]

[35] Q.T. Zhao, J.M. Hartmann, and S. Mantl, "An improved si tunnel field effect transistor with a buried strained sige source", *IEEE Electron Device Lett.,* vol. 32, no. 11, pp. 1480-1482, 2011. [http://dx.doi.org/10.1109/LED.2011.2163696]

[36] B. Dewan, S. Chaudhary, and M. Yadav, "Investigation on the variation effect of gate work function on $$N^+$$ pocket-doped junctionless vertical tunneling FETs", *J. Comput. Electron.,* vol. 24, no. 3, p. 86, 2025. [http://dx.doi.org/10.1007/s10825-025-02320-7]

[37] S. Clara. Atlas device simulation software. IEDM Technical Digest. IEEE International Electron Devices Meeting, 2004., San Francisco, CA, USA, 2004, pp. 195-198.,

[38] T. Nirschl, P-F. Wang, C. Weber, J. Sedlmeir, R. Heinrich, R. Kakoschke, K. Schrufer, J. Holz, C. Pacha, and T. Schulz, "The tunneling field effect transistor (tfet) as an add-on for ultra-low- voltage analog and digital processes", IEDM Technical Digest. IEEE International Electron Devices Meeting, 2004., San Francisco, CA, USA, 2004, pp. 195-198. [http://dx.doi.org/10.1109/IEDM.2004.1419106]

[39] S. Sharma, and B. Kaur, "Performance investigation of asymmetric double□gate doping less tunnel FET with Si/Ge heterojunction", *IET Circuits Dev. Syst.,* vol. 14, no. 5, pp. 695-701, 2020. [http://dx.doi.org/10.1049/iet-cds.2019.0290]

[40] Y.P. Varshni, "Temperature dependence of the energy gap in semiconductors", *physica,* vol. 34, no. 1, pp. 149-154, 1967. [http://dx.doi.org/10.1016/0031-8914(67)90062-6]

[41] S. Saurabh, and M.J. Kumar, *Fundamentals of tunnel field-effect transistors* CRC press, 2016. [http://dx.doi.org/10.1201/9781315367354]

[42] R. Narang, M. Saxena, R.S. Gupta, and M. Gupta, "Impact of temperature variations on the device and circuit performance of tunnel fet: a simulation study", *IEEE Trans. Nanotechnol.,* vol. 12, no. 6, pp. 951-957, 2013. [http://dx.doi.org/10.1109/TNANO.2013.2276401]

[43] K. Boucart, and A.M. Ionescu, "Double-gate tunnel fet with high-k gate dielectric", *IEEE Trans. Electron Dev.,* vol. 54, no. 7, pp. 1725-1733, 2007. [http://dx.doi.org/10.1109/TED.2007.899389]

[44] J. Min, L.D. Wang, J. Wu, and P.M. Asbeck, "Analysis of temperature dependent effects on i–v characteristics of heterostructure tunnel field effect transistors", *IEEE J. Electron Devices Soc.,* vol. 4, no. 6, pp. 416-423, 2016. [http://dx.doi.org/10.1109/JEDS.2016.2614792]

[45] B. Dewan, S. Chaudhary, D. Singh, and M. yadav, "Investigating temperature reliability of RF performance metrics and linearity for double gate doping less TFET", *Engineering Research Express,* vol. 6, no. 1, p. 015308, 2024. [http://dx.doi.org/10.1088/2631-8695/ad23ca]

CHAPTER 11

Optimizing DC Parameters in Hetero-Stacked Source Tunneling Field-Effect Transistor

George Mili[1,*], **Zohmingliana**[1] and **Brinda Bhowmick**[1]

[1] *Department of Electronics and Communication Engineering, National Institute of Technology Silchar, Silchar, Assam, India*

Abstract: This chapter studies a hetero-stacked source n-type TFET (HSS-nTFET) using TCAD simulations. The device uses a stacked-source structure with gate–source overlap to improve tunneling. This design increases the electric field at the source–channel junction and reduces the tunneling width, which supports higher current conduction. The electrical performance is analyzed by varying the source stack materials and key device dimensions. Based on the simulation results, the proposed HSS-nTFET achieves an average subthreshold swing of about 20 mV/dec, and an I_{ON}/I_{OFF} ratio on the order of 10^{12}. The effect of different gate dielectric materials is also examined, showing that high-k dielectrics improve the ON current and switching performance.

Keywords: Heterostack, Drain underlap, Band to band tunneling (BTBT), Stack source engineering, Subthreshold swing.

INTRODUCTION

Over the past few years, there has been a continuous trend toward downsizing CMOS chips, pushing physical boundaries into the realm of nanoelectronic technology [1]. This scaling focuses on improving device performance, reducing power consumption, and increasing IC density. Even with nanoscale scaling, CMOS faces challenges such as short-channel effects, high leakage current, power loss, and threshold voltage drop. Conventional MOSFETs have a subthreshold swing above 60 mV/decade because of thermionic emission [2, 3]. This limitation has led to interest in TFETs as alternative devices. TFETs use band-to-band tunneling (BTBT) to overcome the 60 mV/decade limit at room temperature. They also provide low off-current, making them suitable for ultra-low-power applications [4 - 6]. Conventional TFETs face challenges such as low ON current and unwanted OFF-state current due to ambipolarity [7 - 9]. Improving these

[*] **Corresponding author George Mili:** Department of Electronics and Communication Engineering, National Institute of Technology Silchar, Silchar, Assam, India; E-mail: miligeorge77@gmail.com

Girdhar Gopal, Meena Panchore, Arun Kishor Johar & Tarun Varma (Eds.)

aspects is important for higher drive current and better circuit performance. For comparison, MOSFETs have a minimum subthreshold swing of 60 mV/decade at room temperature and a very low I_{ON}/I_{OFF} ratio [10]. In contrast, TFETs use tunneling to achieve SS values below 60 mV/decade. To improve their performance, several device designs and strategies have been proposed [11 - 15]. Adding low-bandgap materials to heterojunction TFETs can increase ON current and reduce leakage current. Group III-nitrides such as InN, which has a bandgap of 0.7 eV, are commonly used in these devices [16]. By reducing the tunneling width and causing band bending, InN helps increase the ON current. This work studies a lateral n-channel hetero-stacked TFET (HSS-nTFET) using InN to improve device performance. The HSS-nTFET is expected to show a better SS and lower leakage current.

The chapter is divided into five sections. The first section introduces the study. Section 2 shows the cross-sectional view of the proposed TFET and the main parameters. Section 3 presents the simulation results and analysis. Section 4 compares the results with previous works. The final section provides conclusions.

DEVICE STRUCTURE AND PARAMETER SELECTION

Fig. (**1a**) shows the cross-sectional view of the proposed HSS-nTFET. The device has a two-dimensional lateral n-channel structure. It is similar to a conventional TFET, but a stacked source is used instead of a single source region. In the source region of the n-channel HSS-TFET, two different materials are stacked. Si forms the first layer, while InN is used as the second layer. The source region is formed using three layers. The top and bottom layers are Si with a bandgap of 1.1 eV, while the middle layer is InN with a lower bandgap of 0.7 eV. Silicon is also used for the channel and drain regions to limit leakage current. This layered source structure improves band-to-band tunneling and increases the ON current. A higher electric field is observed at the source–channel junction due to the stacked source. Ambipolar current may appear in n-type TFETs under negative gate bias due to hole tunneling from the drain side. So, a gate–drain underlap is applied to reduce this effect. An optimized gate–source overlap is used to increase the effective tunneling region. This overlap increases the local charge density and facilitates a faster change in channel potential [17]. InN is a III–V direct band-gap semiconductor with a bandgap of about 0.7 eV. When used in the source stack, it reduces band bending and narrows the tunneling width, thereby increasing the ON current. Compared with other III–V materials, InN shows strong electron accumulation at the surface. Device simulations were performed using the Synopsys TCAD Sentaurus tool [18]. The parameters used for the proposed device are listed in Table **1**. The P^+ InN source is doped at $1\times10^{21}cm^{-3}$, while the channel and drain doping are set to $1\times10^{17}cm^{-3}$ and 5×10^{18} cm^{-3}, respectively. An undercut length of L_{UC} = 5 nm is applied in the i-Si channel, allowing the gate to overlap the source and underlap the drain. This design supports a steeper average subthreshold slope (SS_{avg}) and a higher I_{ON}. Fermi–Dirac statistics [18] are used to model the carrier distribution in the device. Furthermore, the Shockley-Read-Hall recombination model and the dynamic non-local band- to-band tunneling (BTBT) model are utilized. Throughout the entire tunneling path, a nonlocal BTBT model is applied. As the gate voltage increases, the channel conduction band moves downward. Band-to-band tunneling starts when the channel conduction band falls below the source valence band. For the proposed n-type HSS-TFET, a positive gate voltage places the device in the ON state. This causes band bending near the source–channel junction and allows efficient tunneling. The OFF-state current is evaluated at a gate voltage of 0 V, as shown in the energy band diagrams in Fig. (**1b** and **1c**) . In the ON state, the tunnel barrier width (λ_1) is smaller than in the OFF state (λ_0). This causes band-to-band tunneling to occur mainly along the path from A to A'. The red arrow shows the electron tunneling direction. Fig. (**2a**) and **2b**) display the volumetric electron generation rate in the device under the two bias conditions: i) OFF and ii) ON-state. Fig. (**3**) shows the transfer characteristics, which are compared with experimental data to validate the simulation models [19]. Fig. (**4**). illustrates the main steps in fabricating the proposed HSS-nTFET. A thick Si wafer is used for the channel. The source region is created by etching an L-shaped cut, followed by masking. The drain doping region is then formed through ion implantation. Within the L-shaped cut, the source stack region is developed by sequentially depositing Si in the first layer, InN in the second layer, and again Si in the third layer, with masking and subsequent ion implantation to form a P-doped source region. Subsequently, a thin oxide deposition layer is applied using the Atomic Layer Deposition (ALD) method [20]. This oxide layer is then shaped using selective etching. Finally, metal gate deposition is performed using sputtering technology and defining contacts at appropriate positions.

Table1. Physical parameters and doping of different regions.

Parameters	Values
Gate length(L_g)	25nm
Oxide thickness(T_{OX})	2nm
Channel length undercut(L_{UC})	5nm
P^+ stack source doping(N_A)	$1\times10^{21}cm^{-3}$
N channel doping	$1\times10^{17}cm^{-3}$
N+ drain doping(N_D)	$5\times10^{18}cm^{-3}$

The entry "Thickness of InN (T_InN) 4 nm" is currently appearing outside the table boundary. Kindly place this line properly within the table column along with the other parameters.

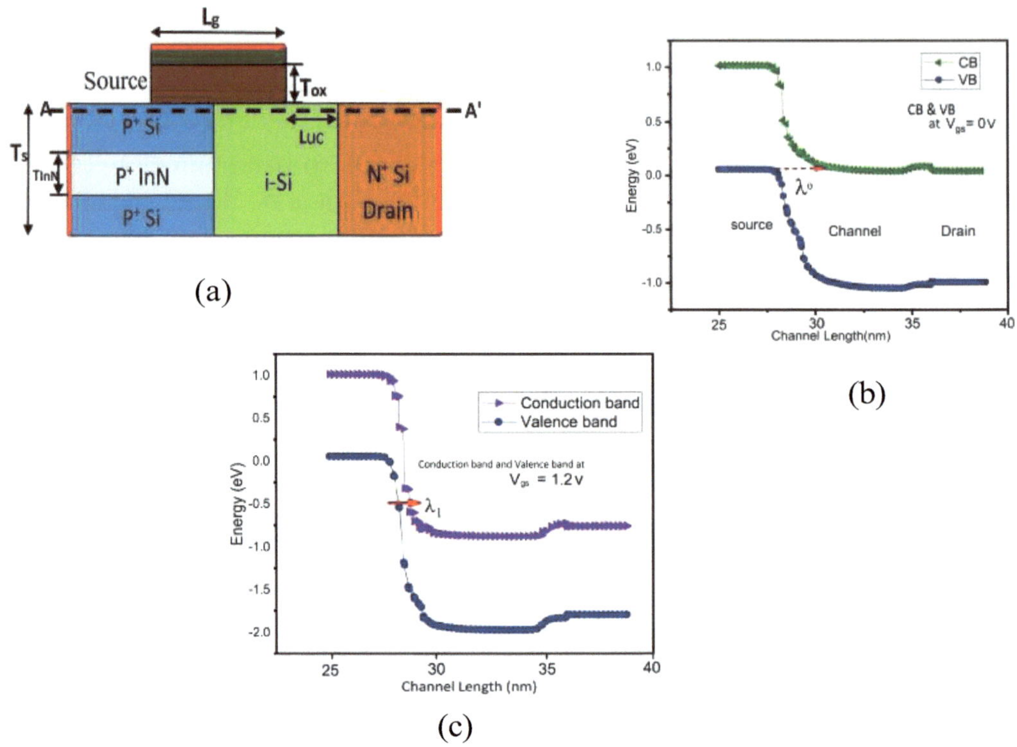

Fig. (1). (a) 2D schematic representation of the proposed device, (b) Energy band diagram at off state at V_{gs}=0V, (c) Energy band diagram at ON state at V_{gs}=1.2V.

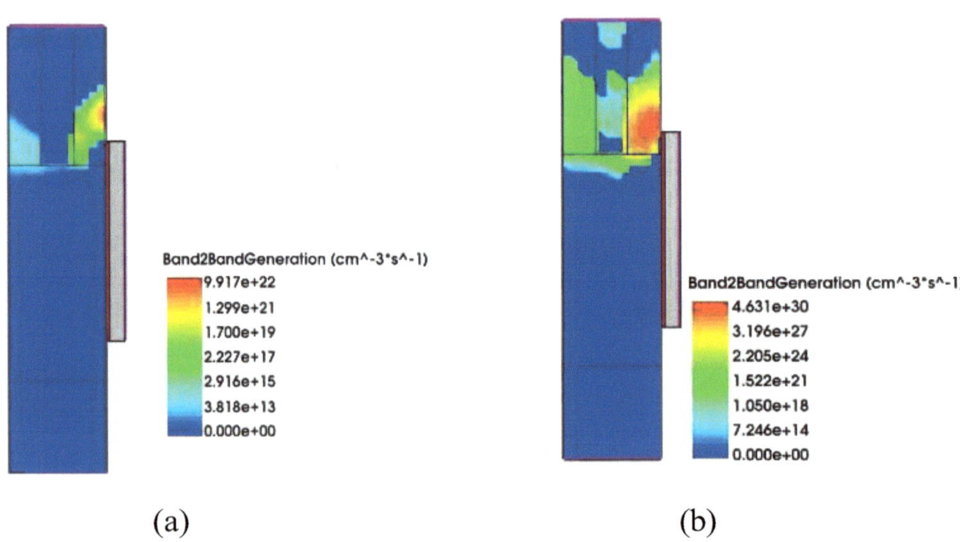

Fig. (2). Electron BTBT generation of the InN source nTFET at (a) OFF (b) ON state.

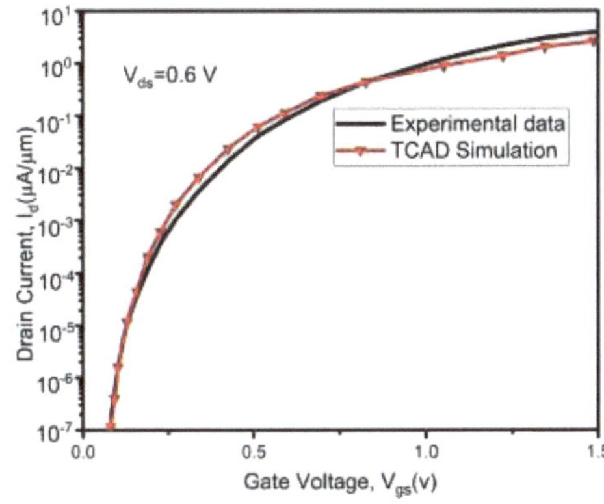

Fig. (3). Model Calibration against the experimental data in [20].

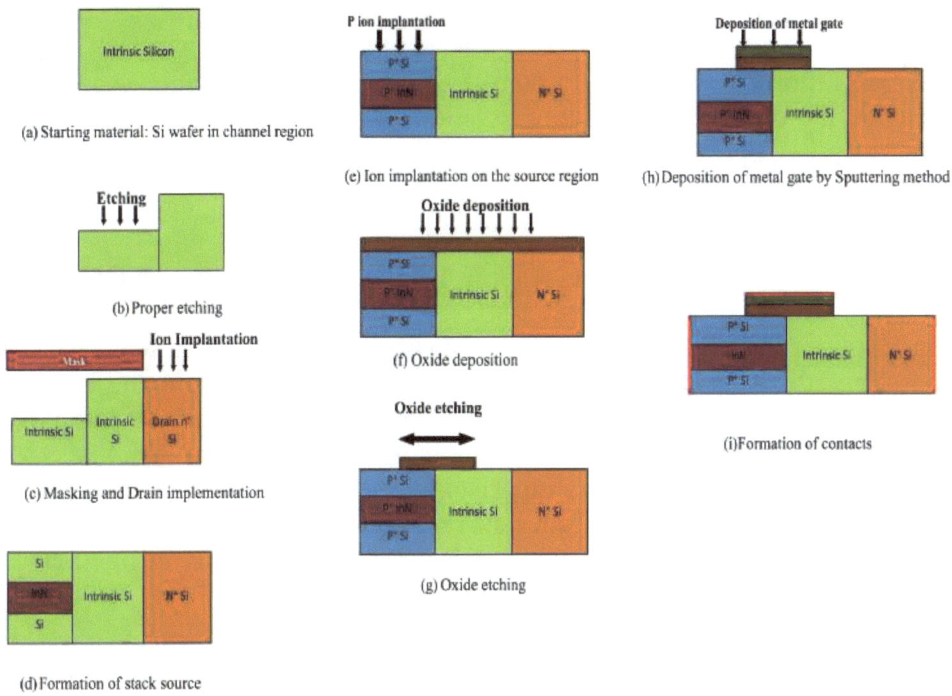

Fig. (4). The proposed process flow for fabricating the device is as.

Here, the fundamental process flow steps for the fabrication of the proposed HSS-nTFET are described and shown in Fig. (**4**). First, a thick Si wafer is formed as the channel region. To produce the source in the proposed device, an L-shaped cut is created by proper etching. After proper masking, the drain-doping region is formed by ion implantation. In that L-shaped cut, the source stack region is formed by depositing Si in the 1st layer, InN in the 2nd layer, and again Si in the 3rd layer, with suitable masking, followed by ion implantation to form a P-doped source region. Next, a thin oxide layer is deposited using Atomic Layer Deposition (ALD) [20]. After that, the oxide layer is shaped by using selective etching. Finally, the metal gate is deposited by sputtering, and the contacts at the appropriate positions are defined.

follows: (a) Start with an intrinsic Si wafer for the channel region. (b) Perform proper etching to create a pattern on the substrate by selectively removing material. (c) Implement drain and deposit ions. (d) Develop the source stack by depositing Si, InN, and Si in the 1st, 2nd, and 3rd layers, respectively, onto the source region. (e) Implement P ions in the source region. (f) Oxide gate deposition by the ALD method. (g) Creating proper oxide etching. (h) Depositing a metal gate using the sputtering technique. (i) Form contacts.

RESULTS AND DISCUSSION

This section reports the simulation results of the proposed device. Sections 3.1 to 3.5 focus on the optimization of key geometrical parameters. The effect of these parameters on I_{ON}, I_{OFF}, I_{ON}/I_{OFF}, and SS is analyzed.

Scaling the Source stack thickness (T_s)

The optimal dimensions for the lateral HSS-nTFET are determined through necessary scaling of device characteristics. Initially, the thickness of Si is scaled for our proposed device, while other device parameters such as T_{OX} (oxide thickness) and the gate length (L_g) are kept constant at T_{OX}= 1 nm and L_g = 25 nm, respectively. Fig. (**5**) shows the transfer curves of the n-TFET for different Si (T_s). The ON current and the ION/IOFF ratio change as the silicon thickness varies. Thinner channels reduce the tunneling distance, which improves the gate's control over the channel. This leads to higher current and a better I_{ON}/I_{OFF} ratio. The SS_{Avg} for each T_s is shown in Fig. (**5b**), where SS_{Avg} is mathematically defined as Eqn.1

$$SS_{AVG} = \frac{V_t - V_{OFF}}{LogI_{V_t} - LogI_{V_{OFF}}}$$

V_{off} = voltage when current is zero

I_{V_t} = current at V_t

$I_{V_{off}}$ = current at V_{off}

Increased gate control over the channel improves the tunneling rate near the gate interface area. Consequently, the drain current is enhanced due to this heightened gate electrostatic control, resulting in improved device performance and a decrease in SS. Specifically, for T_S =9nm, the SS is reduced to 28mV/Dec. Table **2** shows that a superior switching current ratio is achieved at Ts = 9 nm compared to Ts = 2 nm.

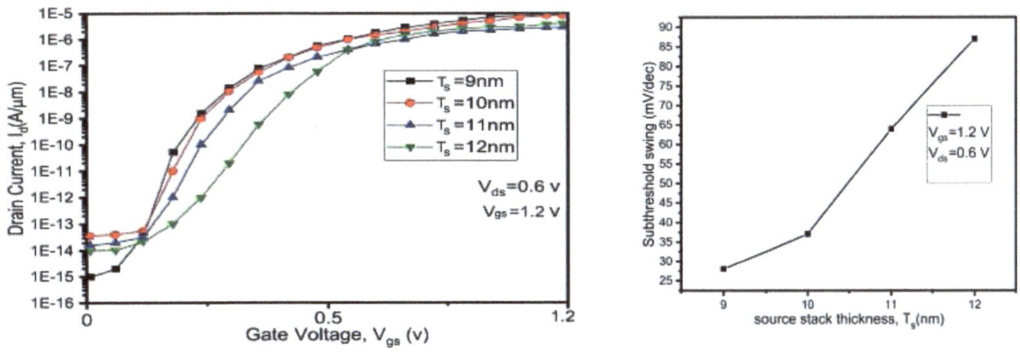

(a) (b)

Fig. (5). Effects of scaling the stack source thickness (T_S) on (a) I_D-V_G characteristics, (b) SS_{Avg}.

Table2. Effect of T_S on I_{ON} and I_{OFF}.

T_S(nm)	I_{on}(A/μm)	I_{off}(A/μm)	I_{on}/I_{off}
9	9.87×10^{-5}	8.17×10^{-15}	1.29×10^{10}
10	8.85×10^{-6}	3.61×10^{-14}	2.45×10^{8}
11	2.91×10^{-6}	1.61×10^{-14}	1.81×10^{8}
12	4.05×10^{-6}	1.02×10^{-14}	3.90×10^{8}

Optimization of channel undercut length

Once the source stack thickness is optimized, it remains constant along with other factors such as T_{OX} (oxide thickness) and the thickness of InN (T_{InN}) set at 1.5nm and 4nm, respectively. Meanwhile, the channel undercut length (L_{UC}) is varied from 2 to 8nm. Fig. (**6a**) shows the V-I curves for different LUC values. Table **3** indicates that L_{UC} has a little effect on the ON current but strongly affects the OFF current. At L_{UC} = 8 nm, the OFF current decreases from 1.5×10^{-10} A/μm (L_{UC} = 2 nm) to 2.96×10^{-14} A/μm, which increases the I_{ON}/I_{OFF} ratio from 5.4×10^{6} to 2.5×10^{10}. Fig. (**6b**) shows the SS_{Avg} for different L_{UC} values, improving from 46 mV/dec at 2 nm to 31.12 mV/dec at 8 nm.

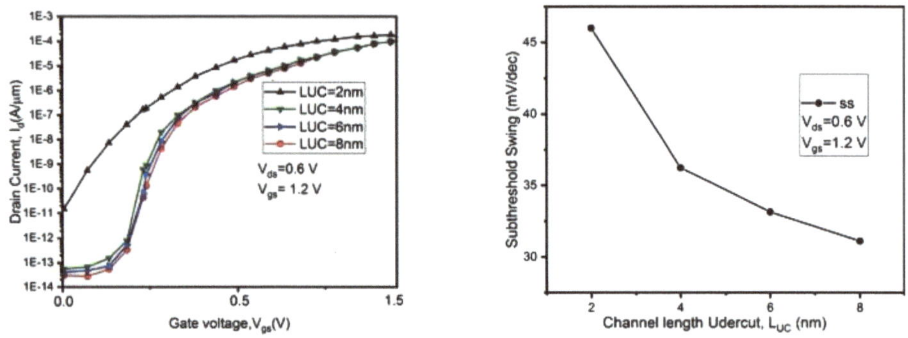

(a) (b)

Fig. (6). Impact of the channel undercut length (L_{UC}) on (a) transfer characteristics, (b) SS.

Table3. Currents measured for different undercut lengths.

L_{UC}(nm)	I_{on}(A/μm)	I_{off}(A/μm)	I_{on}/I_{off}
2	8.24×10^{-4}	1.50×10^{-10}	5.40×10^{6}
4	5.46×10^{-4}	4.47×10^{-14}	1.20×10^{10}
6	7.47×10^{-4}	3.96×10^{-14}	1.80×10^{10}
8	8.52×10^{-4}	2.56×10^{-14}	3.30×10^{10}

Optimization of oxide thickness

To study the effect of oxide thickness (T_{OX}), the channel thickness, channel undercut length, and source thicknesses are kept constant at 27 nm, 5 nm, and 12 nm, respectively. Fig. (**7a**) shows the device behavior for different T_{OX} values. A thinner gate oxide improves the gate's control over the channel, which increases the ON current. At T_{OX} = 1.5 nm, the ON current is higher than at TOX = 4.5 nm. This results in a decrease in the I_{ON}/I_{OFF} ratio (from 9.5 $\times10^{8}$ at T_{OX} = 4.5 nm to 3.01 $\times10^{10}$at T_{OX} =1.5nm) as shown in Table **4**. A SS of 32mV/dec is achieved at T_{OX}=1.5nm, as shown in Fig. (**7b**). Therefore, a dielectric thickness of 1.5nm is deemed optimal for our proposed device.

Table4. Variation of currents and current ratio with gate dielectric oxide thickness.

T_{OX}(nm)	I_{on}(A/μm)	I_{off}(A/μm)	I_{on}/I_{off}
1.5	3.98×10^{-5}	1.32×10^{-15}	3.01×10^{10}
2.5	2.50×10^{-5}	1.76×10^{-14}	1.40×10^{9}
3.5	1.90×10^{-5}	2.58×10^{-14}	7.30×10^{8}
4.5	2.30×10^{-5}	2.40×10^{-14}	9.50×10^{8}

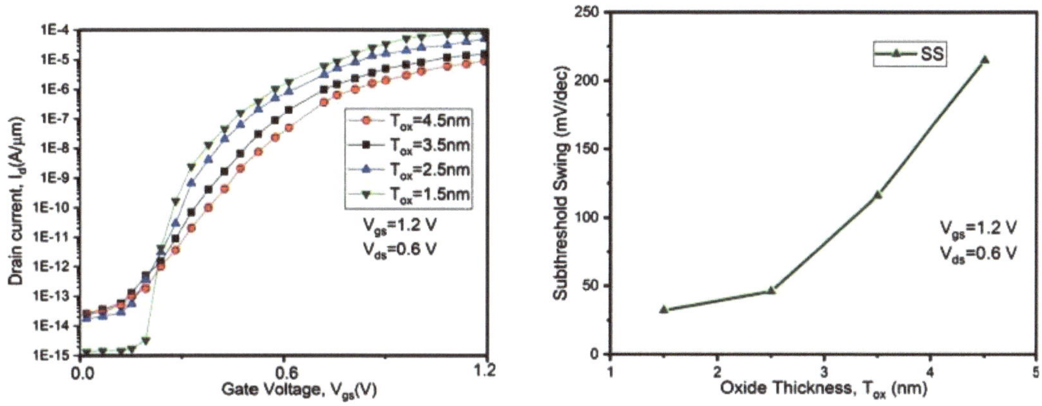

(a) (b)

Fig. (7). Impact of oxide thickness variation (T_{OX}) on device (a) transfer characteristics (b) SS.

Optimization of InN source thickness

This section examines the effect of InN thickness on device performance. The InN-based source region is laterally positioned towards the channel area, with a portion overlapping the gate. With increasing InN thickness, electron tunneling also increases. This results in a light enhancement in the i_{on} current, as depicted in Fig. (**8a**). The I_{ON}/I_{OFF} current ratio is shown to be 2.04×10^{10} at T_{InN} = 4nm, which moderately increases to 4.25×10^{10} at T_{InN} = 6nm. However, as illustrated in Fig. (**8b**), the SS_{Avg} experiences significant improvement at a source thickness of 4nm (InN).

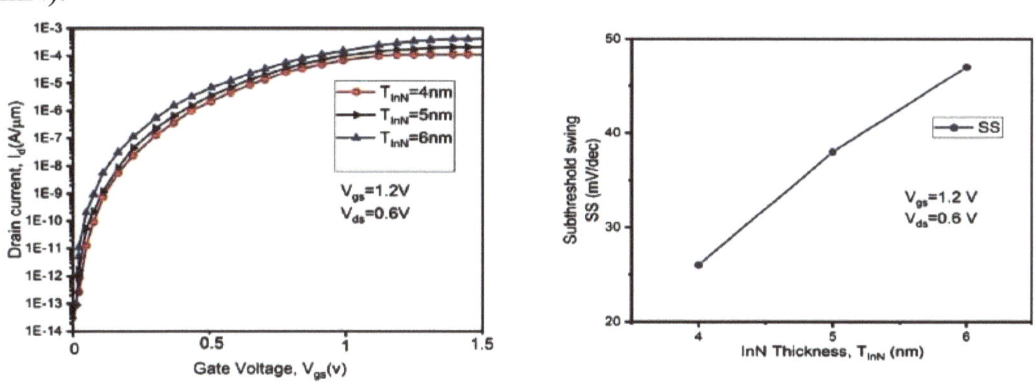

(a) (b)

Fig. (8). T_{InN} thickness variation on (a) V-I characteristics (b) SS_{Avg}.

Impact of different gate oxides

Fig. (**9a**) illustrates the variation in transfer characteristics for different gate oxide materials (HfO_2, Al_2O_3, and SiO_2). The simulation results show that the drain current increases when the dielectric constant is higher. Using a high-κ dielectric improves the coupling between the gate and the channel.

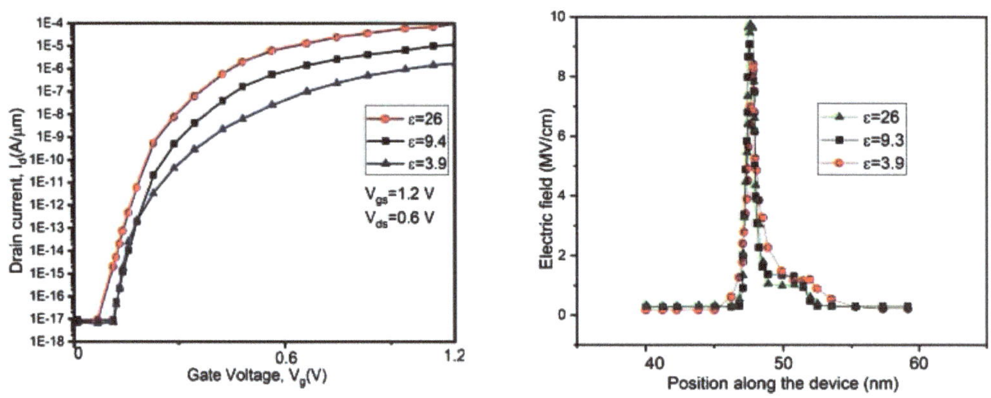

(a) (b)

Fig. (9). Variation in (a) transfer characteristics and (b) electric field for various gate oxides.

A higher dielectric improves the gate's control over the channel, which increases band-to-band tunneling. Fig. (**9b**) shows the electric field for different gate oxide materials. It is evident that high-*κ* gate dielectrics result in larger peak electric fields at the source and channel regions. This increased electric field intensity contributes to higher electron tunneling, as depicted in Fig. (**9b**).

COMPARISON USING DIFFERENT SOURCE STACK MATERIALS

In this section, the proposed device is further analyzed by using different low-band-gap materials in the second source layer. SiGe and InP are considered, leading to Si/SiGe/Si and Si/InP/Si stack source structures, as shown in Fig. (**10**). The simulated results are compared and discussed. SiGe is selected because its band-gap can be controlled by adjusting the Ge content. In $Si_{1-X}Ge_X$, a reduced band-gap of about 0.33 eV is obtained at 40% Ge composition. This reduction increases band bending at the source–channel junction. As a result, the ON current is improved. In addition, InP, a III–V compound semiconductor, is also examined as the source stack material. The results in Fig. (**11**) show that the Si/InN/Si configuration provides better DC performance than the Si/SiGe/Si and

Si/InP/Si structures. A detailed comparison of these parameters is summarized in Table 5.

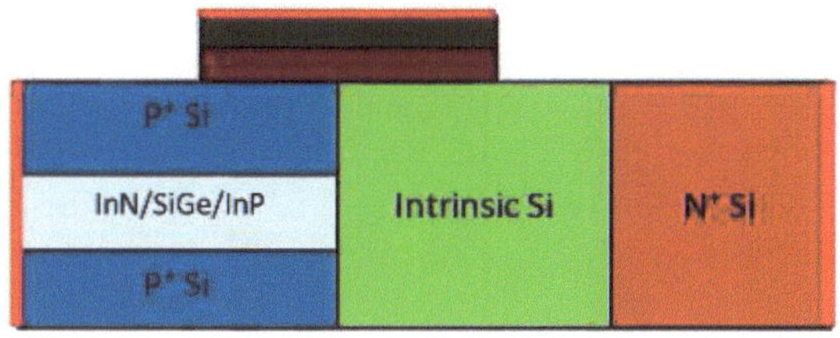

Fig. (10). 2-Dstructure of various source stack materials.

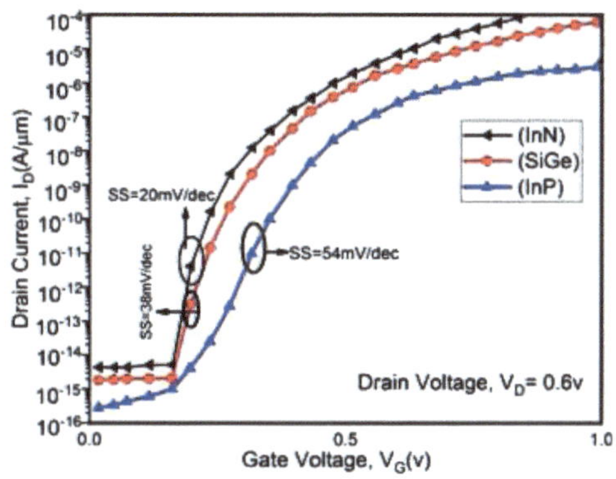

Fig. (11). I_D-V_G Curve of various source stack materials.

Table5. The comparison of DC parameters among Si/InN/Si, Si/SiGe/Si, and Si/InP/Si source stack TFET.

Material	I_{on} (A/μm)	I_{on}/I_{off}	SS (mV/dec)
Si/InN/Si	10^{-4}	10^{11}	20
Si/SiGe/Si	10^{-4}	10^{10}	38
Si/InP/Si	10^{-4}	10^{8}	54

CONCLUSION

In this work, the physical parameters of the HSS-nTFET were optimized, including the stack-source thickness, oxide thickness, channel-undercut length, and gate-oxide thickness. Simulation results show that the introduction of a source stack strengthens the electric field at the source–channel junction and reduces the tunneling width. As a result, the electron tunneling rate and the drive current are improved. The optimized device achieves an I_{ON}/I_{OFF} ratio of the order of 10^{12} with very low leakage current and a steep SS of about 20 mV/dec. These characteristics indicate that the lateral HSS-nTFET is suitable for low-power applications.

AI DISCLOSURE

During the preparation of the work, the author(s) used AI tools and the Grammarly tool to enhance the writing style and readability of the content. The authors clarify that AI was not involved in generating the substantive content and that they take full responsibility for the publication.

REFERENCES

[1] Yuan Taur, D.A. Buchanan, Wei Chen, D.J. Frank, K.E. Ismail, Shih-Hsien Lo, G.A. Sai-Halasz, R.G. Viswanathan, H.J.C. Wann, S.J. Wind, and Hon-Sum Wong, "CMOS scaling into the nanometer regime", *Proc. IEEE,* vol. 85, no. 4, pp. 486-504, 1997. [http://dx.doi.org/10.1109/5.573737]

[2] S. Bangsaruntip, G.M. Cohen, A. Majumdar, and J.W. Sleight, "Universality of short-channel effects in undoped-body silicon nanowire mosfets", *IEEE Electron Device Lett.,* vol. 31, no. 9, pp. 903-905, 2010. [http://dx.doi.org/10.1109/LED.2010.2052231]

[3] K. Roy, S. Mukhopadhyay, and H. Mahmoodi-Meimand, "Leakage current mechanisms and leakage reduction techniques in deep-submicrometer CMOS circuits", *Proc. IEEE,* vol. 91, no. 2, pp. 305-327, 2003. [http://dx.doi.org/10.1109/JPROC.2002.808156]

[4] Qin Zhang, Wei Zhao, and A. Seabaugh, "Low-subthreshold-swing tunnel transistors", *IEEE Electron Device Lett.,* vol. 27, no. 4, pp. 297-300, 2006. [http://dx.doi.org/10.1109/LED.2006.871855]

[5] Byung-Gook Park, B-G. Park, J.D. Lee, and T-J.K. Liu, "Tunneling field-effect transistors(tfets) with subthreshold swing (ss) less than 60mv/dec", *IEEE Electron Device Lett.,* vol. 28, no. 8, pp. 743-745, 2007. [http://dx.doi.org/10.1109/LED.2007.901273]

[6] D. Esseni, M. Pala, P. Palestri, C. Alper, and T. Rollo, "A review of selected topics in physics based modeling for tunnel field-effect transistors", *Semicond. Sci. Technol.,* vol. 32, no. 8, p. 083005, 2017. [http://dx.doi.org/10.1088/1361-6641/aa6fca]

[7] J. Appenzeller, Y.M. Lin, J. Knoch, Z. Chen, and P. Avouris, "Comparing carbon nanotube transistors-the ideal choice: a novel tunneling device design", *IEEE Trans. Electron Dev.,* vol. 52, no. 12, pp. 2568-2576, 2005. [http://dx.doi.org/10.1109/TED.2005.859654]

[8] E.H. Toh, G.H. Wang, G. Samudra, and Y.C. Yeo, "Device physics and design of double-gate tunneling field-effect transistor by silicon film thickness optimization", *Appl. Phys. Lett.,* vol. 90, no. 26, p. 263507, 2007. [http://dx.doi.org/10.1063/1.2748366]

[9] U.E. Avci, D.H. Morris, and I.A. Young, "Tunnel field-effect transistors: Prospects and challenges", *IEEE J. Electron Devices Soc.,* vol. 3, no. 3, pp. 88-95, 2015. [http://dx.doi.org/10.1109/JEDS.2015.2390591]

[10] S. Strangio, P. Palestri, D.A.V.I.D. Esseni, L. Selmi, F. Crupi, S. Richter, Q.T. Zhao, and S. Mantl, "Impact of tfet unidirectionality and ambipolarity on the performance of 6t sram cells", *IEEE J. Electron Devices Soc.,* vol. 3, no. 3, pp. 223-232, 2015. [http://dx.doi.org/10.1109/JEDS.2015.2392793]

[11] B. Bhushan, K. Nayak, and V.R. Rao, "Dc compact model for soi tunnel field- effect transistors", *IEEE Trans. Electron Dev.,* vol. 59, no. 10, pp. 2635-2642, 2012. [http://dx.doi.org/10.1109/TED.2012.2209180]

[12] A. Pal, and A.K. Dutta, "Analytical drain current modeling of double-gate tunnel field-effect transistors", *IEEE Trans. Electron Dev.,* vol. 63, no. 8, pp. 3213-3221, 2016. [http://dx.doi.org/10.1109/TED.2016.2581842]

[13] A.S. Verhulst, W.G. Vandenberghe, K. Maex, S. De Gendt, M.M. Heyns, and G. Groeseneken, "Complementary silicon-based heterostructure tunnel-fets with high tunnel rates", *IEEE Electron Device Lett.,* vol. 29, no. 12, pp. 1398-1401, 2008. [http://dx.doi.org/10.1109/LED.2008.2007599]

[14] M.J. Lee, and W.Y. Choi, "Effects of device geometry on hetero-gate-dielectric tunneling field-effect transistors", *IEEE Electron Device Lett.,* vol. 33, no. 10, pp. 1459-1461, 2012. [http://dx.doi.org/10.1109/LED.2012.2206790]

[15] B. Raad, K. Nigam, D. Sharma, and P. Kondekar, "Dielectric and work function engineered TFET for ambipolar suppression and RF performance enhancement", *Electron. Lett.,* vol. 52, no. 9, pp. 770-772, 2016. [http://dx.doi.org/10.1049/el.2015.4348]

[16] P. Rinke, M. Scheffler, A. Qteish, M. Winkelnkemper, D. Bimberg, and J. Neugebauer, "Band gap and band parameters of InN and GaN from quasiparticle energy calculations based on exact-exchange density-functional theory", *Appl. Phys. Lett.,* vol. 89, no. 16, 2006.. [http://dx.doi.org/10.1063/1.2364469]

[17] K.M. Liu, and C.P. Cheng, "Investigation on the effects of gate-source overlap/underlap and source doping gradient of n-type si cylindrical gate-all-around tunnel field-effect transistors", *IEEE Trans. Nanotechnol.,* vol. 19, pp. 382-389, 2020. [http://dx.doi.org/10.1109/TNANO.2020.2991787]

[18] *S.D.: Version D-2013.03* Synopsys Inc.: Mountain View, CA, USA, 2013.

[19] Q. Huang, R. Huang, C. Wu, H. Zhu, C. Chen, J. Wang, L. Guo, R. Wang, L. Ye, and Y. Wang, "Comprehensive performance re-assessment of tfets with a novel design by gate and source engineering from device/circuit perspective", *2014 IEEE International Electron Devices Meeting,* pp. 13-3, 2014. [http://dx.doi.org/10.1109/IEDM.2014.7047044]

[20] P.O. Oviroh, R. Akbarzadeh, D. Pan, R.A.M. Coetzee, and T.C. Jen, "New development of atomic layer deposition: processes, methods and applications", *Sci. Technol. Adv. Mater.,* vol. 20, no. 1, pp. 465-496, 2019. [http://dx.doi.org/10.1080/14686996.2019.1599694] [PMID: 31164953]

CHAPTER 12

Reliability and Sensitivity Analysis of TFET

Shreyas Tiwari[1,*], **Rajesh Saha**[2] and **Tarun Varma**[3]

[1] *Department of Electronics and Communication Engineering, Pandit Deendayal Energy University, Gandhinagar, Gujarat, India*

[2] *Department of Electronics and Communication Engineering, National Institute of Technology Silchar, Silchar, Assam, India*

[3] *Department of Electronics and Communication Engineering, Malaviya National Institute of Technology, Jaipur, Rajasthan 302017, India*

Abstract: This chapter concentrated on a comprehensive study of TFET with its optical application at three specific wavelengths. Initially, ionic radiation analysis was performed over the Ge-source double gate TFET to evaluate the parameter collected charge (Q_c) and bipolar gain (β). It generated a maximum bipolar gain of 71.6 at LET = 20 MeV.cm^2/mg. After this interface trap analysis, ZHP and hetero stack (HS) TFETs were analyzed, including their effects on the energy band and transfer characteristics. ZHP-TFET reported a major fluctuation in the drain current compared to HSTFET. Further, the impact of noise was evaluated at different frequencies on incorporating various trap effects. Finally, optical assessment of the TFET photo sensor was performed at wavelengths of 300, 500, and 700 nm. Here, HS TFET reported maximum spectral sensitivity (S_n) compared to ZHP-TFET at low gate voltages. The maximum spectral sensitivity of HS-TFET was observed as 421, and ZHP-TFET showed 125 maximum spectral sensitivity at a wavelength of 300 nm. The complete study of TFET was conducted at low supply voltage with minimal power dissipation.

Keywords: Ionic Radiation, Interface trap charge, Noise Analysis, Photo sensor.

INTRODUCTION

Semiconductor businesses are no longer limited to designing mobile devices, chips, and CPUs; they are also venturing into space applications. It is essential to conduct reliability analysis of field-effect transistors to assess their performance under adverse conditions. To address reliability factors, such as radiation [1 - 3], interface trap charges (ITCs) [4, 5], and noise analysis of TFETs [6 - 9] have been performed. Under the radiation effect, high-energy particles enter a semiconductor device, accumulating an extra electron-hole charge carrier. These extra carriers

* **Cooresponding author Shreyas Tiwari:** Department of Electronics and Communication Engineering, Pandit Deendayal Energy University, Gandhinagar, Gujarat, India; E-mail: shreyastiwari21@gmail.com

Girdhar Gopal, Meena Panchore, Arun Kishor Johar & Tarun Varma (Eds.)

cause the device's regular functionality to be perturbed and result in radiation damage. However, the accumulation of trap charges [10] primarily between the semiconductor channel and gate oxide interface inhibits TFET operation. These ITCs arise from a mismatch between two surfaces during fabrication. Since the gate oxide layer follows the amorphous pattern of the layer while the silicon channel is made of crystalline form, the combination of these two layer result an unwanted trap impurities. The final analysis of TFET is application-oriented and includes the optical behavior of the device at a particular wavelength of incident light. Plenty of investigators have already examined the relationship between optical energy and FET-based photo sensors [11 - 14] in their respective fields. They are ideal for photo detectors due to the efficient power consumption, high quantum efficiency, and excellent sensitivity. Consequently, researchers are increasingly focused on FET-based optical detectors for optimal light detection at constant intensity. The literature contains reports on MOSFET-based optical sensors such as the PN Junction [15], Avalanche Photo sensor [16], and PIN Photo sensor [17]. Investigation demonstrates that PN junction and MOSFET devices [18] exhibit minimal photon absorption at excess power dissipation. In this regard, TFET has emerged as an alternative to other FET devices for light detection, offering a low operating voltage and a fast detection rate.

The initial study evaluated the impact of heavy ion radiation on the Ge-sourc--based double gate TFET. Analysis was carried out by calculating the collected charge (Q_C) as well as bipolar gain (β) for different linear energy transfer (LET) [19, 20] values. The impact of various types of trap impurities on the electrical characteristics of horizontal N^+ pocket ZHP-TFET and HS-TFET is further investigated. Subsequently, the optical characteristics of the hetero stack and Z-shape horizontal pocket TFETs were extracted at three different wavelengths in a specific range of the spectrum. The major optical figure of merit (FOMs), such as optical generation rate and spectral sensitivity (Sn) [21, 22], were determined for the corresponding optical device.

This chapter is divided into four sections. Section II provides the device architecture and simulation framework, preceded by the appropriate flow of the fabrication process for these TFETs. Section III describes the trap and optical properties of these TFET devices. Finally, Section IV concludes the chapter.

DEVICE DESIGN AND SIMULATION SETUP

To investigate the effect of heavy ion radiation, a Ge-source based double gate TFET [23, 24] has been considered with its dimensional parameter as illustrated in Fig. (**1a**). The availability of Ge in source side confirms diminished operating voltage of TFET, hence the device executes the fast switching operation and minimum magnitude of leakage current. Further, TFET exhibits tunneling-based carrier conduction, which makes it more resistant to single-event effects (SEE). Fig. (**1b**) depicts the heavy-ion radiation image of the TFET with the corresponding radiation parameters. These essential parameters are: (i) the energy and type of ion, (ii) Angle of penetration, (iii) Penetration Length (L_{Max}), (iv) Radiation width (w), and (v) linear energy transfer (LET).

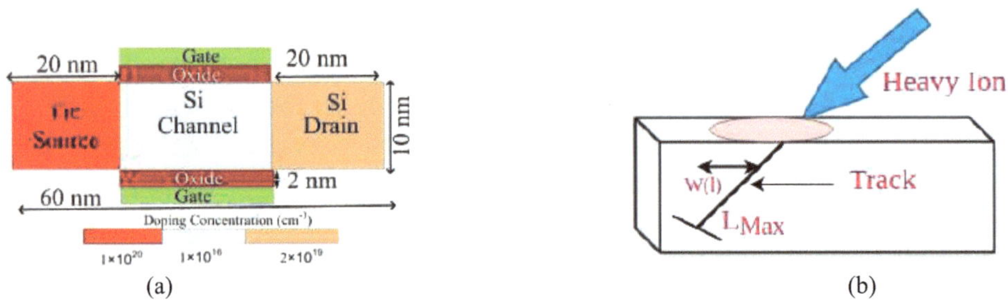

Fig. (1). (a) 2D Schematic of Ge Source Double Gate TFET and (b) Ge source DG-TFET under Heavy ion Radiation effect.

The device architecture of the N^+ pocket-based Z-shaped and stack TFET [24 - 26] as depicted in Fig. (**2a & 2b**). The source side is heavily doped with P-type dopant; the drain region is doped with N-type dopant at 5×1018 cm-3 and the channel region is intrinsically doped at 1×10^{15} cm^{-3}. The ZHP-TFET exhibits one oxide layer on the source side and another silicon oxide region on the bottom side of the gate, confirming both lateral and vertical tunneling along the conductive regime.

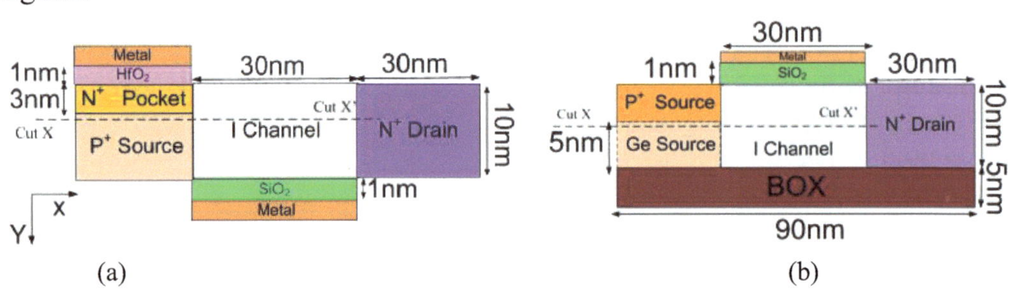

Fig. (2). Schematic of (a) ZHP and (b) HS TFET.

The Hetero stack TFET consists of a Ge source at the source side to operate the device at low V_{GS}. It consists of a silicon layer on the upper side and germanium on the bottom side for improved electrical characteristics. The impact of trap impurities is evaluated on various electrical parameters of TFETs for different trap charge concentrations. To analyze the impact of various ITCs, the acceptor type of traps impurities has been accounted for both uniform and Gaussian types of distribution. For analysis, the trap impurities are considered at the interface between the gate oxide and the silicon channel region of the given TFET. Donor-type impurities are found close to the valence band, whereas acceptor-type trap charges are found close to the conduction band [4]. As a result, the analysis considers acceptor impurities rather than donor charges. In general, Gaussian-type acceptor-trap charge distributions have been considered at the interface between the gate oxide and the silicon channel. The Gaussian trap distribution curve indicates that the maximum trap concentration occurs at the zero-energy level, as illustrated in Fig. (**3**), with the trap distribution parameters provided in Table **1**. The Gaussian distribution of ITCs is mathematically expressed as [27, 28]:

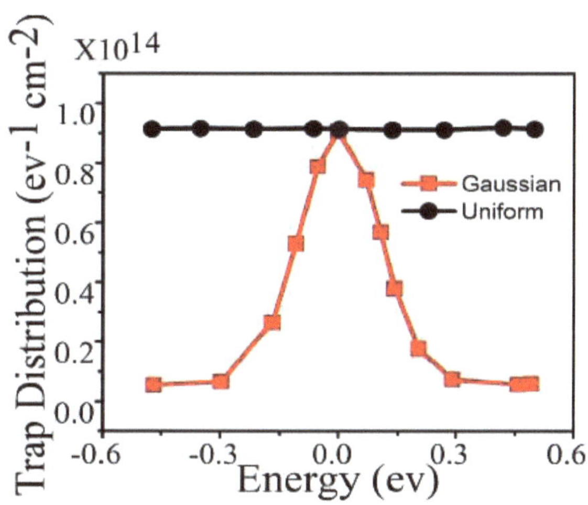

Fig. (3). Trap Distribution [21].

Table 1. Trap Parameters of Various Distributions.

Parameter	Value
Maximum Trap Concentration (N_o)	1×10^{14} eV^{-1} cm^{-2}
Energy Mid (E_o)	0 eV
Energy Sig (E_S)	0.1 eV

$$Q_c = \int_o^t I_d \, dt, \beta = \frac{Q_c}{Q_d} \tag{1}$$

After evaluating the reliability parameter, including heavy-ion radiation, the interface-trap analysis of TFET further explores its application as a photo-sensor. This analysis illustrates their application as photo sensors, performed in a much more proficient way than other FET-based devices.

Fig. (**4**) shows the fabrication steps for ZHP and HS TFETs using different techniques. Initially, an intrinsic silicon wafer is required for device formation; the wafer is then cut to the desired dimensions. Subsequently, the highly doped source and drain regions are allocated using UV-ray photolithography and the ion implantation technique. The next step is the deposition of the high-k and silicon oxide layers using atomic layer deposition (ALD) [28]. Following this, the unwanted oxide layer was etched using rapid ion etching (RIE), and an inductive-charged plasma technique was used. Further fabrication techniques include through pattern recognition using electron beam lithography, followed by polysilicon-metal layer deposition on both the source and channel sides.

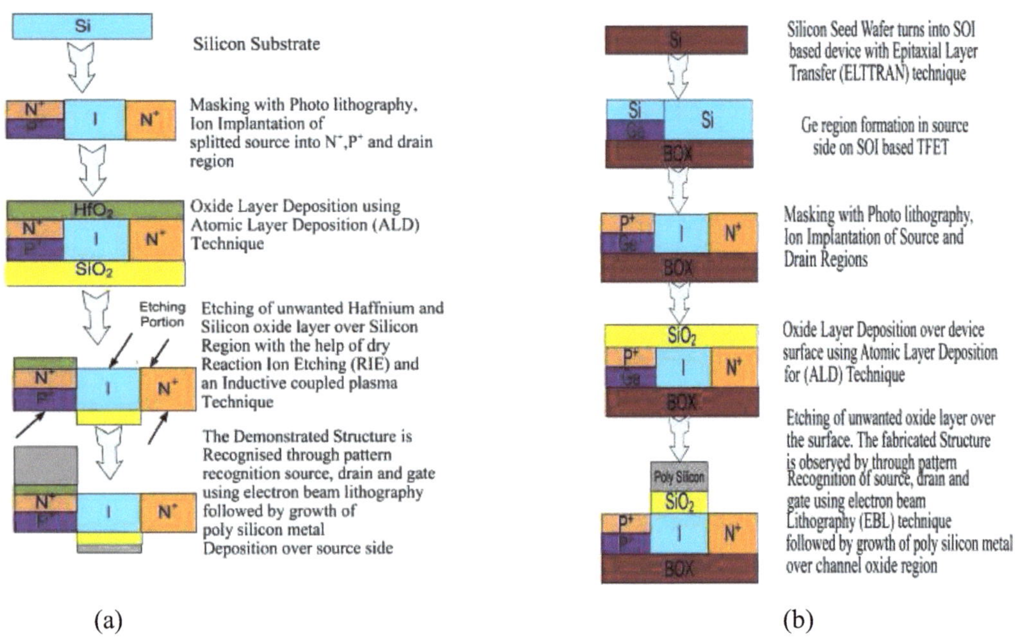

Fig. (4). Fabrication Steps of (a) ZHP and (b) HS TFET.

These TFET-based simulations are performed using the Sentaurus TCAD [29] tool. The Fermi-Dirac distribution model is activated for highly doped source and drain regions. The Shockley-Read-Hall (SRH) recombination and band-to-band tunneling (BTBT) model is applied to describe recombination and tunneling across the energy band. The models' accuracy has been verified by calibrating experimental data with simulation results [30]. To perform the optical operation, an illumination window is considered over the photo-sensitive region of the device, with optical parameters such as light intensity (I), wavelength (λ), polarization angle of incidence, and point of illumination. The non-local tunneling model is used to describe charge-carrier transport in TFETs. Fig. (**5**) shows a strong similarity between the simulated and calibrated data from reference [31].

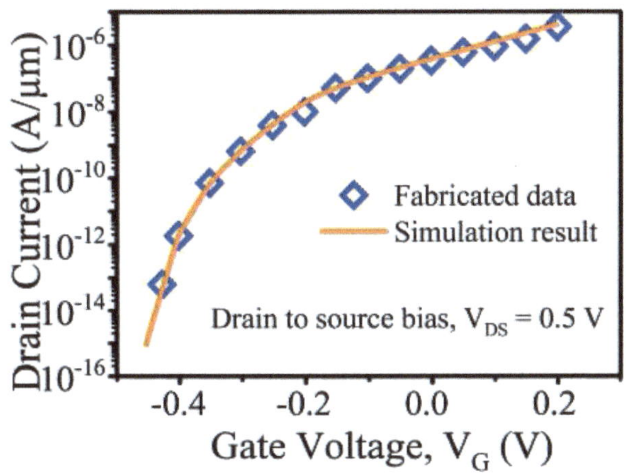

Fig. (5). Calibration of fabricated data from reference [31].

RESULTS AND DISCUSSION

This section presents the final outcomes of the reliability and photo-sensing analyses of TFET at different wavelengths of light. The reliability approach shows the heavy-ion radiation damage to the Ge-source double-gate TFET at various positions along the x-axis.. The analysis extracts radiation parameters, such as collected charge, charge density, and bipolar gain, simulated for different linear energy transfer (LET) values [3, 29]. After radiation analysis, interface-trap charge analysis is performed, and its effects on the electrical parameters of the TFET are evaluated. Finally, photo-sensing analysis of TFETs is executed at different wavelengths of light.

Radiation Analysis

Fig. (**6a**) depicts the ion-radiation location along the x-axis at different positions at constant (LET = 20 MeV-cm^2/mg). Following it, a transient analysis was performed at different positions to signify the most radiation-sensitive area. Subsequently, the electrostatic potential curves of the device before and after ionic radiation at a characteristic time of 10 ps are illustrated in Fig. (**6b**). Under radiation, an accumulated charge lowers the potential barrier of the curve, allowing excess charge conduction in a radiation state.

Fig. (**7a-e**) represents the Gaussian distribution of charges generated by heavy ions at the source side, source and channel junction, channel center, channel-drain junction, and drain terminal, respectively. Fig. (**7f**) shows the greatest heavy-ion generation data for the relevant location. It is observed that as the radiation position shifts from source to drain, the corresponding peak also shifts in the same direction.

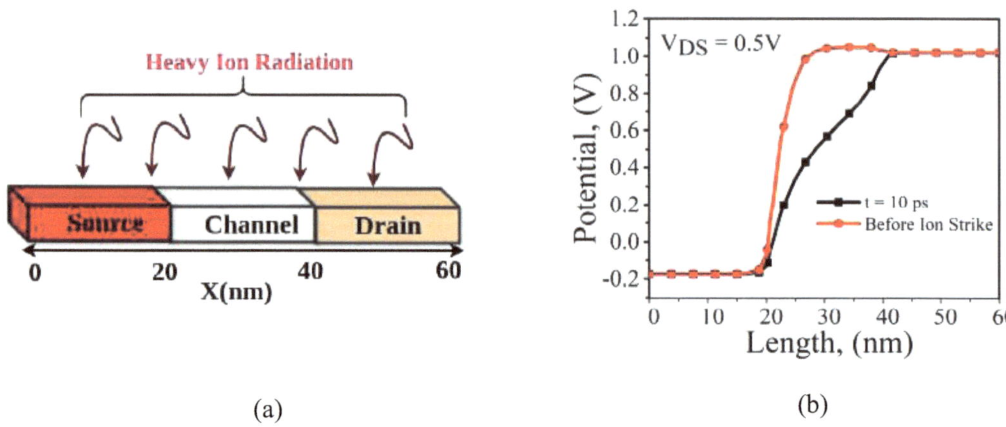

Fig. (6). (a) Heavy-Ion Radiation at different position in X direction and (b) Electrostatic Potential Curve before and after ion strike.

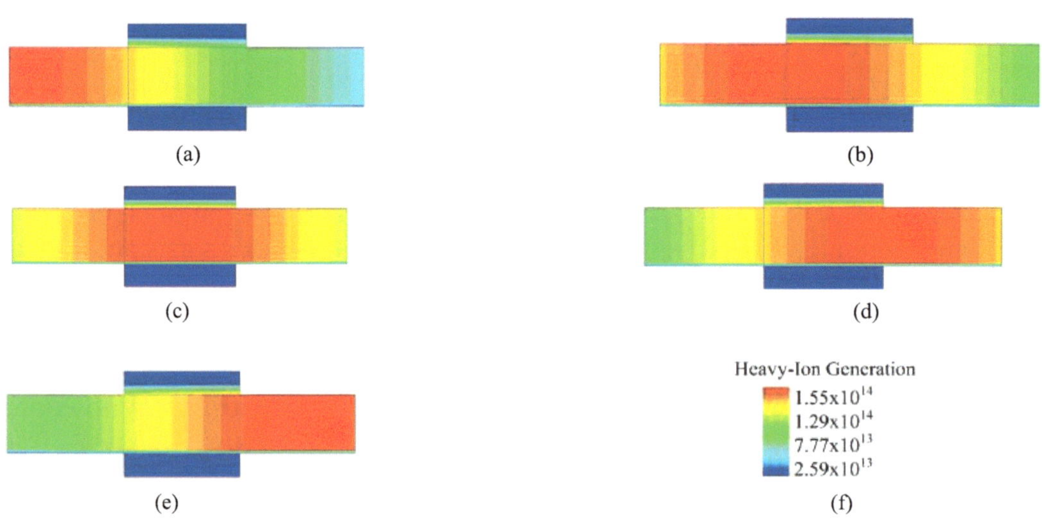

Fig. (7). (a) X = 10 (b) X = 20 (c) X = 30 (d) X = 40 (e) X = 50 and (f) Heavy Ion-Generation at Constant LET = 20 MeV-cm^2/mg.

Fig. (**8a**) illustrates the charge density at the corresponding radiation point. It is observed that the accumulation of ionic charge attains a maximum at the point of radiation. Furthermore, the research demonstrates the device's transient behavior for varied ion hit positions at a constant LET of 20 MeV-cm^2/mg. An ion striking the device causes the transient current to rise for about 12 ps, reaching its maximum peak at 15 ps. I_D is observed to be maximum for position X = 30 nm and minimum for location X = 10 nm as shown in Fig. (**8b**).

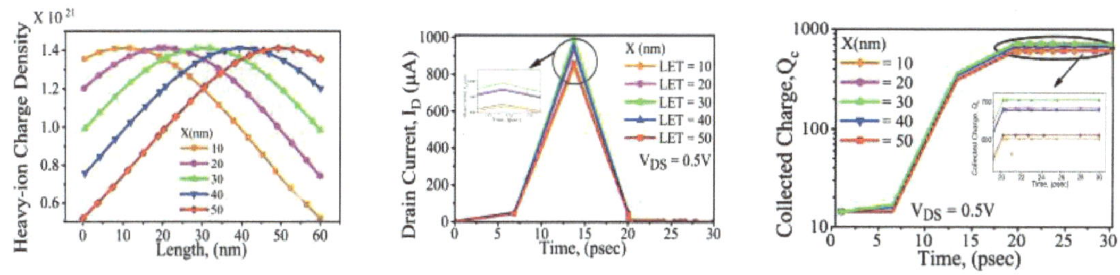

(a) (b) (c)

Fig. (8). (a) The peak variation in Heavy-ion charge density (b) Transient drain current with respect to the Position of ion strikes and (c) Collected Charge (pC/μm) for positions at constant LET = 20 MeV-cm2/mg.

The maximum value of drain current is reported when radiation occurs at the center of the channel region, due to an excess of ionized e-h pairs.

The collected charge at different positions is shown in Fig. (**8c**). Since the maximum magnitude of current is observed at the center, the lowest value of current is observed near the source. Therefore, the lowest accumulation of charge is seen inside the source side at X = 10 nm, whereas the largest collected charge is seen at X = 30 nm, as shown in graph (Fig. **8c**). The maximum radiation impact occurs at x = 30 nm, which is the center of the channel region. Device physics states that the channel region has the highest concentration of charge carriers, the strongest electric field, and the highest density of states. Moreover, ionic particles travel the greatest distance through the channel region, further enhancing the rate of radiation damage. These factors indicate that point x = 30 nm is most prone to radiation damage and most vulnerable, as illustrated in Fig. (**9a** & **b**). The collected charge [2, 3] is given by the mathematical equation as shown in Fig. (**9a**) Heavy Charge density at position X = 30 nm at different LETs and (b) Contour Diagram of Heavy-ion charge density at LET = 20 MeV-cm^2/mg.

Fig. (**9a**) depicts the distribution of ion generation in the device for LET values of 1, 3, 6, 10, 15, and 20 MeV-cm^2/mg [29].

$$D_{GAU} = N_o \exp(\frac{|E-E_O|^2}{2E}) \quad \textbf{(2)}$$

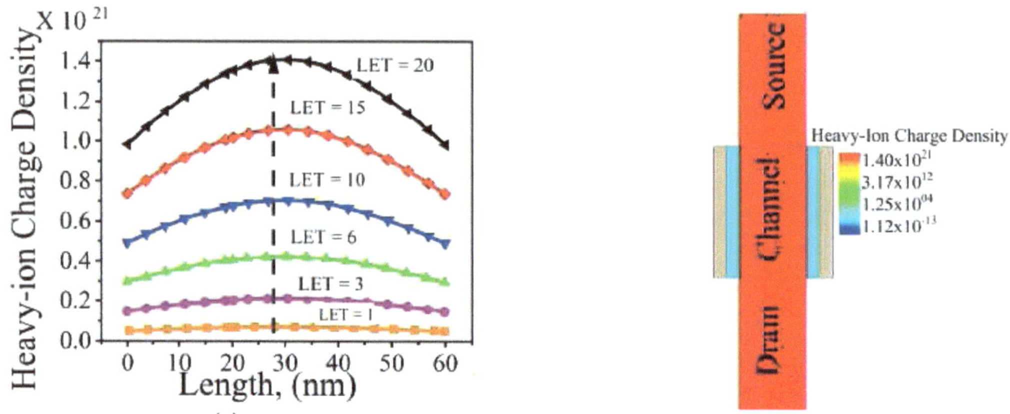

(a) (b)

Fig. (9). (a) Heavy Charge density at position X = 30 nm at different LETs and (b) Contour Diagram of Heavy-ion charge density at LET = 20 MeV-cm2/mg.

Fig. (**9b**) clearly shows that, at the most sensitive position of the device (X = 30 nm), the highest charge density for LET = 20 is approximately 1.4×10^21 cm^{-3}. This is because, as the LET value grows, it also increases the rate of heavy-ion creation. As LET increases, the amount of energy deposited per unit length increases, leading to higher charge density. Moreover, higher LET ions produce more ionizations along their radiation path and generate more electron-hole pairs, which leads to a corresponding increase in charge density.

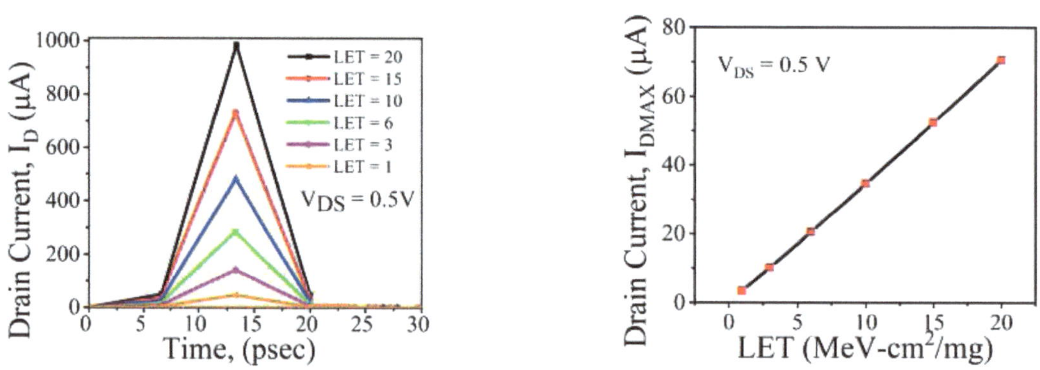

(a) (b)

Fig. (10). (a) The Variation in transient drain current and (b) Collected Charge shown for different LETs with respect to time.

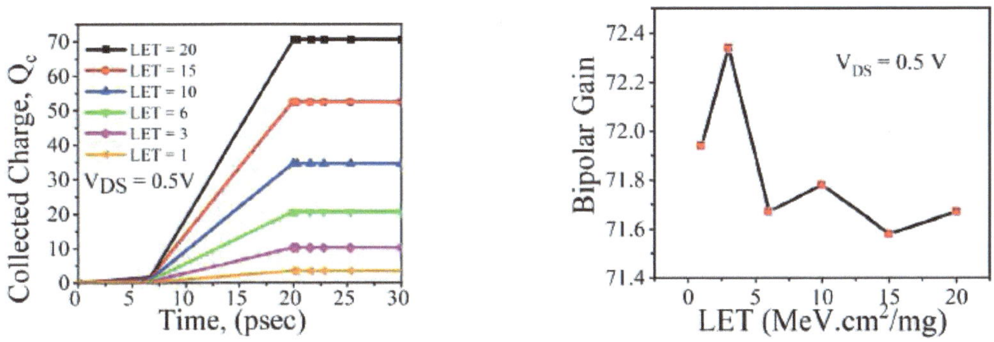

(a) (b)

Fig. (11). (a) Maximum transient Drain current and (b) Bipolar Gain with respect to different LET Value.

Further analysis depicts the (I_D) Transient drain current of the device for different LET values as illustrated in Figs. (**10a, b**). Analysis indicates that the drain current reaches 985 mA, 732 mA, 482 mA, 286 mA, 141 mA, and 46.7 mA at LET = 20, 15, 10, 6, 3, and 1, respectively. Transient current modulation is observed due to ion hits, with higher LET values resulting in stronger transients. Additionally, the collected charge depends directly on the amount of LET delivered to the device after an ion strike. Therefore, radiation under a higher LET value exhibits a larger value of collected charge, as shown in Fig. (**11a**).

The bipolar Gain (β) is an essential metric for analyzing the influence of radiation, as shown in Fig. (**11b**). When ions penetrate the device, they form electron-hole pairs, which boost the transient current and therefore, the charge is collected.

Trap and Noise Analysis

The comparative analysis of DC characteristics for a given ZHP and HS TFET across varying ITCs is presented in this section. The energy band diagrams under ambipolar and OFF states have been extracted as displayed in Fig. (**12**). The investigation states that the impact of trap charges on the energy band diagram in ZHP-TFET is comparatively small compared to HS-TFET in OFF and ambipolar states.

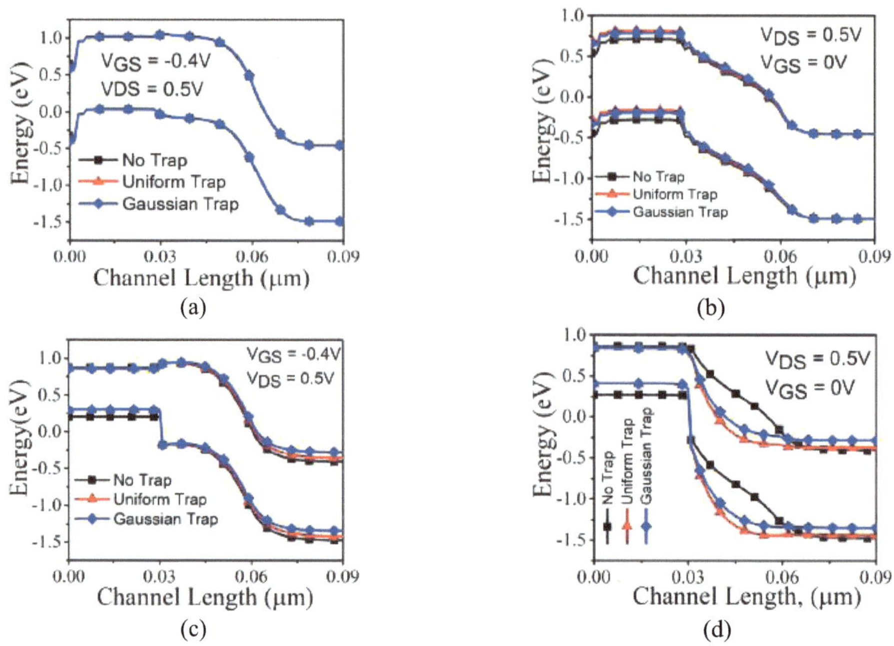

Fig. (12). Energy band diagram (a, b) ZHP TFET at the ambipolar and OFF state, (c, d) HS-TFET at the ambipolar and OFF state.

Fig. (**13a**, **b**) depicts the drain current behavior of ZHP and HS TFETs, respectively. It appears that less drain-current fluctuation is observed in ZHP-TFET than in HS-TFET due to the accounted trap effect. The primary explanation for the low fluctuation in the drain current curve under trap impurities is superior hafnium oxide alignment on the source side and in the channel oxide area.

However, in HS-TFET, the availability of Ge in the source region is responsible for the unnecessary strain effect. As a result, inserting traps over the oxide of HS-TFETs reduces their performance in terms of DC figure of merit (FOM) and makes them less reliable over time.

After analyzing the trap's impact on the device physics and DC behavior of TFET, the analysis further attributes device noise to the devices' impact on the device physics and DC behavior of TFET. Noise is the unwanted behavior generated in the transistor by contaminants at the interface between the gate oxide and the conductive channel. Thus, the noise density of TFET was evaluated at three different frequencies.

The drain current noise spectral density (S_{id}) [32, 33] is an important noise parameter to understand the noise assessment parameter of TFET, primarily in analog and RF applications. Noise sources in TFETs are generated by the tunneling mechanism; these include diffusion noise, flicker noise, and generation-recombination noise. Shot noise mainly arises from tunneling of carriers from the source to the channel region, rather than from the conventional drift and diffusion mechanisms.

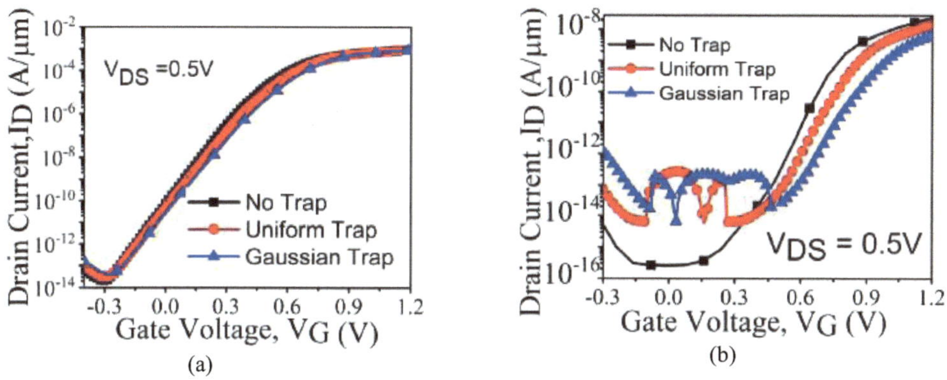

Fig. (13). Transfer Characteristic curves for (a) ZHP and (b) HS TFETs at different impurities.

Fig. (**14**) reveals the S_{id} behavior of ZHP-TFET at three different frequencies. The noise assessment was executed by accounting for both types of trap impurities. The spectral noise density graph shows that the flicker and GR noise components are predominant at 100 Hz, whereas as frequency increases, the impact of diffusion noise dominates. The expression above demonstrates that the relationship between GR noise and frequency square is inverse. It suggests that as frequency increases, GR noise decreases, but simultaneously diffusion noise component increases.

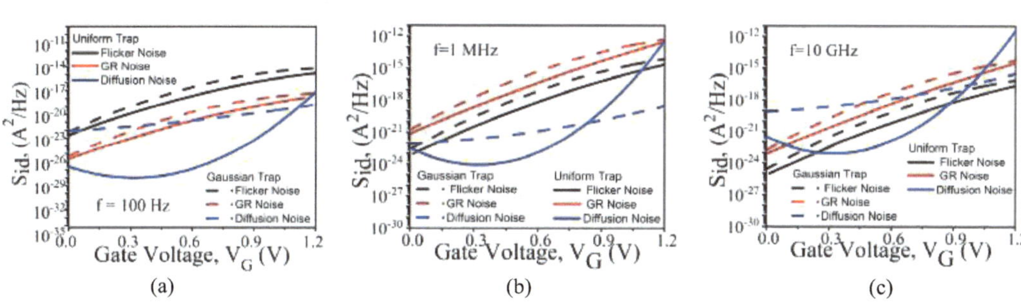

Fig. (14). S_{id} of ZHP-TFET at (a) Low (b) Medium and (c) High frequency under trap impurities.

Further, the Sid noise distribution plot of HS-TFET, extracted by considering trap impurities, is displayed in Fig. (**15**). The Sid plot shows that flicker and Gaussian noise have a major impact on device behavior at low frequencies. Since the availability of Ge at the source side ensures the stack and strain behavior of TFETs, the noise performance of HS devices is improved. Moreover, the device's frequency increases the Gaussian distribution of trap charges, with the highest level occurring at the Sid of HS TFET.

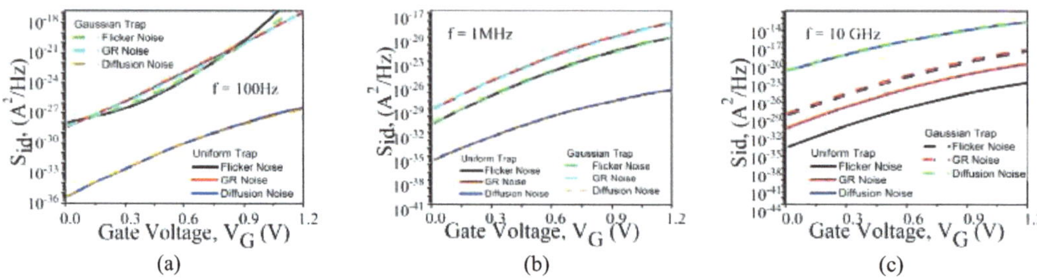

Fig. (15). S_{id} of HS-TFET at (a) Low (b) Medium and (c) High frequency under trap impurities.

Optical Analysis

Fig. (**16**) displays the ZHP TFET-based TFET sensor report excess optical carrier rate, compared with the HS TFET. The primary reason for the excess generation rate is the presence of an oxide layer on both sides of the ZHP-TFET leads, leading to an increased constriction in the energy bands during photo illumination. The light of a particular wavelength incident on the photo-sensitive poly silicon metal region results in an excess generation of charge carriers inside the sensitive region. These extra charge carriers increase the tunneling rate from the source to the channel region, thereby enhancing the drain current of TFET in the wavelength range of (0.3-0.7) m.

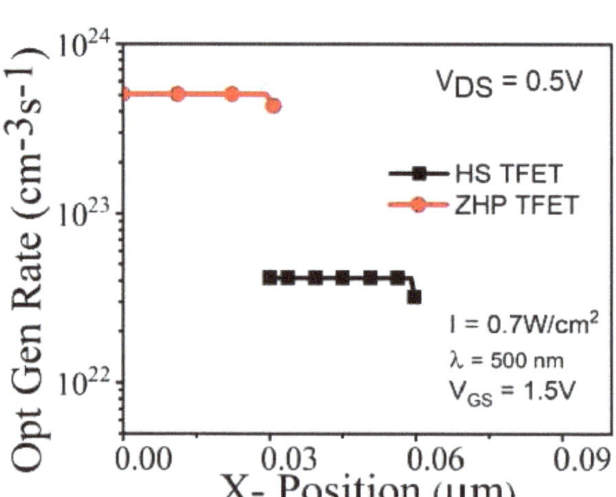

Fig. (16). Photo Generation rate of TFET based Photodiodes.

The energy diagrams for the HS-TFET in both dark and illuminated states are shown in Fig. (**17a** and **b**), respectively. Fig. (**17c** and **d**) show the ZHP-TFET photodiode's energy band diagram under light and dark circumstances, respectively. As shown in these energy bands, under an illuminated state, the electron and hole Fermi levels (Efn, Efp) [14, 21] separate due to the generation of a photo voltage (V_{OP}). While under a dark state, no additional electron-hole pairs are generated inside the device; therefore, the electron Fermi levels coincide.

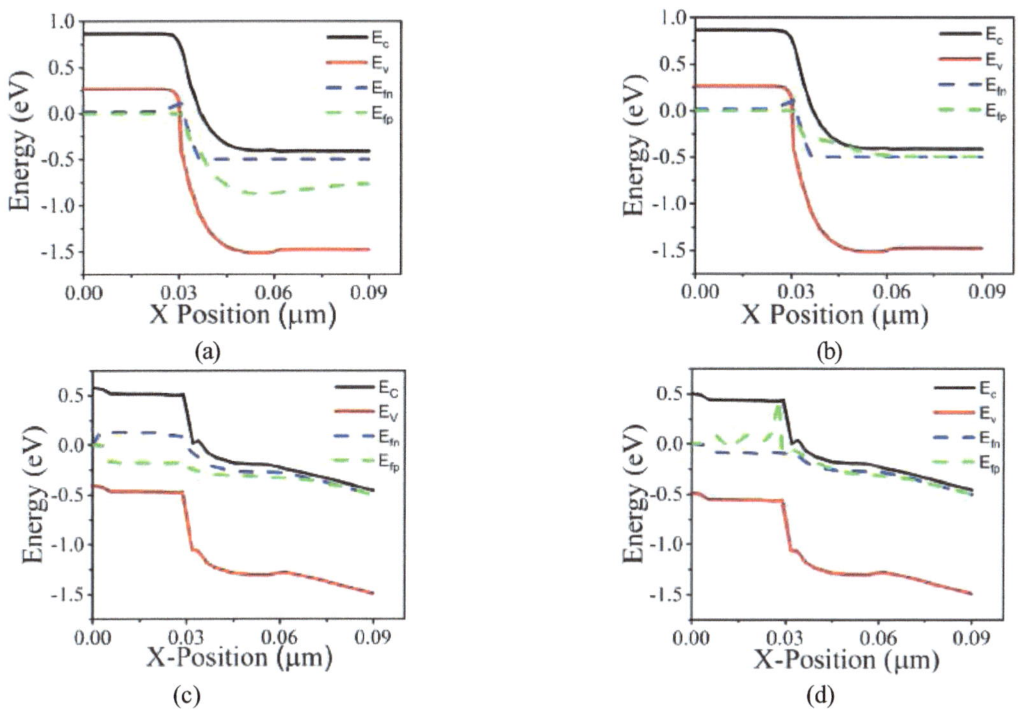

Fig. (17). Energy band diagram for (a, b) HS-TFET Photodiode in illuminated and OFF state (c, d) ZHP TFET Photodiode in illuminated and OFF state at V_{DS} = 0.5 V and V_{GS} = 1.5 V.

Fig. (**18**) shows the drain current behavior of ZHP- and HS-TFET-based photodiodes in the light and dark states. The sensor is considered to be in the light state when the angle of incidence in the simulator's physics file is set to 0^0. A light beam is supposed to move parallel to the light-sensitive region [29] in the +X direction when the angle of incidence is 90^0, which is referred to as the dark state of the device. Additionally, variation in drain current under dark and light conditions is primarily observed at the low end of the gate voltage range because TFET exhibits steepness (SS< 60 mV/dec), thus fluctuations in drain current are observed in the sub-threshold region. Optical analysis shows that the ZHP-based TFET sensor exhibits better current characteristics than the HS-TFET photodiode.

Additionally, it has been observed that the ZHP TFET photodiode exhibits reduced drain current (ID) variance under both light and dark conditions. HS-TFET shows notable fluctuations in drain current under photo operation due to additional charge carriers generated.

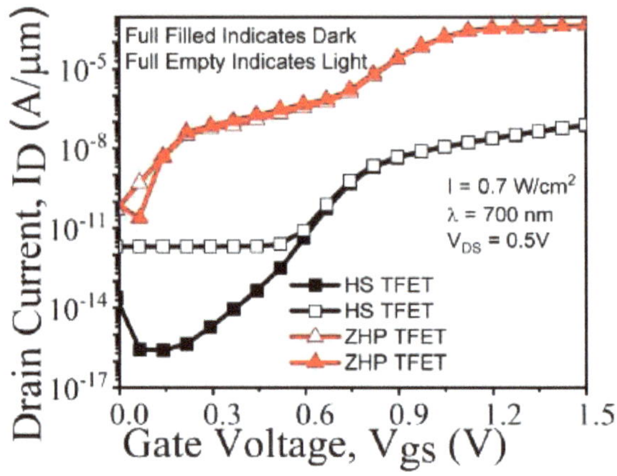

Fig. (18). I_D vs Gate voltage (V_{GS}) characteristics of TFET.

The photodiode's spectral sensitivity (S_n) [21] is an important optical parameter for imaging and sensing applications. It involves materials that respond to different wavelengths of light. The S_n is mathematically expressed as:

$$S_n = \frac{I_D(\lambda_1) - I_D(\lambda_2)}{I_D(\lambda_2)} \tag{3}$$

where $I_D(\lambda_1)$ and $I_D(\lambda_2)$ are the photo current at wavelength λ_1 and λ_2 respectively. As the incident wavelength increases, from λ_1 to λ_2 corresponding light electrical energy decreases, consequently less number of carrier generated at the interface. Fig. (**19a**) depicts the Sn variation across different light wavelengths. The Sn of the HS TFET is about 421, and the ZHP TFET-based photo sensor is 125 at a 300 nm wavelength. Following this, (Fig. **19b**) illustrates the Sn vs. gate voltage (VGS) plot for both TFET-based photo sensors. The photo-current fluctuates more in the case of HS-TFET than in ZHP-TFET; therefore, the sensitivity of HS-TFET is larger than that of ZHP-TFET. The maximum sensitivity of HS-TFET is reported to be 67.5 due to the maximum difference in photocurrent between the dark and illuminated states. Whereas, the ZHP-TFET-based photo sensor achieves a maximum sensitivity of 14.8 due to the minimum difference in photo current between light and dark states.

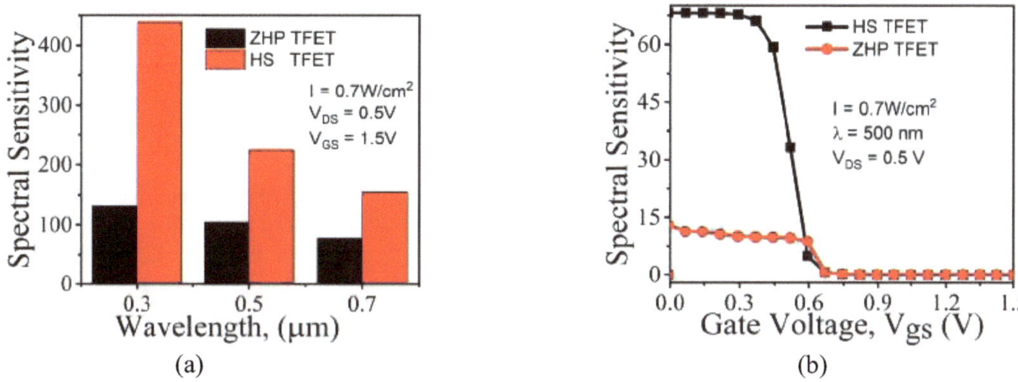

Fig. (19). Comparison of (a) S_n vs. λ and (b) S_n vs. V_{GS} of TFET.

This chapter provides a comprehensive analysis of TFET, including interface-trap charges, noise analysis, and sensing applications. The study states that TFET is the preferred device for researchers to perform operations with minimal time and power dissipation.

CONCLUSION

This section concludes the TFET analysis by simulating heavy-ion radiation, analyzing the impact of trap impurities, performing noise analysis, and applying sensitivity analysis. Ge-source double-gate TFET report a maximum heavy-ion charge density of 1.4×10^{21} at LET = 20 MeV-cm^2/mg. Further, an investigation measures the impact of trap effects on the ZHP and HS TFET. Trap and noise investigations reveal that HS-TFET is more promptly affected by interface impurities than ZHP, which explains the stress & strain effect on the source side of HS-TFET. Finally, the optical application of both TFET carries out at three different wavelengths of light. The HS-TFET reports a (Sn) max of 67.5, and the ZHP-TFET measures a sensitivity of 14.8 at a low gate voltage (V_{GS}). Such analysis decodes the complete information of any device, from reliability to optical application at low supply voltage.

REFERENCES

[1] N. Kamal, M. Panchore, and J. Singh, "3-D Simulation of Junction- and Doping-Free Field-Effect Transistor Under Heavy Ion Irradiation", *IEEE Trans. Device Mater. Reliab,* vol. 18, no. 2, pp. 173-179, 2018. [http://dx.doi.org/10.1109/TDMR.2018.2811493]

[2] A. Maurya, K. Koley, J. Kumar, and P. Kumar, "Investigation of Single-Event-Transient Effects Induced by Heavy-Ion in All-Silicon DG-TFET", *IEEE Access,* vol. 10, pp. 109357-109365, 2022. [http://dx.doi.org/10.1109/ACCESS.2022.3213685]

[3] G.D. Badhwar, and P.M. O'Neill, "Improved model of galactic cosmic radiation for space exploration emission,", *Int. J. Radiat. Appl. Instrum. D. Nucl. Tracks Radiat. Meas.,* vol. 20, no. 3, pp. 403-410,

1992.
[http://dx.doi.org/10.1016/1359-0189(92)90024-P]

[4] X.Y. Huang, G.F. Jiao, W. Cao, D. Huang, H.Y. Yu, Z.X. Chen, N. Singh, G.Q. Lo, D.L. Kwong, and Ming-Fu Li, "Effect of Interface Traps and Oxide Charge on Drain Current Degradation in Tunneling Field-Effect Transistors", *IEEE Electron Device Lett,* vol. 31, no. 8, pp. 779-781, 2010.
[http://dx.doi.org/10.1109/LED.2010.2050456]

[5] Qiu Yingxin, Wang Runsheng, Huang Qianqian, and Huang Ru, "A Comparative Study on the Impacts of Interface Traps on Tunneling FET and MOSFET", *IEEE Trans. Electron Dev.,* vol. 61, no. 5, pp. 1284-1291, 2014.
[http://dx.doi.org/10.1109/TED.2014.2312330]

[6] S. Gupta, Anju, and G.P. Mishra, "Design and Performance Analysis of Z-Shaped Charge Plasma TFET-based label-free Biosensor,", *2021 Devices for Integrated Circuit (DevIC),* pp. 446-449, 2021.
[http://dx.doi.org/10.1109/DevIC50843.2021.9455818]

[7] S. Tiwari, and R. Saha, "Methods to Reduce Ambipolar Current of Various TFET Structures: a Review", *Silicon,* vol. 14, no. 12, pp. 6507-6515, 2022.
[http://dx.doi.org/10.1007/s12633-021-01458-w]

[8] G. Gopal, and T. Varma, "Impact of Temperature on the Reliability of UTB-DG-FE-TFETs and Their RF/Analog and Linearity Parameter Dependence", *J. Electron. Mater.,* vol. 52, no. 9, pp. 6293-6307, 2023.
[http://dx.doi.org/10.1007/s11664-023-10556-x]

[9] R. Molaei Imenabadi, M. Saremi, and W.G. Vandenberghe, "A Novel PNPN-Like Z-Shaped Tunnel Field- Effect Transistor With Improved Ambipolar Behavior and RF Performance", *IEEE Trans. Electron Dev.,* vol. 64, no. 11, pp. 4752-4758, 2017.
[http://dx.doi.org/10.1109/TED.2017.2755507]

[10] M. Dassi, J. Madan, R. Pandey, and R. Sharma, "Impact of interfacial charges on analog and RF performance of Mg2Si source heterojunction double-gate tunnel field effect transistor", *J. Mater. Sci. Mater. Electron.,* vol. 32, no. 19, pp. 23863-23879, 2021.
[http://dx.doi.org/10.1007/s10854-021-06823-4]

[11] J. Talukdar, and B. Muralidharan, "Performance Analysis of Optically Gated MoS2 Photosensor for Visible Light Detection,", *in IEEE Sensors Journal,* vol. 24, no. 15, pp. 23810-23817, 2024.
[http://dx.doi.org/10.1109/JSEN.2024.3416374]

[12] V. Kumari, M. Gupta, and M. Saxena, "TCAD-Based Investigation of Double Gate JunctionLess Transistor for UV Photodetector", *IEEE Trans. Electron Dev.,* vol. 68, no. 6, pp. 2841-2847, 2021.
[http://dx.doi.org/10.1109/TED.2021.3075654]

[13] K. Manikanta, and U. Nanda, "Optical Performance of the Double Gate Reverse T-Shaped Channel TFET in Near Visible Light Photodetector", *Transactions on Electrical and Electronic Materials,* vol. 25, no. 2, pp. 160-172, 2024.
[http://dx.doi.org/10.1007/s42341-023-00493-1]

[14] S. Joshi, P.K. Dubey, and B.K. Kaushik, "Photosensor Based on Split Gate TMD TFET Using Photogating Effect for Visible Light Detection", *IEEE Sens. J.,* vol. 20, no. 12, pp. 6346-6353, 2020.
[http://dx.doi.org/10.1109/JSEN.2020.2966728]

[15] B. Fahs, A.J. Chowdhury, Y. Zhang, J. Ghasemi, C. Hitchcock, P. Zarkesh-Ha, and M.M. Hella, "Design and Modeling of Blue-Enhanced and Bandwidth-Extended PN Photodiode in Standard CMOS Technology", *IEEE Trans. Electron Dev.,* vol. 64, no. 7, pp. 2859-2866, 2017.
[http://dx.doi.org/10.1109/TED.2017.2700389]

[16] M. Moszynski, M. Kapusta, D. Wolski, M. Szawlowski, and W. Klamra, "Energy resolution of scintillation detectors readout with large area avalanche photodiodes and photomultipliers", *IEEE Trans. Nucl. Sci.,* vol. 45, no. 3, pp. 472-477, 1998.
[http://dx.doi.org/10.1109/23.682429]

[17] H.S. Kim, S.H. Park, J.H. Ha, D.H. Lee, and S.Y. Cho, "Characteristics of a Fabricated PIN Photodiode for a Matching With a CsI(Tl) Scintillator", *IEEE Trans. Nucl. Sci.,* vol. 57, no. 3, pp. 1382-1385, 2010.
[http://dx.doi.org/10.1109/TNS.2010.2046334]

[18] C.B. Shaik, and C.K. Pandey, "Enhanced optical performance of a dual-drain vertical TFET photosensor for near-infrared light detection", *Micro and Nanostructures,* vol. 198, p. 208051, 2025.
[http://dx.doi.org/10.1016/j.micrna.2024.208051]

[19] K. Aishwarya, and B. Lakshmi, "TCAD simulation study of heavy ion radiation effects on hetero junctionless tunnel field effect transistor", *Sci. Rep.,* vol. 14, no. 1, p. 7643, 2024.
[http://dx.doi.org/10.1038/s41598-024-58371-6] [PMID: 38561390]

[20] Y. Wu, and Y. Takahash, "The impact of tunnel FET on heavy ion induced transient effect", *Proc. ECS Meeting,* 2017.Chesapeake, VA, USA
[http://dx.doi.org/10.1149/MA2017-02/26/1143]

[21] V. D. Wangkheirakpam, B. Bhowmick, and P. D. Pukhrambam, "Modeling and Simulation of Optically Gated TFET for Near Infra-Red Sensing Applications and Its Low Frequency Noise Analysis", *in IEEE Sensors Journal,* vol. 20, no. 17, pp. 9787-9795, 2020.
[http://dx.doi.org/10.1109/JSEN.2020.2991406]

[22] Y.R. Sipauba Carvalho da Silva, Y. Koda, S. Nasuno, R. Kuroda, and S. Sugawa, "An ultraviolet radiation sensor using differential spectral response of silicon photodiodes,", *2015 IEEE Sensors, Busan, Korea (South),* pp. 1-4, 2015.
[http://dx.doi.org/10.1109/ICSENS.2015.7370656]

[23] O.K. Singh, V. Dhandapani, and B. Kaur, "Investigation of the Electrical Parameters in a Partially Extended Ge-Source Double-Gate Tunnel Field-Effect Transistor (DG-TFET)", *J. Electron. Mater.,* vol. 53, no. 6, pp. 2999-3012, 2024.
[http://dx.doi.org/10.1007/s11664-024-10997-y]

[24] R. Saha, D.K. Panda, R. Goswami, B. Bhowmick, and S. Baishya, "DC and RF /analog parameters in Ge source split drain ZHP TFET : Drain and pocket engineering technique", *Int. J. Numer. Model.,* vol. 35, no. 3, p. e2967, 2022.
[http://dx.doi.org/10.1002/jnm.2967]

[25] R. Molaei Imenabadi, M. Saremi, and W.G. Vandenberghe, "A Novel PNPN-Like Z-Shaped Tunnel Field- Effect Transistor With Improved Ambipolar Behavior and RF Performance", *IEEE Trans. Electron Dev.,* vol. 64, no. 11, pp. 4752-4758, 2017.
[http://dx.doi.org/10.1109/TED.2017.2755507]

[26] K. Vanlalawmpuia, and B. Bhowmick, "Analysis of Hetero-Stacked Source TFET and Heterostructure Vertical TFET as Dielectrically Modulated Label-Free Biosensors", *in IEEE Sensors Journal,* vol. 22, no. 1, pp. 939-947, 2022.
[http://dx.doi.org/10.1109/JSEN.2021.3128473]

[27] L.N. Teja, R. Chaudhary, S. Tiwari, and R. Saha, "Elimination of the Impact of Trap Charges through Heterodielectric BOX in Nanoribbon FET", In: *Nanoelectronic Devices and Applications* Bentham Science Publishers, 2024, pp. 231-245.
[http://dx.doi.org/10.2174/9789815238242124010014]

[28] A.M. Walke, A. Vandooren, R. Rooyackers, D. Leonelli, A. Hikavyy, R. Loo, A.S. Verhuslt, K.H. Kao, C. Huyghebaert, G. Groesken, and V.R. Rao, "Fabrication and Analysis of a Si/Si00.55Ge00.45 Heterojunction Line Tunnel FET. IEEE Tran", *Electron Devices,* vol. 61, pp. 707-715, 2014.
[http://dx.doi.org/10.1109/TED.2014.2299337]

[29] T.C.A.D. Sentaurus User Guide, *Synopsys Inc*Mountain View, CA, USA, 2020.

[30] K.H. Kao, A.S. Verhulst, W.G. Vandenberghe, B. Sorée, G. Groeseneken, and K. De Meyer, "Direct and indirect band-to-band tunneling in germanium-based TFETs", *IEEE Trans. Electron Dev.,* vol. 59,

no. 2, pp. 292-301, 2012.
[http://dx.doi.org/10.1109/TED.2011.2175228]

[31] S.H. Kim, H. Kam, C. Hu, and T-J.K. Liu, "Ge-source tunnel field effect transistors with record high ION/IOFF", *VLSI Symp. Tech. Dig.,* pp. 178-179, 2009.

[32] V.D. Wangkheirakpam, B. Bhowmick, and P.D. Pukhrambam, "Noise behavior of vertical tunnel FETs under the influence of interface trap states", *Microelectronics J.,* vol. 114, p. 105124, 2021.
[http://dx.doi.org/10.1016/j.mejo.2021.105124]

[33] R. Goswami, B. Bhowmick, and S. Baishya, "Effect of scaling on noise in Circular Gate TFET and its application as a digital inverter", *Microelectronics J.,* vol. 53, pp. 16-24, 2016.
[http://dx.doi.org/10.1016/j.mejo.2016.04.009]

CHAPTER 13

Dielectric-Modulated Zinc-Oxide Nanostructured Thin Film Field Effect Transistor

Dasari Srikanya[1,*], **Chitrakant Sahu**[2], **Girdhar Gopal**[3], **Arun kishor Johar**[4] and **Tarun Varma**[5]

[1] *Department of Electronics and Communication Engineering, Dayannda Sagar University, Harohalli, Karnataka, India*

[2] *Department of Electronics and Communication Engineering, National Institute of Technology Raipur, Raipur, Chhattisgarh, India*

[3] *Department of Electronics and Communication Engineering, National Institute of Technology Patna, Patna, Bihar, India*

[4] *Department of Electronics and Communication Engineering, Manipal University Jaipur, Jaipur, Rajasthan, India*

[5] *Department of Electronics and Communication Engineering, Malaviya National Institute of Technology Jaipur, Jaipur, Rajasthan, India*

Abstract: This chapter discusses the design, simulation, and performance evaluation of a dielectric-modulated zinc oxide thin-film transistor (DM-ZnO-TFT) biosensor aimed at label-free biomolecule detection. The sensor exploits the dielectric modulation caused by biomolecules with different dielectric constants located within a nanocavity near the TFT's drain region. To improve stability and enhance molecular binding, a biocompatible Al_2O_3 layer along with a dielectric SiO_2 stack is incorporated. Through 2D TCAD simulations, key electrical parameters such as threshold voltage shift, drain current, surface potential, and transconductance were analyzed. The findings reveal a significant change in device behavior in response to biomolecules with dielectric constants ranging from 1 to 15, demonstrating the sensor's high sensitivity and real-time detection capabilities. This DM-ZnO-TFT biosensor offers a cost-effective, CMOS-compatible, and promising approach for future biomedical and environmental sensing technologies.

Keywords: Dielectric-modulated biosensor, Zinc oxide (ZnO), Thin-film transistor (TFT), TCAD simulation label-free detection, Biomolecule sensing.

[*] **Corresponding author Dasari Srikanya:** Department of Electronics and Communication Engineering, Dayannda Sagar University, Harohalli,Karnataka, India; E-mail: srikanya99@gmail.com

Girdhar Gopal, Meena Panchore, Arun Kishor Johar & Tarun Varma (Eds.)

INTRODUCTION

BioFETs, or Field-effect transistor (FET)-based biosensors, have been a significant innovation in the biosensor field since their realization in 1980 [1]. They have broad applications in the agriculture, environment, food, and medical industries due to their signal amplification and unique biological recognition abilities. BioFETs consist of two primary components: the biological recognition element and the signal transduction. The biorecognition element uniquely recognizes and binds specific analytes. Signal transduction translates the biochemical interactions between the analyte and bioreceptor into readable signals. These components enable the creation of highly selective, sensitive, and reliable biosensors by integrating various bioactive materials. They offer fast response times, small size, low cost, lightweight, high reliability, and the ability to incorporate on-chip biosensor arrays for portable microanalysis systems [2, 3]. In BioFETs, the conventional gate metal is replaced with a biologically specific material. This modification enables the detection of target molecules present in the biofilm layer. The biofilm layer achieves this by modulating the channel conductivity of the FET. Ion-sensitive field-effect transistors (ISFETs), introduced in the 1970s, are a pioneering type of BioFET because of their small size, high input impedance, rapid response, and high sensitivity [4].

FET-based biosensors have seen widespread exploration, such as the catalyzed FET biosensors, which use biocatalysts like enzymes for binding and recognizing biological molecules [5, 6]. Enzyme-based ISFETs (ENFETs) are commonly used, where enzymes are stabilized on a gate insulator for specific substrate binding. The extended-gate FET (EGFET) introduced improvements over traditional ISFETs, offering advantages such as flexible gate area and ease of packaging [7, 8]. MOSFET (metal-oxide-semiconductor FET) biosensors are highly sensitive and operate on the principle of electrical biomolecule detection, modifying the electric field and affecting the drain current [9]. They have been used for detecting DNA-protein interactions and diagnosing diseases like Alzheimer's [10, 11]. Label-free detection techniques are common in various FET-based biosensors, offering advantages in simplicity and cost-effectiveness by integrating with CMOS technology.

TFT-based Biosensors

Thin-film transistor (TFT)-based biosensors are a significant class of biosensors that enable label-free detection of biomolecules. TFT biosensors consist of a thin film of semiconductor material deposited on an insulating substrate. TFTs do not require additional labeling of biomolecules, reducing the complexity and cost of the detection process [12]. TFT biosensors have been used to detect biomarkers for diseases like birth asphyxia, sickle cell disease (SCD), traumatic brain injury, and stroke. Their high sensitivity and rapid detection capabilities make them suitable for monitoring critical health conditions. The large-area coverage and fast response times of TFT biosensors enable real-time monitoring of low-concentration protein analytes [13]. This capability is crucial in medical diagnostics, where timely information is critical for patient care. Additionally, solution-processed TFTs enable low-cost fabrication. These TFT-based biosensors are versatile and can be fabricated on a variety of substrates, including flexible materials, which broadens their potential applications in wearable and portable medical devices.

TFTs utilize various materials as the active semiconductor layer. Traditional materials include amorphous silicon and polycrystalline silicon, but these have limitations in terms of mobility and uniformity. Consequently, researchers have explored alternative materials such as carbon nanotubes, organic semiconductors, cadmium selenide, and metal halide perovskites. Among these, metal oxide-based semiconductors have gained significant attention due to their high carrier mobility, transparency, and stability. Zinc oxide (ZnO) is a widely used oxide semiconductor for TFTs [14]. ZnO offers outstanding electrical properties, such as high optical transparency, a large exciton binding energy of 60 meV, and a wide bandgap of 3.37 eV. Its ability to be deposited at low temperatures further enhances its suitability as an active layer material for thin-film transistor (TFT)-based biosensors. Moreover, ZnOs biocompatibility broadens its potential applications. ZnO-based TFT biosensors hold significant promise in diverse fields, including environmental monitoring, clinical diagnostics, and on-site detection [15]. Their high sensitivity, real-time monitoring capabilities, and cost-effectiveness make them ideal for various applications, providing efficient and reliable sensing solutions.

Dielectric Modulated Biosensors

Among FET-based biosensors, dielectric-modulated (DM) FETs have demonstrated superior sensitivity and are widely preferred for biosensing applications. The inclusion of nanocavities in DM-FETs provides ample space for effective biomolecule conjugation, thereby enhancing overall sensitivity. Choi *et al.* introduced the first DM-FET biosensor with nano-cavities for label-free biomolecule detection [16]. Since then, numerous dielectric-modulated FET biosensors with improved sensitivity have been developed [17]. Dielectric modulation is a technique that involves altering the dielectric properties within a device to control its electrical behavior. By incorporating biomolecules with varying dielectric constants into the cavity regions, a gradient is created, which influences the electric field distribution and the overall device response. When a voltage is applied, the differing dielectric constants of biomolecules affect the electric field and charge distribution, leading to changes in key performance parameters like threshold voltage, on-state current, and biomolecule sensitivity in biosensors. Technology Computer-Aided Design (TCAD) simulations are instrumental in studying the effects of dielectric modulation on device performance. Through TCAD, researchers can define device geometry, material properties, and operating conditions to analyze how dielectric modulation affects electrical characteristics [18, 19]. These simulations enable the optimization of device design and allow predictions of device behavior under varying conditions, facilitating the identification of optimal designs for specific biosensing applications.

This chapter presents the development and implementation of a dielectric-modulated ZnO thin-film transistor (TFT) biosensor for biomolecule detection. ZnO TFT-based biosensors are highly attractive due to their simple, cost-effective fabrication processes, along with high sensitivity, stability, and biocompatibility. These qualities make ZnO TFTs an excellent platform for diverse biosensing applications. This work explores integrating dielectric modulation into ZnO TFTs to further enhance their sensing performance. TCAD simulations were utilized to model and analyze the effects of dielectric modulation, enabling the optimization of device design and the prediction of its behavior under various conditions.

DIELECTRIC MODULATED ZNO TFT-BASED BIOSENSOR

The cross-sectional structure of the dielectric-modulated ZnO TFT (DM-Zn--TFT) biosensor is illustrated in Fig. (**1**). key dimensions of the device include a ZnO channel layer with a thickness (t_{ch}) of 20 nm, a gate length (L_G) of 3 μm, a gate metal electrode work function (ϕ_m) of 4.6 eV, and source/drain metal electrode work functions (ϕ_{SD}) of 4.33 eV. The lengths of the source and drain regions are equal to each other, and the device width (W) is set to 100 μm. The gate dielectric is designed as a stacked SiO2/Al2O3 structure. To protect the ZnO active layer from direct contact with biomolecules, which could degrade device performance due to corrosion, a 5 nm thick Al_2O_3 oxide layer is deposited over the ZnO surface. This Al_2O_3 layer not only serves as a protective barrier but also provides a high surface density, improving biomolecule adhesion by absorbing more −OH groups, thereby enhancing the device's sensitivity [20, 21].

On top of the Al_2O_3 layer, a 15 nm SiO_2 layer is deposited. A cavity, intended for biomolecule placement, is created by etching into the SiO_2 layer near the drain region. This cavity has a length (L_{Cavity}) of 1.5 μm and a thickness (t_{Cavity}) of 15 nm. When biomolecules are placed in the cavity, they form a tightly bound layer of ions directly adjacent to the Al_2O_3 surface, called the Stern layer [22]. This layer creates a localized electric field that significantly impacts the overall charge distribution within the device. Specifically, ions in the Stern layer interact closely with charges in the ZnO active layer, altering the local electric field and influencing the device's threshold voltage and on-state current. In simulations, the stern layer is represented as an additional SiO_2 layer with an approximate thickness (t_{Stern}) of 1 nm and its dielectric constant of 2.25 [23, 24].

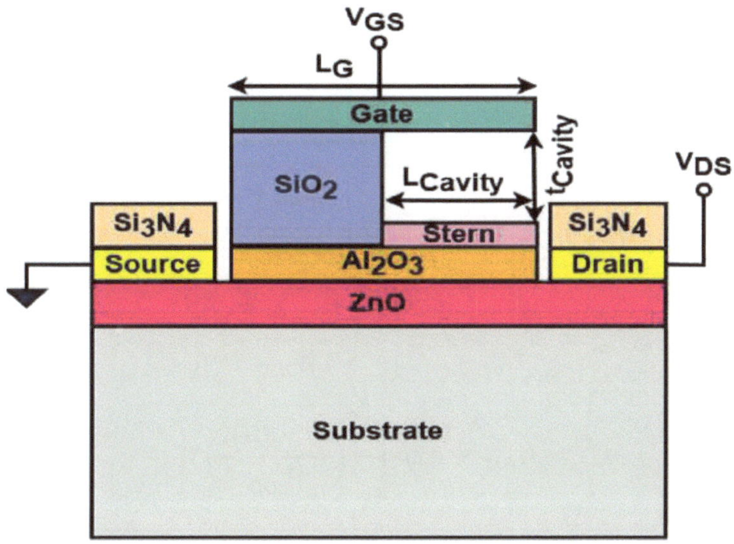

Fig. (1). 1: 2-D Schematic illustration of the Dielectric-Modulated ZnO Nanostructured TFT (DM-Zn--TFT) biosensor.

Physical Models

The 2D TCAD simulations were conducted using the following physical models: The Shockley-Read-Hall (SRH) model is applied to account for recombination processes. The FERMI model was used for carrier statistics. The CVT mobility model was employed to capture mobility behavior accurately. Oxide-based semiconductor materials typically exhibit numerous defect states within their bandgap. To model devices composed of such materials accurately, a continuous density of states (DOS) must be applied in the simulations, including the physical models mentioned above. This is achieved using the defect statement, which defines the defect states as a combination of exponentially decaying band tail states and Gaussian-distributed mid-gap states [25]. The overall DOS, denoted as g(E), consists of four distinct bands: two tail bands, one donor-like near the valence band, and one acceptor-like near the conduction band; and two mid-gap deep-level bands, one acceptor-like and the other donor-like, represented by Gaussian distributions.

$$n(E) = n_{TA}(E) + n_{TD}(E) + n_{GA}(E) + n_{GD}(E) \quad \textbf{(1)}$$

For the tail states, the DOS is characterized by the densities at the conduction and valence band edges (N_{TA} and N_{TD}) along with the characteristic decay energies (W_{TA} and W_{TD}). Meanwhile, the Gaussian distributions for the deep-level bands are defined by the total defect densities (N_{GA} and N_{GD}), characteristic decay energies (W_{GA} and W_{GD}), and the energies where the distributions peak (E_{GA} and E_{GD}).

$$n_{TA}(E) = N_{TA} \times exp\left[\frac{E - E_C}{W_{TA}}\right] \quad \textbf{(2)}$$

$$n_{TD}(E) = N_{TD} \times exp\left[\frac{E_V - E}{W_{TD}}\right] \quad \textbf{(3)}$$

$$n_{GA}(E) = N_{GA} \times exp\left[-\left(\frac{E_{GA} - E}{W_{GA}}\right)^2\right] \quad \textbf{(4)}$$

$$n_{GD}(E) = N_{GD} \times exp\left[-\left(\frac{E - E_{GD}}{W_{GD}}\right)^2\right] \quad \textbf{(5)}$$

This chapter investigates the influence of various neutral biomolecules on the performance of the DM-ZnO-TFT by introducing a cavity, represented as a nanogap in simulations, beneath the metal gate toward the drain side. The cavity serves as a user-defined dielectric region, with its dielectric constant varying according to the specific dielectric properties of the incorporated biomolecules. The absence of biomolecules is denoted by air (k = 1). Table **1** provides a list of biomolecules detected by this approach.

Table 1. Dielectric constant values of biomolecules examined in this work.

Biomolecule	Dielectric constant (k)	Reference
Uricase	1.54	[26]
Urease	1.64	[27]
Streptavidin	2.10	[16]
Biotin	2.63	[28]
Cholesterol Oxidase	3.28	[29]
Glucose Oxidase	3.46	[30]
APTES	3.57	[31]

Biomolecule	Dielectric constant (k)	Reference
Zein	5	[32]
kertain	7.5	[33]
Peptide	15	[34]

Process Flow

The process flow for fabricating the DM-ZnO-TFT is depicted in Fig. (**2**), outlining each critical step necessary for constructing the device. The fabrication process begins with thoroughly cleaning the substrate, which can either be a glass substrate or an oxidized silicon wafer, to ensure a pristine surface. Photolithography is then performed to create a mask for the deposition and patterning of the ZnO layer. Afterward, ZnO deposition is carried out using atomic layer deposition (ALD) or radio frequency (RF) sputtering techniques to create a high-quality semiconductor layer (active layer). The ZnO layer is then patterned by a lift-off technique to remove ZnO from undesired areas, such as regions outside the channel region.

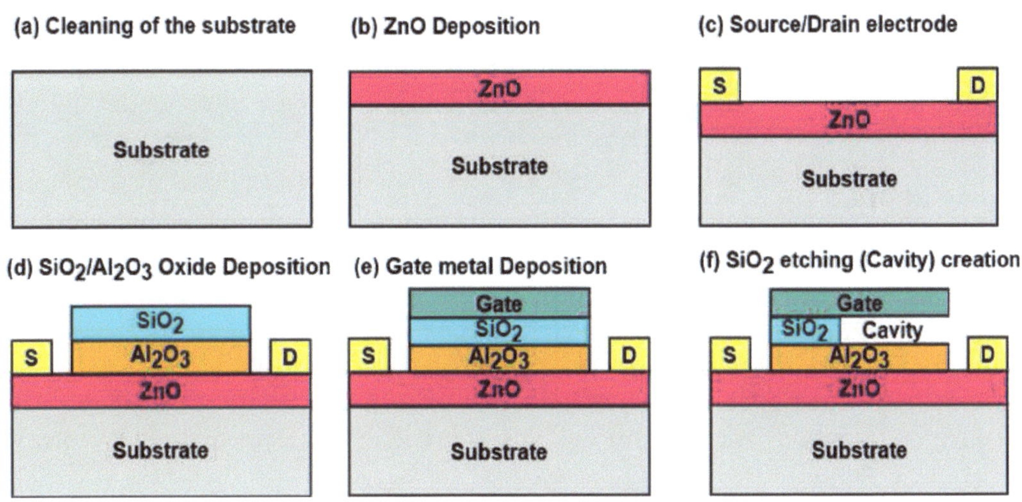

Fig. (2). Proposed process flow of DM-ZnO-TFT biosensor.

Next, source and drain electrodes are deposited onto the ZnO layer via electron-beam (e-beam) evaporation, a preferred method for achieving accurate thin-film growth of conductive materials. These electrodes are patterned using lithography, followed by a lift-off process to retain only the designated electrode regions, thus forming electrical contacts for the device. Subsequently, the gate insulator stack

(Al_2O_3/SiO_2) is formed, beginning with the deposition of Al_2O_3 through sputtering. This layer is patterned by creating a mask using photo-lithography, followed by etching to remove Al_2O_3 from undesired locations. After the Al_2O_3 layer is defined, SiO_2 is deposited, typically through thermal oxidation. It is then patterned using photo-lithography to create a mask for selective etching, ensuring SiO_2 is removed from the unwanted regions, leaving the SiO_2 layer in the required locations. Following this, gate metal is deposited and patterned accordingly to form the gate electrode. Finally, selective SiO_2 etching is implemented to generate the cavity or nanogap under the gate electrode, which is essential for immersing biomolecules to detect and analyze. This step is achieved through techniques such as reactive ion etching (RIE), spacer etching, and controlled lateral etching to precisely define the cavity beneath the gate.

RESULTS AND DISCUSSION

Detecting these biomolecules is crucial in the medical industry, as it enables early diagnosis of various health conditions. Prompt identification of disease biomarkers leads to better prognoses and improved patient care. This rapid and accurate detection allows for timely intervention, accelerating treatment and significantly enhancing patient recovery outcomes. With these objectives, this section provides a detailed analysis of the simulation results, focusing on the device's response to different biomolecules immersed in the cavity. The biomolecules have distinct dielectric constants based on their type, which directly impacts the device's electrical characteristics.

Contour Study

The simulations show significant changes in the device's electrical performance when exposed to different biomolecules. Initially, these changes are examined through contour plot studies, including the energy profile and surface potential. The contour plots are drawn along the device channel, following a horizontal line set 1 nm below the SiO_2/ZnO interface. As shown in Fig. (**3**), changes in the dielectric constants of these biomolecules significantly affect the energy profile and surface potential displayed in the contour plots.

These contour plots reveal that the dielectric properties of biomolecules cause variations in the oxide capacitance. When the dielectric value of the biomolecules increases, the oxide capacitance also rises. This increase is due to improved gate control, which enhances the electric field's effect on the channel region. As the oxide capacitance increases, the conduction band energy levels decrease, as illustrated in Fig. (**3a**). This reduction leads to a rise in surface potential beneath the cavity area. The shift in surface potential due to changes in biomolecule dielectric constants is clearly shown in Fig. (**3b**), highlighting the device's improved capacity to detect biomolecular interactions through variations in its electrical response.

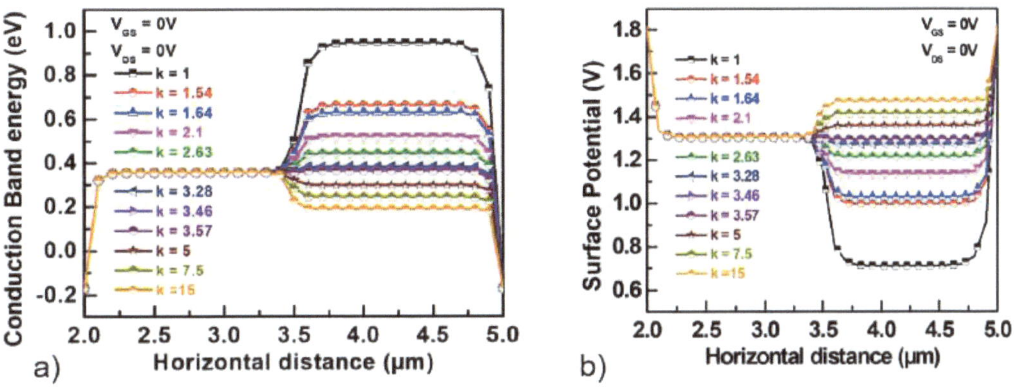

Fig. (3). (a) Conduction band energy, (b) Surface potential profiles across the ZnO channel for different biomolecules.

Electrical Characteristics

The transfer characteristics ($I_{DS}-V_{GS}$) of the device, when exposed to various biomolecules, are presented in Fig. (**4a**). During this analysis, the gate voltage (V_{GS}) was varied from −1 V to 5 V, with a fixed drain bias (VDS) of 1 V. It is observed that as the dielectric value of the biomolecule increases, a reduction in the device's threshold voltage is observed, along with a significant increase in the output drain current (I_{DS}). The I_{DS} value was found to rise significantly from 16.89 mA/mm for a cavity filled with air (k = 1) to 34.78 mA/mm when the cavity was filled with biomolecules having a dielectric constant of k = 15, at a gate-source voltage (V_{GS}) of 5 V. To quantify the shift in threshold voltage (ΔV_{th}), the difference in threshold voltage (V_{th}) was measured in comparison to the reference Vth, where the cavity is air-filled. The Vth was determined using the constant current method [35]. The maximum observed shift in Vth reached up to 589 mV for biomolecules having a dielectric constant of k = 15. This shift underscores the enhanced sensitivity of the device to variations in the biomolecule's dielectric constant, which is critical for detecting biomolecular interactions with high accuracy.

Fig. (**4 (b)**) illustrates the output characteristics ($I_{DS}-V_{DS}$) of the device for various biomolecules at a gate bias of 5 V. The data clearly show that the output drain current increases as the biomolecule's dielectric constant increases. Notably, the highest output drain current achieved was 56.18 mA/mm when the cavity was filled with a peptide biomolecule (k = 15). This value is approximately three times larger than the current value observed for an air-filled cavity, which was 18.19

mA/mm. The observed increase in the output drain current (ΔIDS) can be attributed to the enhanced dielectric constant of the biomolecules, which improves the device's gate control and overall conductivity.

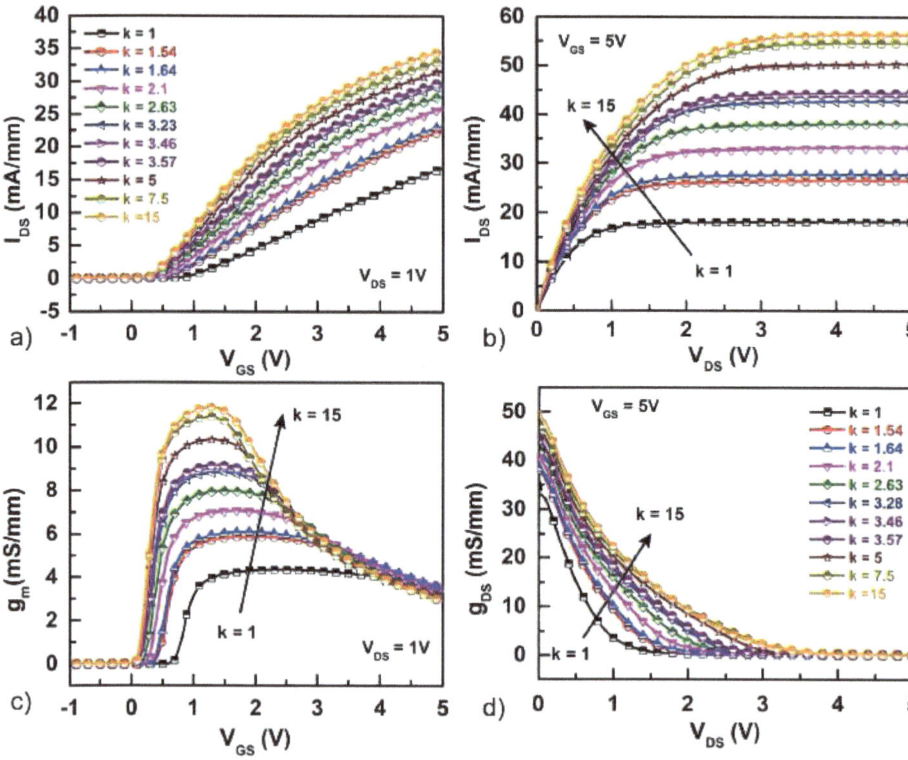

Fig. (4). Transfer (I_{DS} $-V_{GS}$), (b) Output (I_{DS} $-V_{DS}$), (c) Transconductance ($g_m$$-V_{GS}$), and (d) Output Conductance ($g_{DS}$$-V_{DS}$) characteristics of the DM-ZnO-TFT for various biomolecules.

The transconductance ($g_m = \partial I_{DS}/\partial V_{GS}$) of the DM-ZnO-TFT is also observed for biomolecules, as shown in Fig. (**4c**). The transconductance (g_m) is the most important performance metric to consider when analyzing any biosensor device's radio frequency (RF) performance. It represents the relationship between the change in the transistor's output current (IDS), and the change in its input gate voltage (V_{GS}) at a constant bias condition. In radio RF applications, gm is a critical parameter, as it directly impacts several key factors such as gain, linearity, and noise performance of the device. The device's transconductance (g_m) enhances with the biomolecule's dielectric constant, as previously discussed; a higher k value raises the current and lowers the threshold voltage, thereby further enhancing transconductance. The gm value increased from 4.34 to 11.78 mS/mm as the dielectric value, k, of biomolecules increased from 1 to 15. In Fig. (**4d**), the output conductance characteristics (g_{DS}–V_{DS}) of the proposed DM-ZnO-TFT are analyzed for various neutral biomolecules at a V_{GS} of 5 V. The output conductance ($g_{DS} = \partial I_{DS}/\partial V_{DS}$) demonstrates variations in the dielectric values of the biomolecules in the cavity region. The highest output conductance (g_{DS}) is observed when a biomolecule with the dielectric constant of (k = 15) is present, reaching 47.44 mS/mm compared to 33.66 mS/mm at (k = 1).

Sensitivity

In this section, a sensitivity analysis is conducted. Sensitivity is a critical parameter in biosensors, reflecting their ability to detect and respond to low concentrations of target analytes, such as biomolecules. The reliability and utility of a biosensor largely depend on its sensitivity, as a highly sensitive biosensor can accurately identify even minute amounts of a substance. This capability is essential for early detection and precise diagnostics. The impact of the biomolecule's dielectric constant on different sensitivities is analyzed, including threshold voltage sensitivity (SV_{th}), transconductance sensitivity (Sg_m), output drain current sensitivity (SI_{DS}), and output conductance sensitivity (Sg_{DS}). The performance of the device, DM-ZnO-TFT, was thoroughly assessed by evaluating its sensitivity. Sensitivity was calculated using a specific formula [36], where the sensitivity (S_x) for a parameter "x", such as (V_{th}, g_m, I_{DS}, or g_{DS}), was determined as follows:

$$S_x = \frac{x_{Ref} - x_{Bio}}{x_{Ref}} \times 100\% \qquad \textbf{(6)}$$

Here, x_{Ref} represent the reference values that are measured with respect to an air-filled cavity, while x_{Bio} denotes the parameter value in the presence of the biomolecule or analyte being detected. The sensitivity calculation quantitatively assesses key device parameters based on changes in the dielectric constant of biomolecules. This sensitivity analysis provides insights into the device's ability to detect and respond to biomolecular interactions. Fig. (**5 (a)**) shows the threshold voltage sensitivity (SV_{th}) of the device as a function of k. It is observed that as the dielectric constant of the biomolecule increases, there is a corresponding increase in the device's on-current (I_{ON}), which results in lowering the V_{th} value. This reduction in V_{th} results in a further increase in the difference between the Vth of a biomolecule-filled cavity and that of an air-filled cavity. When the cavity is filled with air, the DM-ZnO-TFT exhibits a higher Vth compared to when it is filled with biomolecules having a dielectric value of 1. When the cavity is filled with air, without any biomolecule immersion, the threshold voltage is measured at 636 mV, noted as the reference threshold voltage ($V_{th(Ref.)}$). Consequently, higher dielectric values correlate with greater sensitivity.

The device demonstrates a maximum SV_{th} of 92.59% at k = 15, indicating its enhanced performance in detecting biomolecules with higher dielectric properties.

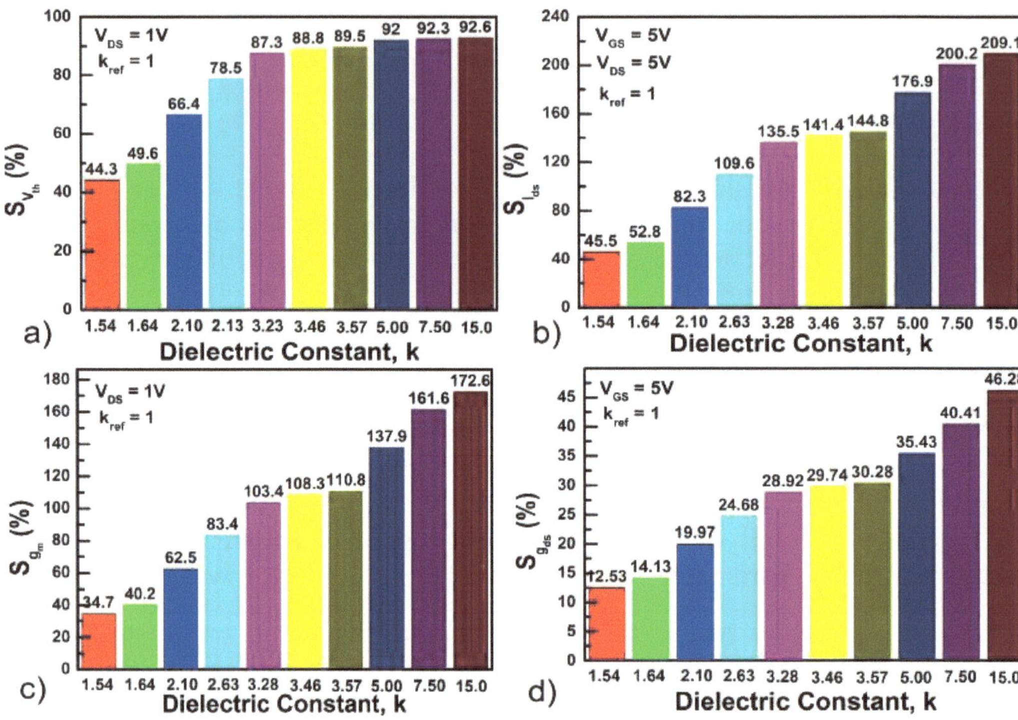

Fig. (5). Proposed devices relative (a) Threshold voltage sensitivity (SV_{th}), (b) Output current sensitivity (SI_{DS}), (c) Trans-conductance sensitivity (Sg_m), and (d) Output conductance sensitivity (Sg_{DS}) for various biomolecules.

Additionally, Fig. (**5b**) presents the drain current sensitivity (SI_{DS}) across different dielectric biomolecules. When compared to a cavity filled with air, the SI_{DS} increases from 45.48% at k = 1.54 to 209.09% at k = 15 as the dielectric constant of the biomolecules rises. This enhancement is primarily due to the corresponding increase in the I_{DS}. The results indicate that the device exhibits increased sensitivity as the cavity is filled with biomolecules with higher dielectric constants, demonstrating its potential for effective biosensing applications.

This chapter extends the analysis of the device's performance by evaluating its transconductance sensitivity (Sg_m), as illustrated in Fig. (**5c**). Transconductance (g_m) is a critical parameter for assessing the RF performance of any biosensor. The transconductance plot shows that gm increases with the dielectric constant of the biomolecules, peaking at 11.87 mS/mm at a VDS of 1 V when the cavity is filled with peptide biomolecules. Moreover, the device achieves a maximum Sg_m of approximately 172.59% at k = 15 compared to air-filled cavities.

In Fig. (**5d**), the relative output conductance sensitivity (Sg_{DS}) is analyzed for different neutral biomolecules compared to a cavity filled with air. The relative output conductance sensitivity (Sg_{DS}) demonstrates a notable increase, ranging from 12.52% at k = 1.54 to 46.31% at k = 15, showcasing a significant shift with the dielectric constant. This enhancement in sensitivity aligns with the observed rise in output conductance. The results demonstrate a significant enhancement in sensitivity as the cavity of the DM-ZnO-TFT is immersed in neutral biomolecules possessing a higher dielectric constant. These findings show that the variations in SV_{th}, SI_{DS} Sg_m, and Sg_{DS} with the dielectric constant (k) provide an understanding of the device's response to different biomolecular environments. Furthermore, these sensitivity values increase with increasing k, demonstrating the role of the dielectric properties of biomolecules in analyzing the device's sensing performance.

Impact of Cavity Thickness

The influence of cavity thickness (t_{Cavity}) on the device was investigated with respect to key sensing parameters (ΔV_{th}, ΔI_{DS}, Δg_m, and Δg_{DS}) for different neutral biomolecules, including uricase (k = 1.54), biotin (k = 2.63), glucose oxidase (k = 3.46), and zein (k = 5). Cavity thicknesses of 10, 15, 20, and 25 nm have been considered in this analysis. An increase in cavity thickness results in an inadequate channel formation beneath the cavity due to the increased effective thickness of the gate oxide (t_{oxeff}). Consequently, the device's output current (I_{DS}) decreases with the increment in cavity thickness. The device demonstrated maximum drain currents of 36.98, 26.46, 20.26, and 16.16 mA/mm for cavity thicknesses of 10, 15, 20, and 25 nm, respectively, at a constant k (1.54). As the t_{Cavity} increases, the t_{oxeff} also increases, leading to a decrease in the I_{ON}.

Additionally, the device's threshold voltage (V_{th}) increases with the cavity thickness. At a cavity thickness of 10 nm, the device demonstrates an I_{ON} of 36.29 mA/mm, which has increased by a factor of 2.4 compared to the value at a 25 nm cavity thickness device. For cavity thicknesses of 10, 15, 20, and 25 nm, the device's Vth values were 165, 352, 542, and 730 mV, respectively. The shift in threshold voltage (ΔV_{th}) concerning an air-filled cavity is shown in Fig. (**6a**). As discussed earlier, an increase in cavity thickness corresponds to a higher threshold voltage (V_{th}). It is noted that the ΔV_{th} increases from 182 mV to 484 mV when the cavity thickness increases from 10 to 25 nm at k = 1.54. At k = 5, the shift intensifies, escalating from 371 mV to 1015 mV, due to the higher Vth at greater cavity thickness in an air-filled cavity (k = 1.5). Similarly, Fig. (**6b**) depicts the impact of cavity thickness on the shift in the output drain current (ΔI_{DS}). In contrast to the effects on (ΔV_{th}), the value of ΔI_{DS}decreases with an increase in cavity thickness. This reduction in I_{DS} with cavity thickness is a notable aspect, aligning with the trends observed in ΔI_{DS}. The shift in output drain current (ΔI_{DS}) also follows this pattern, decreasing from 9.96 mA/mm at 10 nm to 6.09 mA/mm at 25 nm when the cavity is filled with uricase (k = 1.54) biomolecules. At the zein-filled cavity (k = 5), the ΔI_{DS} value decreased from 34.24 to 26.23 mA/mm. This decreasing trend signifies the diminishing change in output drain current as the cavity thickness increases.

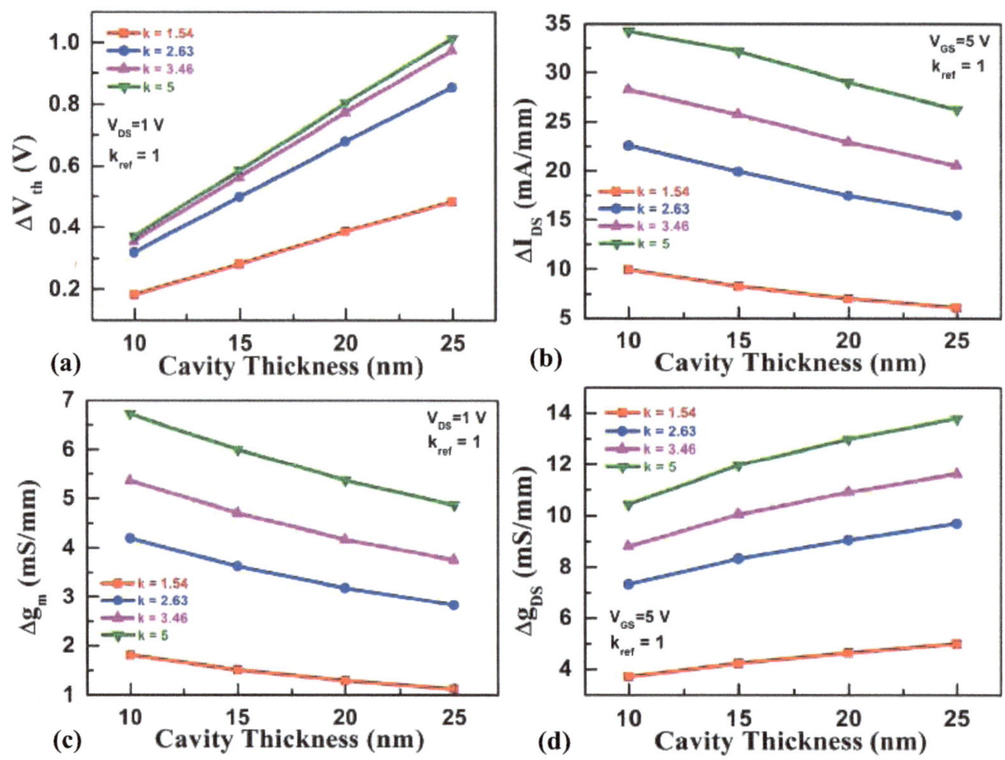

Fig. (6). The impact of cavity thickness on sensing parameters: (a) Shift in threshold voltage (ΔV_{th}), (b) Shift in drain current (ΔI_{DS}), (c) Shift in transconductance (Δg_m), and (d) Shift in output conductance (Δg_{DS}) for various biomolecules with dielectric constants.

Additionally, Fig. (**6c**and **6d**) delve into the impact of cavity thickness on the shifts in transconductance (Δgm) and output conductance (Δg_{DS}), respectively. Analogous to the trend observed in ΔI_{DS}, the Δg_m decreases with an increase in cavity thickness. The Δgm exhibits a decreasing trend from 1.81 mS/mm at 10 nm to 1.12 mS/mm at 25 nm for the uricase-filled cavity (k = 1.54). At the zein-filled cavity (k = 5), Δgm decreases from 6.73 mS/mm to 4.86 mS/mm. This trend reflects the device's response to changes in cavity thickness, which is crucial for optimizing its transconductance performance. Turning attention to the shift in output conductance (Δg_{DS}), the behavior of Δg_{DS} is opposite to ΔI_{DS}, showcasing an increasing trend from 3.72 mS/mm at 10 nm to 5 mS/mm at 25 nm for uricase-filled cavities. A similar pattern is observed at the zein-filled cavity (mS/mm), with Delta Δg_{DS} increasing from 10.44 mS/mm to 13.79 mS/mm. This trend underscores the intricate interplay of cavity thickness with the output conductance of the DM-ZnO-TFT.

Impact of Full Cavity

Fig. (7) presents a comprehensive analysis of the device, considering the full cavity length, with the gate length equal to the cavity length. This analysis explores the impact of full cavity length on the device's electrical performance through various characteristics. In Fig. (**7a**), the transfer characteristics demonstrate the influence of the full cavity length. As the cavity length increases, the on-current continues to increase, while the threshold voltage decreases. This observed trend is attributed to the extended cavity length, which enhances the device's performance. Fig. (**7b**) illustrates the impact of a full cavity on the output characteristics, revealing that the output current increases with increasing cavity length. Specifically, at k=15, the output current is 1.2× higher at V_{GS} = 5 V compared to the current at the half-filled cavity. This enhancement can be attributed to increased gate control capability.

Similarly, Fig. (**7c**) highlights the impact of the full cavity on the transconductance characteristics, which represent the rate at which the output current changes with gate voltage. The transconductance increases with the growing cavity length, leading to higher output current and reduced threshold voltage. Consequently, the increased cavity length significantly influences the transconductance, with the g_m value at k = 15 being 1.2×higher than at half cavity length. The output conductance characteristics, depicted in Fig. (**7d**), exhibit a similar trend to the transconductance characteristics. The rate at which the output current changes with drain voltage increases with the cavity length, emphasizing the influence of cavity length on device performance. This increased output conductance is significant for the sensitivity of the biosensor, as it directly affects the device's responsiveness to changes in the dielectric properties of biomolecules. The higher output conductance enhances the device's sensitivity, making it more effective in biosensing applications.

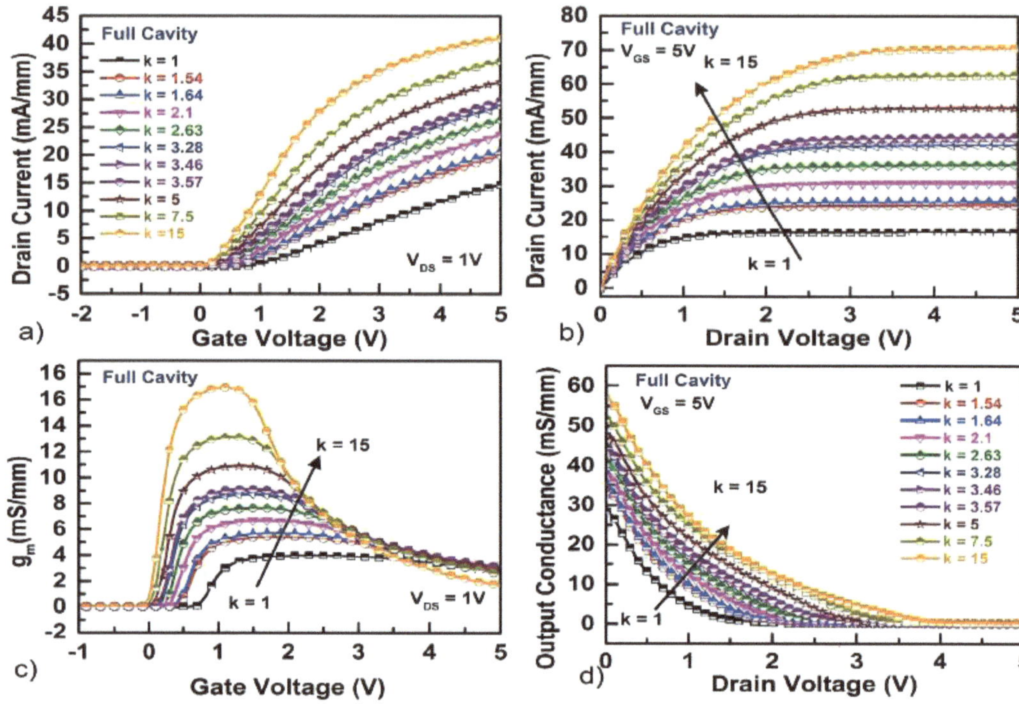

Fig. (7). (a) Transfer (b) Output, (c) Transconductance, and (d) Output Conductance characteristics of DM-ZnO-TFT for various biomolecules at full cavity length ($L_G = L_{Cavity}$).

Fig. (**8**) illustrates the influence of full cavity length on the relative sensitivities of threshold voltage, output current, transconductance, and output conductance. Fig. (**8a**) shows the impact of full cavity length on threshold voltage sensitivity. As the cavity length increases, there is a slight increase in sensitivity. The full cavity length demonstrates an SV_{th} 1.1 × higher than the half-filled cavity at k = 1. This is due to the threshold voltage variation with respect to k being nearly identical at full and half cavity lengths. Fig. (**8b**) portrays the impact of full cavity length on output current sensitivity. With the increased cavity length, the sensitivity of the output current experiences a notable enhancement. At k = 15, the full cavity length has SI_{DS} 1.55 ×higher than the half-filled cavity. This improvement can be attributed to heightened gate control, leading to a more responsive biosensor. The influence of full cavity length on transconductance sensitivity is illustrated in Fig. (**8c**). As the cavity length increases, the transconductance sensitivity increases sharply. At k = 15, the full cavity length exhibits an Sg_m =1.87× higher than the half-filled cavity. The heightened transconductance, resulting from the extended cavity, contributes to a more sensitive response to changes in gate voltage. Fig.

(**8d**) demonstrates the impact of full cavity length on output conductance sensitivity. Similar to the other sensitivities, there is a substantial improvement, indicating that the extended cavity enhances the device's responsiveness to changes in drain voltage. At k = 15, the full cavity length has $Sg_{DS} = 2.16\times$ higher than the half-filled cavity. This heightened sensitivity is pivotal for the overall effectiveness of the biosensor in detecting variations in the dielectric properties of biomolecules.

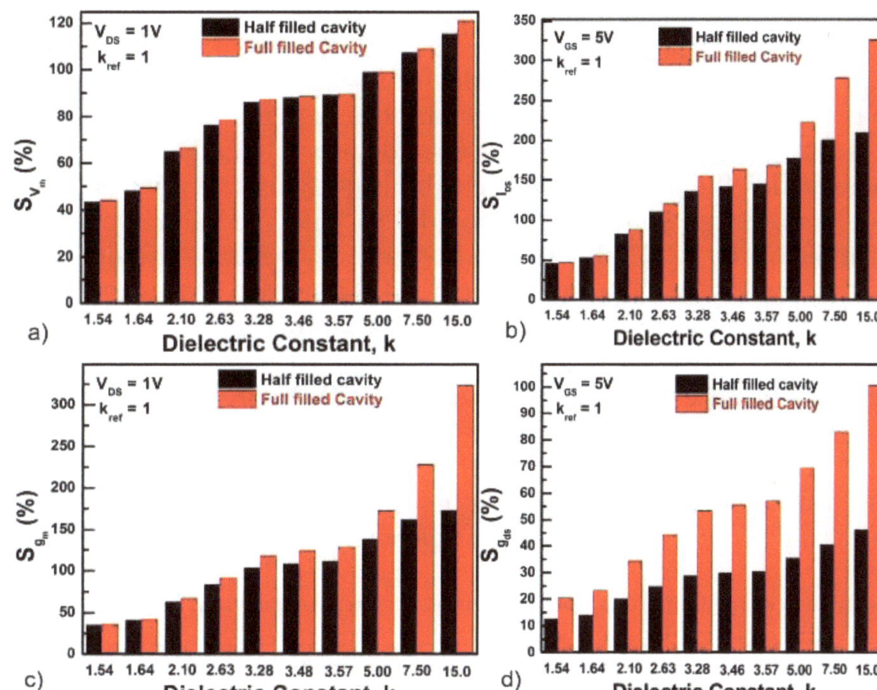

Fig. (8). Comparison of sensitivities of half-filled and fully-filled cavities: (a) Relative threshold voltage sensitivity (SV_{th}), (b) Relative output current sensitivity (SI_{DS}), (c) Relative trans-conductance sensitivity (Sg_m), and (d) Relative output conductance sensitivity (Sg_{DS}).

Comparative Analysis

The comparative analysis of this work with previously reported studies is presented in this section. The key sensing parameter, ΔVth, defines the shift in threshold voltage when a biomolecule is present, relative to an air-filled cavity. A larger ΔV_{th} indicates a stronger interaction between the biomolecules and the device. It is defined as:

$$\Delta V_{th} = V_{th(Ref)} - V_{th(Bio)} \tag{7}$$

where Vth(Ref) is the threshold voltage of the device for an air-filled cavity (k = 1), and $V_{th(Bio)}$ is the threshold voltage of the device when a biomolecule is present. Another key sensing parameter, ΔI_{DS}, represents the change in drain current due to biomolecule adsorption in the cavity compared to an air-filled cavity. A higher ΔI_{DS} signifies an increased modulation of the channel current, leading to improved sensitivity. It is defined as:

$$\Delta I_{DS} = I_{DS(Ref)} - I_{DS(Bio)} \quad \textbf{(8)}$$

where $I_{DS(Ref.)}$ is the drain current of the device for an air-filled cavity, and $I_{DS(Bio)}$ is the drain current of the device when a biomolecule is present. Table **2** compares these key sensing parameters of the proposed device with earlier studies. For k = 5 (zein biomolecule), the proposed device exhibits a significant ΔV_{th} of 585 mV and ΔI_{DS} of 32.16 mA/mm, which are 2.3× and 3.6× than the values reported earlier [32]. When the cavity is filled with keratin (k= 7.5), the proposed device exhibits a 5.2× improvement in ΔV_{th} compared to the reported study at k = 10 [31]. Additionally, the DM-ZnO-TFT demonstrates a 2-order increase in ΔI_{DS} even at k = 7.5, highlighting its enhanced sensitivity at lower dielectric constants. Notably, at k = 2.63, the proposed device achieves a V_{th} value ≈ 7.68 times that of earlier reported devices [28, 37], while at k = 2.1, it exhibits a 2.76 times increment compared to [37]. These results highlight the superior sensitivity of the DM-Zn--TFT biosensor, demonstrating its potential for advanced biosensing applications with improved detection accuracy and efficiency.

Table 2. Sensitivity comparison: proposed versus previously reported devices.

S.No	Sensitivity Parameter	k value	Sensitivity Value	This Work
1	ΔVth(mV)	k = 5	250 [32]	585 (k = 5)
2	$\Delta I_{DS}(\mu A/\mu m)$	k = 5	2 [32]	32.16 (k = 7.5)
3	ΔV_{th}(mV)	k ≈ 10	113 [31]	587 (k = 7.5)
4	$\Delta I_{DS}(\mu A/\mu m)$	k ≈ 10	0.375 [31]	36.39 (k = 7.5)
5	ΔV_{th}(mV)	k =2.63	65 [28]	499 (k =2.63)
6	ΔV_{th}(mV)	k =2.63	60 [37]	499 (k =2.63)
7	ΔV_{th}(mV)	k =2.1	180 [37]	422 (k =2.1)

CONCLUSION

This chapter introduces a dielectric-modulated Zinc-Oxide nanostructured thin film transistor (DM-ZnO-TFT) featuring a cavity under the gate near the drain side. Through 2D TCAD Atlas Silvaco simulations, various neutral biomolecules were considered, each characterized by its respective dielectric constant, including Uricase, Urease, Streptavidin, Biotin, ChOX, GOx, APTES, Zein, keratin, and Peptide. The electrostatic effect of these biomolecules on the cavity was analyzed by studying contour cut line plots of conduction band energy and surface potential. The electrical characteristics of the devices were assessed by evaluating key sensing parameters, including threshold voltage shift (ΔV_{th}), shift in drain current (ΔI_{DS}), shift in transconductance (Δg_m), and shift in output conductance (Δg_{DS}). Results indicate a maximum ΔV_{th} of 589 mV, a maximum ΔI_{DS} of 38.01 mA/mm, a maximum Δg_m of 7.49 mS/mm, and a maximum Δg_{DS} of 15.62 mS/mm when the cavity is immersed with the high dielectric constant biomolecules (peptide) compared to an air-filled cavity.

The sensitivity analysis of the proposed DM-ZnO-TFT was further examined, revealing a threshold sensitivity (SV_{th}) of 92.6%, output drain current sensitivity (SI_{DS}) of 172.6%, transconductance sensitivity (Sg_m) of 209.1%, and an output conductance sensitivity (Sg_{DS}) of 46.28% at k = 15. The sensitivity values were observed to rise with the dielectric constant of the biomolecules. Additionally, the impact of t_{Cavity} thickness on the device characteristics was analyzed. As cavity thickness increases from 10 to 25 nm, there is a considerable increase in ΔV_{th} by 173.58%, while ΔI_{DS} decreases by 23.10%, Δgm decreases by 27.79%, and Δg_{DS} increases by 24.29%. Finally, the effect of full cavity length on device sensitivity was analyzed and compared with that of half-length cavities. All relative sensitivity parameters showed a notable increase with increasing cavity length. These findings highlight the enhanced sensitivity of the DM-ZnO-TFT, providing a more sensitive range for detecting neutral biomolecules.

AI DISCLOSURE

During the preparation of the work, the author(s) used AI tools and the Grammarly tool to enhance the writing style and readability of the content. The authors clarify that AI was not involved in generating the substantive content and that they take full responsibility for the publication.

REFERENCES

[1] M.J. Schöning, and A. Poghossian, "Recent advances in biologically sensitive field-effect transistors (BioFETs)", *Analyst (Lond.)*, vol. 127, no. 9, pp. 1137-1151, 2002. [http://dx.doi.org/10.1039/B204444G] [PMID: 12375833]

[2] S.V. Dzyadevich, Y.I. Korpan, V.N. Arkhipova, M.Y. Alesina, C. Martelet, A.V. El'Skaya, and A.P. Soldatkin, "Application of enzyme field-effect transistors for determination of glucose concentrations in blood serum", *Biosens. Bioelectron.*, vol. 14, no. 3, pp. 283-287, 1999. [http://dx.doi.org/10.1016/S0956-5663(99)00007-X] [PMID: 10230028]

[3] B.K. Sohn, and C.S. Kim, "A new pH-ISFET based dissolved oxygen sensor by employing electrolysis of oxygen", *Sens. Actuators B Chem.*, vol. 34, no. 1-3, pp. 435-440, 1996. [http://dx.doi.org/10.1016/S0925-4005(97)80017-2]

[4] P. Bergveld, "Development of an ion-sensitive solid-state device for neurophysiological measurements", *IEEE Trans. Biomed. Eng.,* vol. BME-17, no. 1, pp. 70-71, 1970. [http://dx.doi.org/10.1109/TBME.1970.4502688] [PMID: 5441220]

[5] A. Girard, O. De Sagazan, F. Le Bihan, F. Geneste, T. Mohammed-Brahim, S. Dauphas, P. Brissot, and C. Guguen-Guillouzo, "Electrical quantification of low concentrations of specific protein in complex biological solutions", In: *in SENSORS, 2008 IEEE*, 2008, pp. 638-641. [http://dx.doi.org/10.1109/ICSENS.2008.4716522]

[6] Y. Ishige, S. Takeda, and M. Kamahori, "Direct detection of enzyme-catalyzed products by FET sensor with ferrocene-modified electrode", *Biosens. Bioelectron.,* vol. 26, no. 4, pp. 1366-1372, 2010. [http://dx.doi.org/10.1016/j.bios.2010.07.053] [PMID: 20728332]

[7] S.A. Pullano, C.D. Critello, I. Mahbub, N.T. Tasneem, S. Shamsir, S.K. Islam, M. Greco, and A.S. Fiorillo, "Egfet-based sensors for bioanalytical applications: A review", *Sensors (Basel),* vol. 18, no. 11, p. 4042, 2018. [http://dx.doi.org/10.3390/s18114042] [PMID: 30463318]

[8] D.S. Kim, J.E. Park, J.K. Shin, P.K. Kim, G. Lim, and S. Shoji, "An extended gate FET-based biosensor integrated with a Si microfluidic channel for detection of protein complexes", *Sens. Actuators B Chem.,* vol. 117, no. 2, pp. 488-494, 2006. [http://dx.doi.org/10.1016/j.snb.2006.01.018]

[9] D. Sadighbayan, M. Hasanzadeh, and E. Ghafar-Zadeh, "Biosensing based on field-effect transistors (FET): Recent progress and challenges", *Trends Analyt. Chem.,* vol. 133, p. 116067, 2020. [http://dx.doi.org/10.1016/j.trac.2020.116067] [PMID: 33052154]

[10] N. Nikkhoo, P.G. Gulak, and K. Maxwell, "Rapid detection of E. coli bacteria using potassium-sensitive FETs in CMOS", *IEEE Trans. Biomed. Circuits Syst.,* vol. 7, no. 5, pp. 621-630, 2013. [http://dx.doi.org/10.1109/TBCAS.2013.2276013] [PMID: 24107977]

[11] S. S. Nemati, G. Dehghan, N. Sheibani, and Y. Abdi, "Ion-Sensitive Field-Effect Transistor-Based Biosensors: The Sources Of Gates And Their Sensitive Layers A Review", *IEEE Sens. J.,* vol. 24, no. 11, pp. 17324-17336, 2024. [http://dx.doi.org/10.1109/JSEN.2024.3389680]

[12] A. Kumar, A.K. Goyal, and N. Gupta, "thin-film transistors (tfts) for highly sensitive biosensing applications: a review", *ECS J. Solid State Sci. Technol.,* vol. 9, no. 11, p. 115022, 2020. [http://dx.doi.org/10.1149/2162-8777/abb2b3]

[13] F. Yan, and H. Tang, "Application of thin-film transistors in label-free DNA biosensors", *Expert Rev. Mol. Diagn.,* vol. 10, no. 5, pp. 547-549, 2010. [http://dx.doi.org/10.1586/erm.10.50] [PMID: 20629501]

[14] Y.S. Rim, H. Chen, B. Zhu, S.H. Bae, S. Zhu, P.J. Li, I.C. Wang, and Y. Yang, "Interface engineering of metal oxide semiconductors for biosensing applications", *Adv. Mater. Interfaces,* vol. 4, no. 10, p. 1700020, 2017. [http://dx.doi.org/10.1002/admi.201700020]

[15] P.I. Reyes, C.J. Ku, Z. Duan, Y. Lu, A. Solanki, and K.B. Lee, "ZnO thin film transistor immunosensor with high sensitivity and selectivity", *Appl. Phys. Lett.,* vol. 98, no. 17, p. 173702, 2011. [http://dx.doi.org/10.1063/1.3582555]

[16] H. Im, X.J. Huang, B. Gu, and Y.K. Choi, "A dielectric-modulated field-effect transistor for biosensing", *Nat. Nanotechnol.,* vol. 2, no. 7, pp. 430-434, 2007. [http://dx.doi.org/10.1038/nnano.2007.180] [PMID: 18654328]

[17] P. Dwivedi, and A. Kranti, "Dielectric modulated biosensor architecture: Tunneling or accumulation based transistor?", *IEEE Sens. J.,* vol. 18, no. 8, pp. 3228-3235, 2018. [http://dx.doi.org/10.1109/JSEN.2018.2808948]

[18] R. Narang, M. Saxena, and M. Gupta, "Comparative analysis of dielectric-modulated fet and tfet-based biosensor", *IEEE Trans. Nanotechnol.,* vol. 14, no. 3, pp. 427-435, 2015. [http://dx.doi.org/10.1109/TNANO.2015.2396899]

[19] S. Kumar, Y. Singh, B. Singh, and P. K. Tiwari, "Simulation Study Of Dielectric Modulated Dual Channel Trench Gate TFET-Based Biosensor", *IEEE Sens. J.,* vol. 20, no. 21, pp. 12565-12573, 2020. [http://dx.doi.org/10.1109/JSEN.2020.3001300]

[20] J.C. Chou, C.Y. Weng, and H.M. Tsai, "Study on the temperature effects of Al2O3 gate pH-ISFET", *Sens. Actuators B Chem.,* vol. 81, no. 2-3, pp. 152-157, 2002. [http://dx.doi.org/10.1016/S0925-4005(01)00945-5]

[21] Ajay, R. Narang, M. Saxena, and M. Gupta, "Novel junctionless electrolyte-insulator-semiconductor field-effect transistor (JL EISFET) and its application as pH/biosensor", *Microsyst. Technol.,* vol. 23, no. 8, pp. 3149-3159, 2017. [http://dx.doi.org/10.1007/s00542-016-3013-1]

[22] M.A. Brown, A. Goel, and Z. Abbas, "Effect of electrolyte concentration on the stern layer thickness at a charged interface", *Angew. Chem. Int. Ed.,* vol. 55, no. 11, pp. 3790-3794, 2016. [http://dx.doi.org/10.1002/anie.201512025] [PMID: 26880184]

[23] P.K. Bera, R.R. Sahoo, R. Kar, and D. Mandal, "Design and sensitivity investigation of dielectric modulated and electrolyte-based ph sensing of vertical tfet biosensor", *IEEE Trans. Nanotechnol.,* vol. 22, pp. 537-544, 2023. [http://dx.doi.org/10.1109/TNANO.2023.3307632]

[24] A. Bandiziol, P. Palestri, F. Pittino, D. Esseni, and L. Selmi, "A tcad-based methodology to model the site-binding charge at isfet/electrolyte interfaces", *IEEE Trans. Electron Dev.,* vol. 62, no. 10, pp. 3379-3386, 2015. [http://dx.doi.org/10.1109/TED.2015.2464251]

[25] SILVACO International, ATLAS User's Manual – Device Simulation Software, Santa Clara, CA, USA, 2004.,

[26] P. Dwivedi, and A. Kranti, "Applicability of Transconductance-to-Current Ratio (g_m/I_ds) as a Sensing Metric for Tunnel FET Biosensors", *IEEE Sens. J.,* vol. 17, no. 4, pp. 1030-1036, 2017. [http://dx.doi.org/10.1109/JSEN.2016.2640192]

[27] C. Chong, H. Liu, S. Wang, and S. Chen, "Simulation and performance analysis of dielectric modulated dual source trench gate tfet biosensor", *Nanoscale Res. Lett.,* vol. 16, no. 1, p. 34, 2021. [http://dx.doi.org/10.1186/s11671-021-03486-2] [PMID: 33576891]

[28] N. Shafi, C. Sahu, and C. Periasamy, "Analytical modeling of surface potential and drain current for virtually doped symmetrical dielectric modulated biofet", *IEEE Sens. J.,* vol. 20, no. 9, pp. 4749-4757, 2020. [http://dx.doi.org/10.1109/JSEN.2020.2964625]

[29] A. Paliwal, M. Tomar, and V. Gupta, "Complex dielectric constant of various biomolecules as a function of wavelength using surface plasmon resonance", *J. Appl. Phys.,* vol. 116, no. 2, p. 023109, 2014. [http://dx.doi.org/10.1063/1.4890027]

[30] A.M. Bhat, A. Varghese, N. Shafi, and C. Periasamy, "A dielectrically modulated gan/aln/algan moshemt with a nanogap embedded cavity for biosensing applications", *J. Inst. Electron. Telecommun. Eng.,* vol. •••, pp. 1-10, 2021.

[31] S.N. Mishra, and K. Jena, "A dielectric-modulated normally-off algan/gan moshemt for bio-sensing application: Analytical modeling study and sensitivity analysis", *J. Korean Phys. Soc.,* vol. 74, no. 4, pp. 349-357, 2019. [http://dx.doi.org/10.3938/jkps.74.349]

[32] R. Ajay, R. Narang, M. Saxena, and M. Gupta, "Narang, M. Saxena, and M. Gupta, “Drain current

model of a four-gate dielectric modulated mosfet for application as a biosensor,"", *IEEE Trans. Electron Dev.,* vol. 62, no. 8, pp. 2636-2644, 2015.
[http://dx.doi.org/10.1109/TED.2015.2441753]

[33] R. Goswami, and B. Bhowmick, "Comparative analyses of circular gate tfet and heterojunction tfet for dielectric-modulated label-free biosensing", *IEEE Sens. J.,* vol. 19, no. 21, pp. 9600-9609, 2019.
[http://dx.doi.org/10.1109/JSEN.2019.2928182]

[34] G. Löffler, H. Schreiber, and O. Steinhauser, "Calculation of the dielectric properties of a protein and its solvent: theory and a case study", *J. Mol. Biol.,* vol. 270, no. 3, pp. 520-534, 1997.
[http://dx.doi.org/10.1006/jmbi.1997.1130] [PMID: 9237916]

[35] W. Huo, Z. Mei, M. Zhao, Y. Sui, B. Zhao, Y. Zhang, T. Wang, S. Cui, H. Liang, H. Jia, and X. Du, "Flexible zno thin-film transistors on thin copper substrate", *IEEE Trans. Electron Dev.,* vol. 65, no. 9, pp. 3791-3795, 2018.
[http://dx.doi.org/10.1109/TED.2018.2859277]

[36] D. Singh, S. Pandey, K. Nigam, D. Sharma, D.S. Yadav, and P. Kondekar, "A charge-plasma-based dielectric-modulated junctionless tfet for biosensor label-free detection", *IEEE Trans. Electron Dev.,* vol. 64, no. 1, pp. 271-278, 2017.
[http://dx.doi.org/10.1109/TED.2016.2622403]

[37] H. Im, X.-J. Huang, B. Gu, and Y.-K. Choi, "A Dielectric-Modulated Field-Effect Transistor For Biosensing", *Nat. Nanotechnol.,* vol. 2, no. 7, pp. 430-434, 2007.
[http://dx.doi.org/10.1038/nnano.2007.180]

CHAPTER 14

Dielectric-Modulated Negative Capacitance Tunnel FET for Highly Sensitive Biosensing Applications

Anil Kumar Pathakamuri[1], **Chandan Kumar Pandey**[1,*] and **Girdhar Gopal**[2]

[1] *School of Electronics Engineering, VIT-AP University, Amaravati, Andhra Pradesh, India*

[2] *Department of Electronics & Communication Engineering, National Institute of Technology Patna, Patna, Bihar, India*

Abstract: The biosensing potential of a ferroelectric negative capacitance biosensor tunnel field-effect transistor (FENCB-TFET) is investigated in this chapter using a dielectric modulation technique. The design features nanoscale pores near the source and drain regions to enhance the surface area available for biomolecule immobilisation, therefore directly enhancing electrostatic coupling. Adjustments in turn-on voltage, drain current sensitivity, and current switching ratio affect performance. We evaluated biomolecules with neutral, positive, and negative charges at the control layer and at the insulator edge under a range of dielectric conditions. The study takes into account practical limitations, such as partially filled holes caused by steric hindrance, in addition to perfect sensing. The findings show that the FENCB-TFET is a strong, extremely sensitive competitor for low-power, label-free biosensing.

Keywords: Biosensor, Dielectric modulation Sensitivity, Nano-cavity.

INTRODUCTION

Modern diagnostics, environmental monitoring, and biochemical sensing all require the ability to identify biomolecules at the nanoscale. Traditional methodologies often fail to deliver high reliability in these dimensions. Nanoscale electronic devices provide an alternative because of their tiny size, superior electrostatic control, and ability to directly convert biological interactions into electrical impulses [1]. Modern diagnostics, environmental monitoring, and biochemical sensing all demand the capacity to identify biomolecules on a nanoscale. Traditional methodologies usually fail to provide high reliability at these dimensions. Nanoscale electronic devices provide an alternative because of their tiny size, superior electrostatic control, and ability to transform biological

* **Corresponding author Chandan Kumar Pandey:** School of Electronics Engineering, VIT-AP University, Amaravati, Andhra Pradesh, India; E-mail: chandankumarpandey@gmail.com

interactions directly into electrical impulses [1]. However, traditional ion-sensitive FETs (ISFETs) have a blind spot: they have trouble detecting neutral molecules. This gap led to the development of dielectric-modulated FETs (DM-FETs). By employing nanocavities to detect variations in the dielectric constant within the cavity, these devices are able to recognise both charged and neutral objects [2 - 5]. Since then, tunnel FETs (TFETs) have become a powerful alternative to traditional CMOS. Their key feature is that they are suitable for low-power sensing as they can achieve a subthreshold swing below the thermal limit of mV/dec [6, 7].

Combining nanogaps with high-k dielectrics in TFETs sharpens sensitivity by increasing band-to-band tunnelling at the source. Tunnelling efficiency improves when biomolecules inhabit high-k cavities, allowing detection at extremely low concentrations [6 - 8]. According to a study, geometry is what determines performance. Targets near the source lower the tunnelling barrier, directly increasing sensitivity, as established by Narang *et al.* [8]. While others employed structural alterations, such as Gate-All-Around (GAA) designs, to improve electrostatic control [10, 11], Kanungo *et al.* further optimised these architectures [9]. A dual-metal gate method was successfully employed by Maradan *et al.* to control drain-induced effects [12]. Charge plasma approaches, which minimise random dopant oscillations without compromising performance, have been used to junctionless TFETs in order to streamline production [13]. Dwivedi *et al.* emphasised the transconductance-to-current ratio as an important factor [14], while Verma demonstrated that dual-metal designs enhance on-currents [15]. Noor *et al.* demonstrated this vertical TFET, proving that vertical TFETs outperform lateral ones because they have a larger tunnelling area [16]. Despite this, altering the dielectric constant remains a fundamental way for label-free detection [17, 18], which is frequently stabilised against temperature changes using charge plasma methods. Core-shell junctionless nanotube TFETs are another low-cost and scalable solution [19]. Recently, the emphasis has shifted to maximising efficiency metrics such as the current sensing ratio (CSR) and subthreshold swing (SS). Published SS values currently range from as low as mV/dec [20 - 24] to about mV/dec [25], validating TFETs' ultra-low-power capability [26].

The most recent innovation includes the use of ferroelectric materials. Ferroelectric negative capacitance provides better gate control and lower voltage operation, ushering in a new era of biosensor performance [26]. Building on previous advances, this chapter studies a ferroelectric negative capacitance dielectric-modulated TFET with nanocavities at the source and drain. To demonstrate its suitability for practical, label-free biosensing, we rigorously analyse this design under a variety of situations, including changing dielectrics, biomolecular charges, and non-uniform distributions.

2D CONFIGURATION AND DEVICE DESCRIPTION

Fig. (**1**) depicts the FENCB-TFET's two-dimensional design. We designed the shape and materials to improve electrostatic control and biosensing capacity. In the proposed device, the conventional gate oxide is replaced with a ferroelectric layer, which plays a crucial role in improving electrostatic control. As a result, due to the negative-capacitance effect, the subthreshold swing is lowered below the theoretical thermal limit of 60 mV/decade at room temperature [27 - 35]. To further address the concern about the leakage, a split-source architecture is used. This configuration has the advantage of suppressing leakage current while increasing band-to-band tunnelling. By strengthening the tunnelling junction at the source-channel interface, the device can generate a faster, more efficient electrical response.

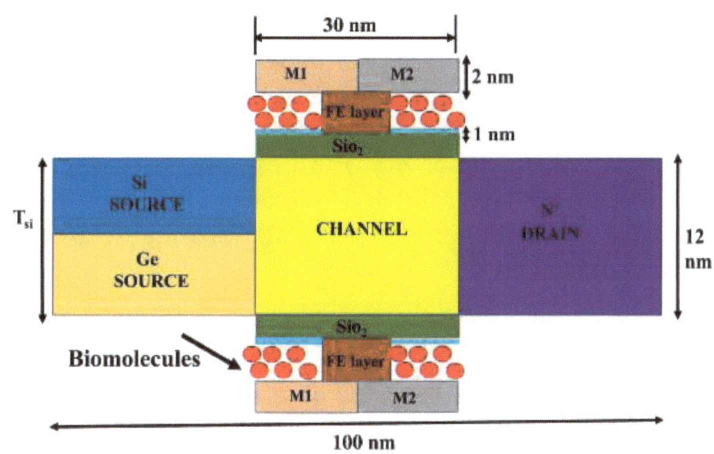

Fig. (1). The construction of the proposed device (FENCB-TFET).

The total device size is kept to 100 nm, and the device layout is based on standard fabrication-compatible protocols owing to the practical feasibility [36]. A thin buffer layer of 1 nm is put between the ferroelectric oxide and the silicon substrate. This layer enhances interfacial stability and reduces lattice mismatch and strain-related effects. During the fabrication process, a thin layer of silicon dioxide (1 nm thick) is deposited and then etched to produce nano-scale cavities at the source and drain regions. These cavities can act as active sensing areas with stable sites for biomolecule immobilisation during sensing operation. Dielectric modulation within these cavities is the basis for the sensing technology. Gate-t--channel coupling affects tunnelling probability and sensitivity as the cavity contents change their dielectric constant. We tested Sentaurus TCAD [37] in dry conditions to replicate real-world conditions. The simulation framework uses complex doping-dependent mobility models and high doping in both the source and drain to induce bandgap narrowing. To adequately represent the physics, we employed a nonlocal band-to-band tunnelling (BTBT) model for interband events and the Shockley-Read-Hall (SRH) model for carrier recombination. Biomolecules were modelled by varying dielectric constants and charge densities within the cavities, with empty (air-filled) cavities serving as a reference for sensitivity comparisons [17, 18]. Finally, combining ferroelectrics with dielectric-modulated cavities results in a very precise and efficient system. These characteristics distinguish the FENCB-TFET as a strong contender for real-time, label-free biosensing.

RESULTS AND DISCUSSION

We examined the FENCB-TFET's reactivity to a wide variety of biomolecules, namely acids, DNA, proteins, RNA, and staphylococcal nucleases. Because the exact dielectric constant of a target determines how it interacts with the sensing zone, we swept across values comparable with existing literature [38]. The findings show a clear difference in how positively and negatively charged molecules behave inside the nano-cavity. Furthermore, we extended ideal settings to simulate non-uniform and chaotic hybridisation. This phase was critical in ensuring that the device remained robust and dependable even in the messy realities of practical biosensing.

The Influence of Uncharged Biomolecules on Detection Sensitivity

The FENCB-TFET (Fig. **1**) was first tested against neutral biomolecules. With air (K=1) serving as the baseline, the simulation sweeps the dielectric constant through values of and assuming a fully filled nanocavity. At VGS and VDS, operating conditions were set. Raising causes both ON-state and ambipolar currents to clearly spike, but the OFF-state current stays mostly constant, as shown in the transfer characteristics in Fig. (**2a**). Stronger gate-to-channel capacitive coupling reduces the tunnelling barrier, according to simple physics. The polarisation field is amplified by a larger dielectric constant, increasing band bending close to the drain and opening the way for more effective electron tunnelling. In Fig. (**2b** and **2c**), we plotted drain current, charge sensitivity, and turn-on voltage sensitivity to measure this. According to the findings, sensitivity consistently rises with increases. This demonstrates that dielectric modulation is the critical determinant in performance, supporting the device's design for dependable, label-free biosensing.

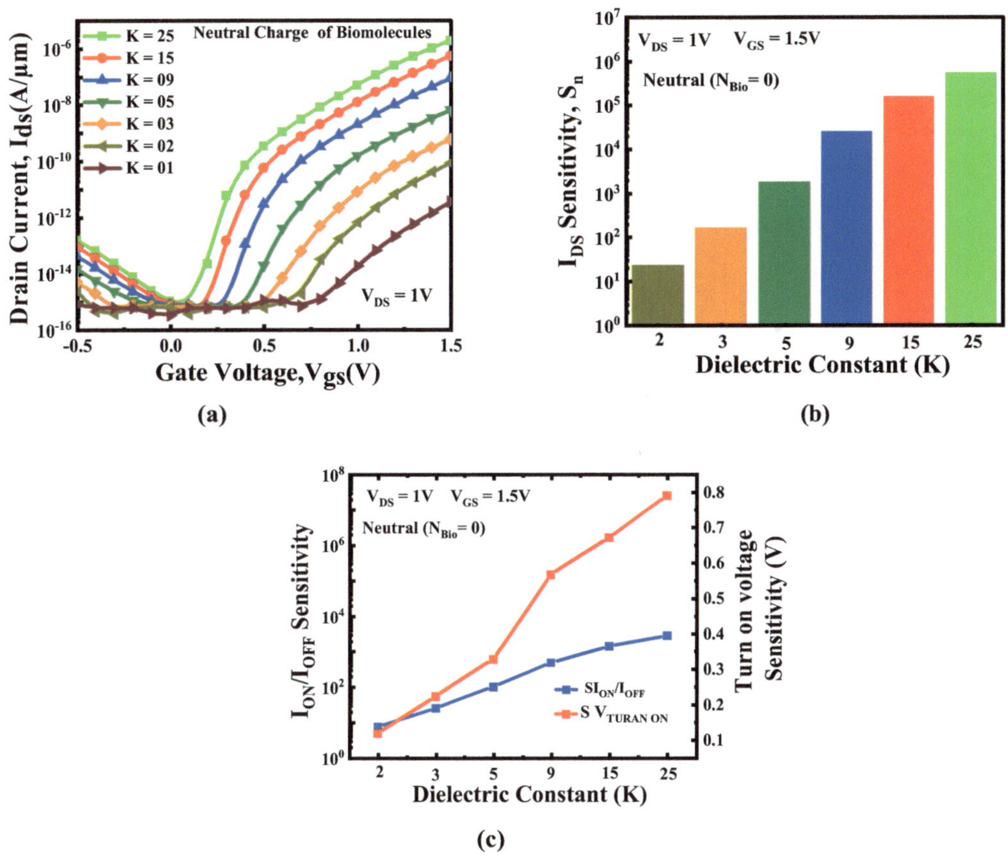

Fig. (2). (a) Transfer characteristics of FENCB-TFET. (b) I_D Sensitivity (c) CSR and $V_{turn\text{-}ON}$ Sensitivity.

The drain current sensitivity is defined by Equation (1) as:

S_{Id}

$$\frac{I_{d\varepsilon k>1}(bio)-I_{d\varepsilon k=1}(air)}{I_{d\varepsilon k=1}(air)} \tag{1}$$

CSR of SI_{ON}/I_{OFF}

$$\frac{I_{ON}}{I_{OFF}} = \frac{\frac{I_{ON}}{I_{OFF}}|_{\varepsilon k>1} - \frac{I_{ON}}{I_{OFF}}|_{\varepsilon k=1}}{\frac{I_{ON}}{I_{OFF}}|_{\varepsilon k=1}} \tag{2}$$

$V_{Turn\text{-}ON}$ sensitivities

$$V_{Turn-ON} = V_{Turn-ON\,\varepsilon k>1}(\text{bio}) - V_{Turn-ON\,\varepsilon k=1}(\text{air}) \qquad \textbf{(3)}$$

The relation between the charge sensitivity ratio (CSR) and the dielectric constant, K, is shown in Fig. (**2c**). A significant upward trend in CSR is evident with increasing value of (K).

The underlying mechanism for this phenomenon could be inferred from Fig. (**2a**), where the ON bout current shows a conspicuous increase, while the current in the OFF bout is quite steady. This difference in behavior directly enhances the CSR and results in an overall increase in the sensor sensitivity. In addition to modulating the current, the dielectric constant also affects the turn-on voltage sensitivity. Increased dielectric values provide for a stronger response in the turn-on voltage, indicating enhanced sensitivity to modulation of the dielectric in the Nano cavity region. This strengthens the electrostatic control of the channel and significantly alters the tunnelling conditions. Further substantiating evidence of this behaviour is shown in Fig. (**2b**). The addition of the ferroelectric layer increases the electric field near the tunnelling junction, facilitating carrier transfer and further band bending. Consequently, both the CSR and turn on voltage sensitivity are enhanced simultaneously. Owing to this synergistic effect, the FENCB-TFET can reliably detect biomolecules even under low dielectric-contrast conditions. At higher polarization strengths, CSR conceals its theoretical maximum, and hence confirms the dominant role played by the polarization field in enhancing sensitivity. These results help understand the exact mechanisms that control performance, thereby offering a roadmap to optimize FENCB-TFETs for practical detection applications.

Negative-Charged Biomolecules Influence Detection Sensitivity

Fig. (**3a**) presents the influence of different negative biomolecule concentrations ($-N_{bio}$) on the drain current (I_D) and sensing response of the FENCB-TFET when the nanocavity is completely filled. To capture the role of dielectric modulation, multiple dielectric constants (εk = 1, 2, 3, 5, 9, 15, and 25) are considered. (Fig. **3a**) specifies the impact of disparate concentrations of negatively charged biomolecules ($-N_{bio}$) over drain current (I_D) and consequent sensing response of the FENCB-TFET in the case of full saturation of the nanocavity. In order to conclude the role of dielectric modulation, a suite of dielectric constants (εk = 1, 2, 3, 5, 9, 15, and 25) is introduced into the analysis. The injection of negatively charged biomolecules strongly affects the drain current because the surface-accumulated charge alters the local electrostatic environment in the nanocavity. As the density of negative biomolecules increases, conspicuous deviations in I_D occur, indicating robust coupling between the biomolecules' charge and the device's electrical features. Concurrently, changes in dielectric constant also modify this response, highlighting the interrelated nature of biosensor performance within the context of charge density and also dielectric properties. The simulations are done for negative biomolecule surface charge densities of -$3\text{x}10^{11}$ cm^{-2}, -$6\text{x}10^{11}$ $cm^{-2,}$ and -$9\text{x}10^{11}$ cm^{-2}. The corresponding electrical characteristics for values of k equal to 2, 3, 5, 9, 15 and 25 are shown in Fig. (**3a**). These results provide unambiguous information about how changes in the dielectric environment, in parallel with the increasing negative charge density, control the sensing behaviour of the device. Fig. (**3b**) shows the variation in the sensitivity with respect to -N_{bio}. As the concentration of negative biomolecules increases, a progressive reduction in sensitivity is found. Larger values of the dielectric constant tend to dampen the fluctuations in sensitivity, while values of smaller values of εk engender a more pronounced response. This loss of sensitivity at elevated charge concentrations can be mostly attributed to changes in the local electrostatic potential in the nanocavity, which, in turn, reduces the channel's modulating effect. A similar trend can be seen in the charge - sensitivity ratio (CSR), as shown in Fig. (**3c**). The CSR decreases with higher values of -Nbio, mainly due to the decreasing ON - state current and a nearly constant OFF - state current. This behaviour suggests that the CSR is dominantly controlled by changes in the ON-current. The trend is further accentuated in Fig. (**3d**), but especially at lower values of negative charge densities where the effects of current modulation are the strongest.

Overall, these results expound a complicated game of biomolecular charge density, dielectric modulation, and device sensitivity. The observed behaviour is intimately related to limitations imposed by the electrostatic equilibrium in the nanocavity, which in the end determine sensing performance of the FENCB-TFET biosensor.

$$V_{gs} = \emptyset_s + \emptyset_{ms} + \left\{\frac{-q(\pm N_{bio})}{\Delta_c}\right\} \quad \textbf{(4)}$$

Equation (4) is the general electrical model of the FENCB-TFET that considers the applied bias to the gate, work functions of the metal, and change in capacitance, as well as the charge imposed by the biomolecules. This term describes the fundamental electrostatics by which changes caused by the biomolecules affect the functioning of a device. Specifically, it describes how a difference in surface potential controls carrier transport and, hence, sensing response.

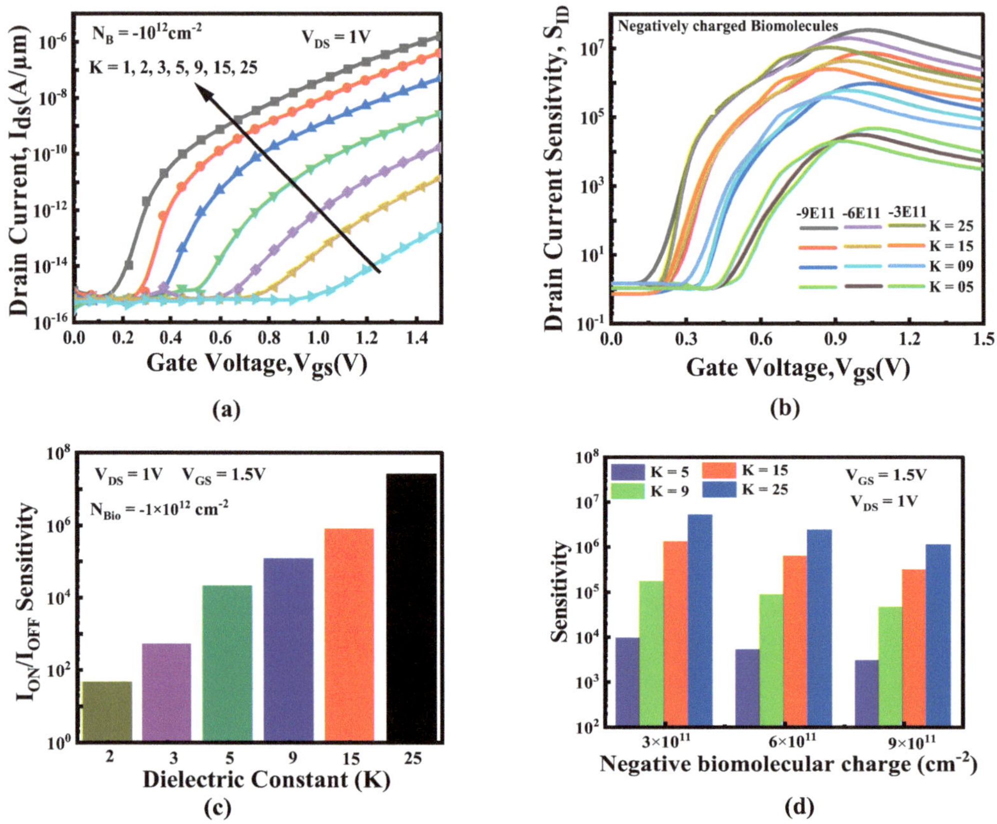

Fig. (3). ID-VG of (a) $-N_{bio}$ (b) $-N_{bio}$ sensitivity to various $-N_{bio}$ (SI_D). Various charge concentration sensitivity (c) $-N_{bio}$ CSR sensitivity (d) $-N_{bio}$.

The change in the concentration of the negatively charged biomolecules ($-N_{bio}$) at constant gate voltage (V_{GS}) and capacitance change results in a greater value of the term q(±N_bio)ΔC. This additional charge contributes to a lower surface potential (φs), thereby reducing carrier transport across the channel and leading to a lower drain current. Subsequently, the biosensor's overall sensitivity decreases with increasing negative charge concentration.

Conversely, maintaining a constant negative concentration of biomolecules and gate bias, the higher the dielectric constant (εk), the larger the effective capacitance of the nanocavity. The increase in ΔC, in turn, partially overcomes the effect of biomolecular charge to increase the surface potential and enhance carrier transport. This contradicting action justifies the fact that the sensing performance of the FENCB-TFET relies on the ratio of biomolecular charge density and dielectric modulation in the nanocavity.

This communication assists in explaining the trends of the present modulation and sensitivity. It stresses the importance of the dielectric environment and the biomolecular charge to the device performance. The larger the effective capacitance (C), the stronger the contribution of the term $q(+N_{bio})\Delta C$ and/or $q(-N_{bio})\Delta C$ in the equation, which causes the surface potential (φs) to increase further. The increased φs enhances carrier movement through the channel, thereby increasing the drain current (I_D). This enhancement in I_D is directly proportional to the better sensing responsiveness of the FENCB-TFET.

The interaction between the dielectric constant (εk) and the capacitance change, together with the surface potential, strongly influences the device's sensing behaviour. As long as biomolecular condition changes to a negatively charged one at a concentration of $N_{bio} = -3x10^{-11}$ cm^{-1}, it is observed that the sensitivity index (S_{ID}) changes considerably. In the case of a dielectric constant of k = 5, there is a relatively small increment of 9% in the S_{ID}. However, at a higher dielectric constant of k = 25, the enhancement is much greater with a near improvement of 62%. The result of this comparison is that the greater the dielectric constant, the more the device responded to changes in the biomolecular charge.

Positive-Charged Biomolecules Influence Detection Sensitivity

Fig. (**4a**) demonstrates the I_d-V_g characteristics of the FENCB-TFET as a function of dielectric constant and as a function of positively charged biomolecule concentration of the $+N_{bio}$. An increase in $+N_{bio}$ leads to a larger ON-state current, which is in line with the behavior that was predicted by Equation (4). The increase in $+N_{bio}$ at fixed gate voltage and dielectric constant increases the surface potential. Its electron tunneling from the source to the channel improves. Effects of dielectric modulation are visible when εk is increased at a constant $+N_{bio}$. This increases the drain current, surface potential, and capacitance. Figs. (**4b** and **4c**) show the corresponding sensitivities of the ON/OFF current ratio and drain current, confirming the strong dependence of sensor response on both the dielectric constant and biomolecule concentration. SI_{ON}/I_{OFF} increases with εk as a result of the ON-state current enhancement, as shown in Fig. (**4d**). The drain current sensitivity increases by approximately 91% for εk = 5 and 29% for εk = 25, when the biomolecule condition changes from neutral to $+N_{bio} = 3 \times 10^{11}$ cm^{-2}. These findings show the combination effect of dielectric properties and positive biomolecular charge on FENCB-TFET sensing performance.

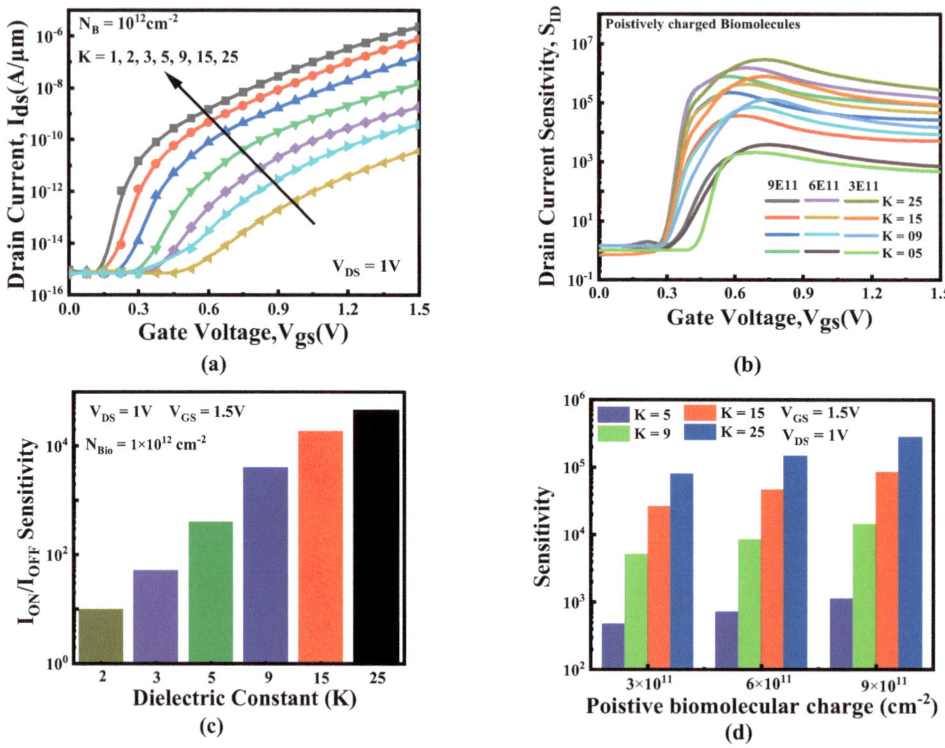

Fig. (4). ID-VG of (a) N_{bio} (b) N_{bio} sensitivity to various N_{bio} (SI_D). Various charge concentration sensitivity (c) N_{bio} CSR sensitivity (d) N_{bio}.

Symmetrical Hybridization of Biomolecules and the Effect of Probe Positioning in Nanocavities

This section studies the FENCB-TFET biosensor's performance in the presence and absence of charged biomolecules, considering actual sensing conditions in which steric hindrance leads to non-uniform biomolecule distribution within the nanocavity. Uneven hybridisation is widespread in practical biosensing and has a significant impact on sensor performance. To reflect this behaviour, we select four sample biomolecule distribution profiles: declining, rising, concave, and convex, as shown in Fig. (**5a-d**).

The influence of these profiles on sensing performance is evaluated for a partially filled nano-cavity with a fill factor (FF) of 55%, under both negatively charged biomolecules ($-Nbio = 9 \times 10^{11}$ cm^{-2}) and positively charged biomolecules ($+Nbio = 9 \times 10^{11}$ cm^{-2}), as shown in Figs. (**5** and **6**). The sensitivity study, shown in Fig. (**6a** and **6b**), demonstrates the biosensor's responsiveness to negatively charged biomolecules under different hybridisation profiles. These findings reveal that the sensitivity of the biomolecules toward the tunnelling junction is higher as the biomolecules are drawn nearer to the location, and for declining and concave spatial profiles, at a fill factor of 55. In such structures, the local electrostatic field at the junction is more perturbed, and hence the differences in the drain current are larger. Conversely, rising and convex profiles are less sensitive since they have less effect in the tunnelling region. These findings underscore the importance of the geometric location of biomolecules in the nanocavity and show that hybridization geometry plays a central role in the performance of a sensor in practical operating conditions.

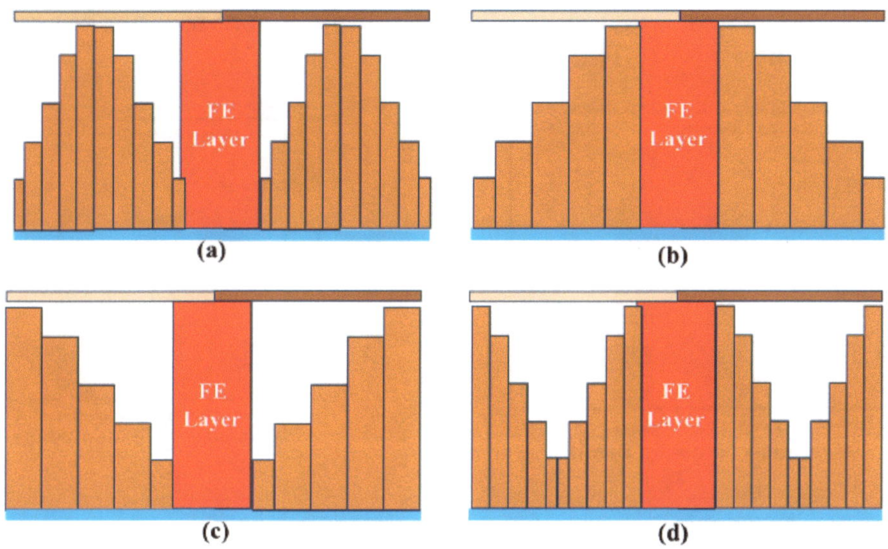

Fig. (5). Depicts the structure of the partially occupied nanocavity, highlighting (a) Convex, (b) Declining, (c) Increasing, and (d) Concave profiles.

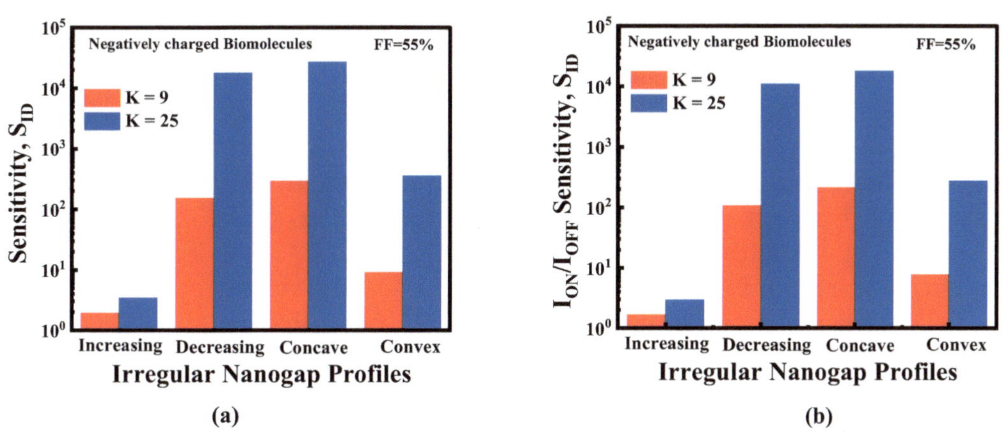

Fig. (6). FF = 55%, (a) SI_d (drain current), and (b) SCSR -N_{bio}.

These features, which lead to a greater surface potential (φs) compared to the higher step profiles, are directly linked with the subordinate step lines of outlines with negative biomolecules charged close to the excavation junction. These results indicate that nanocavity shape is important for optimizing the biosensor's sensitivity and performance. The effect of the distribution profile of biomolecules under conditions where positively charged biomolecules are present has a great influence on the sensing mechanism.

Fig. (**7a** and **7b**) depict the reaction of concave and declining profiles, which provide a heightened reaction in the tunnelling junction, thus, an increased sensitivity of the FENCB-TFET.

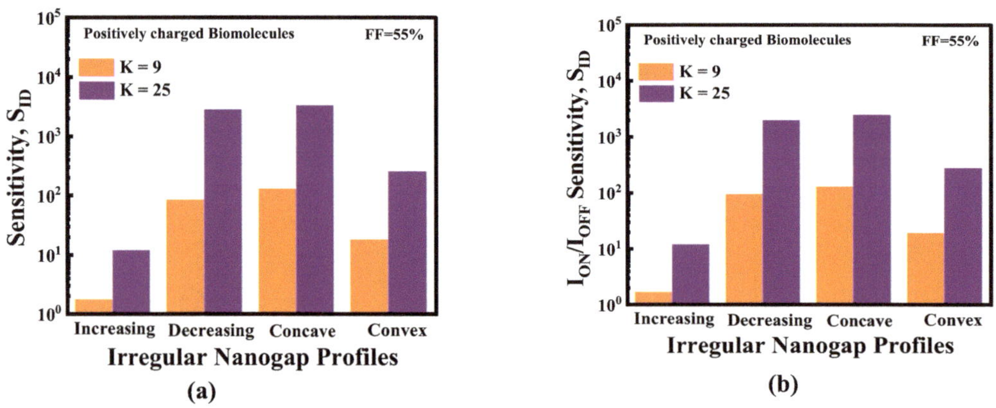

Fig. (7a, b). FF = 55%, showcasing (a) SI_d (drain current) and (b) SI_{ON}/I_{OFF} N_{bio}.

This improvement results from biomolecules being brought closer to the tunnelling area, thereby augmenting local electrostatic modulation. Probe and receptor location effects are examined more deeply under non-ideal hybridization conditions. The probes' and target biomolecules' spatial positioning within the nano-cavity determines the degree of hybridisation between them, as this positioning directly affects the effective sensing area. Consequently, the probe's position affects the biosensor's performance and sensitivity to a considerable degree. Table **1** presents a detailed comparison of various probe designs. Regions that are shaded indicate the presence of biomolecules (k > 1), while the unshaded parts are associated with air-filled cavities (k = 1). The results clearly show that biomolecule placement significantly affects sensitivity. For instance, Cases 2 and 4 have the same fill factor (50%), but Case 4 has a much better sensitivity because the biomolecules are closer to the tunnelling junction. This type of comparative analysis thus shows that the distribution of biomolecules in space may have a greater effect than just the fill factor. Earlier investigations involving dielectric-modulated FETs and TFET-based biosensing platforms have, in a similar way, shown these outcomes and therefore confirm the more general characteristics of these things.

Table 1. The sensitivity analysis of the FENCB-TFET biosensor in various probe placement configurations within the nanocavity, which demonstrates the effect of the location of the biomolecule and fill factor of the cavity on the device response.

Case	Biomolecules are present within the shaded area.										Fill Factor	Sensitivity		
												K=09	K=15	K=25
1	■	■	■			■		■	■		60%	3.5×10^{3}	0.9×10^{4}	3.4×10^{4}
2		■			■	■		■		■	50%	2.6×10^{1}	1.2×10^{2}	2.5×10^{3}
3	■		■		■	■			■		50%	2.6×10^{1}	3.2×10^{2}	6.3×10^{3}
4			■		■		■		■		40%	2.5×10^{4}	4.4×10^{4}	5.6×10^{5}
5		■					■				30%	1.5×10^{4}	3.4×10^{4}	1.6×10^{5}

CONCLUSION

In the current chapter, a systematic investigation into the enhancement of sensitivity in a FENCB-TFET biosensor has been conducted by deliberately modulating the dielectric properties and incorporating ferroelectric negative capacitance within embedded nanocavities, with the overarching objective of facilitating label-free biosensing endeavours. The assessment of sensing performance is performed through a set of key performance indicators such as drain-current sensitivity, charge-sensitivity ratio, and turn-on voltage sensitivity, which collectively provide the ability to fully evaluate the electrical behaviour of the device. The device's response is interrogated for biomolecules in neutral, negatively charged, and positively charged states over a range of dielectric constants. The data obtained show unequivocally that the biosensor exhibits the highest sensitivity in the presence of negatively charged biomolecules. On the other hand, biomolecules that carry a positive charge give a moderate response, and neutral species give the lowest sensitivity. This systematic trend highlights the dominant influence of biomolecular charge on the sensor's electrical response. To elucidate this phenomenon further, the sensing characteristics are scrutinised over a wide range of dielectric constants for each of the three charge conditions. The findings steadily show increased sensitivity for negatively charged biomolecules, followed by positive and neutral ones. This is a consistent observation, which supports that both biomolecular charge and modulation of the dielectric charge are used in concert with one another to govern the sensing mechanism and overall performance of the biosensor. In addition, practical sensing scenarios are dealt with by an examination of the effects induced by non-uniform probe and receptor localisation in the nanocavity. Empirical results show that the cavity fill factor and the spatial positioning of the biomolecules have a significant impact on the sensor's performance. These observations underscore the critical role of including the realistic configuration of biomolecules when assessing biosensor response. Overall, the present study shows that the proposed FENCB - TFET architecture represents a reliable and efficient platform to achieve high-sensitivity biosensing. The analysis provided herein provides a lucid understanding of device behaviour under actual working conditions, thereby addressing the credibility of the biosensor for biological sensing applications.

AI DISCLOSURE

During the preparation of the work, the author(s) used AI tools and the Grammarly tool to enhance the writing style and readability of the content. The authors clarify that AI was not involved in generating the substantive content and that they take full responsibility for the publication.

REFERENCES

[1] V. Eskandari, H. Sahbafar, E. Karooby, M.H. Heris, S. Mehmandoust, D. Razmjoue, and A. Hadi, "Surface-Enhanced Raman scattering (SERS) filter paper substrates decorated with silver nanoparticles for the detection of molecular vibrations of Acyclovir drug", *Spectrochim. Acta A Mol. Biomol. Spectrosc.,* vol. 298, p. 122762, 2023. [http://dx.doi.org/10.1016/j.saa.2023.122762] [PMID: 37130482]

[2] M. Ghodrati, A. Farmani, and A. Mir, "Nanoscale sensor-based tunneling carbon nanotube transistor for toxic gases detection: a first-principle study", *IEEE Sens. J.,* vol. 19, no. 17, pp. 7373-7377, 2019. [http://dx.doi.org/10.1109/JSEN.2019.2916850]

[3] J. Eveness, L. Cao, J. Kiely, and R. Luxton, "Equivalent circuit model of a non-faradaic impedimetric ZnO nano-crystal biosensor", *J. Electroanal. Chem. (Lausanne),* vol. 906, p. 116003, 2022. [http://dx.doi.org/10.1016/j.jelechem.2021.116003]

[4] X. Chen, Z. Guo, G-M. Yang, J. Li, M-Q. Li, J-H. Liu, and X-J. Huang, "Electrical nanogap devices for biosensing", *Mater. Today,* vol. 13, no. 11, pp. 28-41, 2010. [http://dx.doi.org/10.1016/S1369-7021(10)70201-7]

[5] M. Ghodrati, A. Mir, and A. Farmani, *"Carbon nanotube field effect transistors–based gas sensors." Nanosensors for smart cities* Elsevier, 2020, pp. 171-183.

[6] J.M. Kinsella, and A. Ivanisevic, "Taking charge of biomolecules", *Nat. Nanotechnol.,* vol. 2, no. 10, pp. 596-597, 2007. [http://dx.doi.org/10.1038/nnano.2007.313] [PMID: 18654379]

[7] P. Ghosh, and B. Bhowmick, "Performance enhancement of a FET device with ferroelectric tunnel junction and its application as a biosensor", *J. Comput. Electron.,* vol. 21, no. 6, pp. 1416-1424, 2022. [http://dx.doi.org/10.1007/s10825-022-01935-4]

[8] R. Narang, K.V.S. Reddy, M. Saxena, R.S. Gupta, and M. Gupta, "A dielectric-modulated tunnel-FET-based biosensor for label-free detection: Analytical modeling study and sensitivity analysis", *IEEE Trans. Electron Dev.,* vol. 59, no. 10, pp. 2809-2817, 2012. [http://dx.doi.org/10.1109/TED.2012.2208115]

[9] S. Kanungo, S. Chattopadhyay, P.S. Gupta, and H. Rahaman, "Comparative performance analysis of the dielectrically modulated full-gate and short-gate tunnel FET-based biosensors", *IEEE Trans. Electron Dev.,* vol. 62, no. 3, pp. 994-1001, 2015.

[http://dx.doi.org/10.1109/TED.2015.2390774]

[10] D.B. Abdi, and M.J. Kumar, "Dielectric modulated overlapping gate-on-drain tunnel-FET as a label-free biosensor", *Superlattices Microstruct.,* vol. 86, pp. 198-202, 2015. [http://dx.doi.org/10.1016/j.spmi.2015.07.052]

[11] R. Narang, K.V.S. Reddy, M. Saxena, R.S. Gupta, and M. Gupta, "A dielectric-modulated tunnel-FET-based biosensor for label-free detection: Analytical modeling study and sensitivity analysis", *IEEE Trans. Electron Dev.,* vol. 59, no. 10, pp. 2809-2817, 2012. [http://dx.doi.org/10.1109/TED.2012.2208115]

[12] S. Kanungo, S. Chattopadhyay, P.S. Gupta, K. Sinha, and H. Rahaman, "Study and analysis of the effects of SiGe source and pocket-doped channel on sensing performance of dielectrically modulated tunnel FET-based biosensors", *IEEE Trans. Electron Dev.,* vol. 63, no. 6, pp. 2589-2596, 2016. [http://dx.doi.org/10.1109/TED.2016.2556081]

[13] D. Singh, S. Pandey, K. Nigam, D. Sharma, D.S. Yadav, and P. Kondekar, "A charge-plasma-based dielectric-modulated junctionless TFET for biosensor label-free detection", *IEEE Trans. Electron Dev.,* vol. 64, no. 1, pp. 271-278, 2017. [http://dx.doi.org/10.1109/TED.2016.2622403]

[14] P. Dwivedi, and A. Kranti, "Applicability of Transconductance-to-Current Ratio (g_m/I_ds) as a Sensing Metric for Tunnel FET Biosensors", *IEEE Sens. J.,* vol. 17, no. 4, pp. 1030-1036, 2017. [http://dx.doi.org/10.1109/JSEN.2016.2640192]

[15] A.K. Pathakamuri, and C.K. Pandey, "A novel extended back-gate negative capacitance TFET for improved device performance", *2022 IEEE International Conference of Electron Devices Society Kolkata Chapter (EDKCON),* IEEE, 2022. [http://dx.doi.org/10.1109/EDKCON56221.2022.10032855]

[16] S.L. Noor, and Z.R.K. Md, *"Application of nanocavity embedded dual metal double gate TFET in biomolecule detection." 2017 IEEE Region 10 Humanitarian Technology Conference (R10-HTC)* IEEE, 2017.

[17] P. Venkatesh, K. Nigam, S. Pandey, D. Sharma, and P.N. Kondekar, "A dielectrically modulated electrically doped tunnel FET for application of label free biosensor", *Superlattices Microstruct.,* vol. 109, pp. 470-479, 2017. [http://dx.doi.org/10.1016/j.spmi.2017.05.035]

[18] M. Ghodrati, A. Mir, and A. Farmani, "Non-destructive label-free biomaterials detection using tunneling carbon nanotube-based biosensor", *IEEE Sens. J.,* vol. 21, no. 7, pp. 8847-8854, 2021. [http://dx.doi.org/10.1109/JSEN.2021.3054120]

[19] A. Chakraborty, and D. Singha, *"Sensitivity study of staggered heterojunctions based SRG TFET sensor." 2017 Devices for Integrated Circuit (DevIC)* IEEE, 2017.

[20] A. Anam, S. Anand, and S.I. Amin, "Design and performance analysis of tunnel field effect transistor with buried strained Si 1− x Ge x source structure based biosensor for sensitivity enhancement", *IEEE Sens. J.,* vol. 20, no. 22, pp. 13178-13185, 2020. [http://dx.doi.org/10.1109/JSEN.2020.3004050]

[21] V.D. Wangkheirakpam, B. Bhowmick, and P.D. Pukhrambam, "N+ pocket doped vertical TFET based dielectric-modulated biosensor considering non-ideal hybridization issue: A simulation study", *IEEE Trans. Nanotechnol.,* vol. 19, pp. 156-162, 2020. [http://dx.doi.org/10.1109/TNANO.2020.2969206]

[22] Anil Kumar Pathakamuri, Chandan Kumar Pandey, and Tammisetti Ashok, "Performance Analysis Of Highly Sensitive Vertical Tunnel FET For Detecting Light In Near-IR Range", *J. Mater. Sci. Mater. Electron.,* vol. 35, no. 18, p. 1238, 2024. [http://dx.doi.org/10.1007/s10854-024-12953-2]

[23] W.V. Devi, B. Bhowmick, and P.D. Pukhrambam, "N+ pocket-doped vertical TFET for enhanced

sensitivity in biosensing applications: modeling and simulation", *IEEE Trans. Electron Dev.,* vol. 67, no. 5, pp. 2133-2139, 2020.
[http://dx.doi.org/10.1109/TED.2020.2981303]

[24] S. Mukhopadhyay, D. Sen, B. Goswami, and S.K. Sarkar, "Performance evaluation of dielectrically modulated extended gate single cavity InGaAs/Si HTFET based label-free biosensor considering non-ideal issues", *IEEE Sens. J.,* vol. 21, no. 4, pp. 4739-4746, 2021.
[http://dx.doi.org/10.1109/JSEN.2020.3033576]

[25] S. Kim, R. Lee, D. Kwon, T-H. Kim, T.J. Park, S-J. Choi, H-S. Mo, D.H. Kim, and B-G. Park, "Multiplexed silicon nanowire tunnel FET-based biosensors with optimized multi-sensing currents", *IEEE Sens. J.,* vol. 21, no. 7, pp. 8839-8846, 2021.
[http://dx.doi.org/10.1109/JSEN.2021.3054052]

[26] A.M. Ionescu, L. Lattanzio, G.A. Salvatore, L. De Michielis, K. Boucart, and D. Bouvet, "The hysteretic ferroelectric tunnel FET", *IEEE Trans. Electron Dev.,* vol. 57, no. 12, pp. 3518-3524, 2010.
[http://dx.doi.org/10.1109/TED.2010.2079531]

[27] S. Ghosh, K. Koley, and C.K. Sarkar, "Impact of the lateral straggle on the Analog and RF performance of TFET", *Microelectron. Reliab.,* vol. 55, no. 2, pp. 326-331, 2015.
[http://dx.doi.org/10.1016/j.microrel.2014.10.008]

[28] A.K. Pathakamuri, and C.K. Pandey., "Impact Of Back Gate-Drain Overlap On DC And Analog/HF Performance Of A Ferroelectric Negative Capacitance Double Gate TFET", *Phys. Scr.,* vol. 98, no. 12, p. 124001, 2023.
[http://dx.doi.org/10.1088/1402-4896/acf5ae]

[29] P. Ghosh, and B. Bhowmick, "Analysis of kink reduction and reliability issues in low-voltage DTD-based SOI TFET", *Micro & Nano Lett.,* vol. 15, no. 3, pp. 130-135, 2020.
[http://dx.doi.org/10.1049/mnl.2019.0427]

[30] M. Ghodrati, A. Mir, and A. Naderi, "Proposal of a doping-less tunneling carbon nanotube field-effect transistor", *Mater. Sci. Eng. B,* vol. 265, p. 115016, 2021.
[http://dx.doi.org/10.1016/j.mseb.2020.115016]

[31] Maryam Ghodrati, and Ali Mir, "Improving The Performance Of A Doping-Less Carbon Nanotube FET With Dual Junction Source And Drain Regions: Numerical Studies", *J. Circuits Syst. Comput.,* vol. 31, no. 10, p. 2250182, 2022.
[http://dx.doi.org/10.1142/S0218126622501821]

[32] A.K. Pathakamuri, and C.K. Pandey, "Performance investigation of ferroelectric L-shaped tunnel FET with suppressed corner tunneling for low power applications", *AEU Int. J. Electron. Commun.,* vol. 179, p. 155314, 2024.
[http://dx.doi.org/10.1016/j.aeue.2024.155314]

[33] M. Ghodrati, A. Mir, and A. Naderi, "New structure of tunneling carbon nanotube FET with electrical junction in part of drain region and step impurity distribution pattern", *AEU Int. J. Electron. Commun.,* vol. 117, p. 153102, 2020.
[http://dx.doi.org/10.1016/j.aeue.2020.153102]

[34] Ali Naderi, and Maryam Ghodrati, "Improving Band-To-Band Tunneling In A Tunneling Carbon Nanotube Field Effect Transistor By Multi-Level Development Of Impurities In The Drain Region", *Eur. Phys. J. Plus,* vol. 132, no. 12, p. 510, 2017.
[http://dx.doi.org/10.1140/epjp/i2017-11784-1]

[35] A. Naderi, M. Ghodrati, and S. Baniardalani, "The use of a Gaussian doping distribution in the channel region to improve the performance of a tunneling carbon nanotube field-effect transistor", *J. Comput. Electron.,* vol. 19, no. 1, pp. 283-290, 2020.
[http://dx.doi.org/10.1007/s10825-020-01445-1]

[36] A.K. Pathakamuri, C.K. Pandey, and P. Ghosh, "Negative capacitance split source tunnel FET as dielectric modulated highly sensitive biosensor", *Microsyst. Technol.,* vol. 31, no. 8, pp. 1975-1986,

2025.
[http://dx.doi.org/10.1007/s00542-024-05824-7]

[37] *TCAD Sentaurus Device Version T-2022.03*. http://www.synopsys.com

[38] S. Anand, A. Singh, S.I. Amin, and A.S. Thool, "Design and performance analysis of dielectrically modulated doping-less tunnel FET-based label free biosensor", *IEEE Sens. J.,* vol. 19, no. 12, pp. 4369-4374, 2019.
[http://dx.doi.org/10.1109/JSEN.2019.2900092]

CHAPTER 15

Design and Analysis of a Junction-less Double Gate Tunnel Field Effect Transistor (JL-DG-TFET) Based Inverter for Low-Power, High-Performance Digital Circuits

Arun Kumar Sharma[1,*], **Ritik Kumar**[1] and **Tarun Varma**[1]

[1] *Department of Electronics and Communication Engineering, Malaviya National Institute of Technology Jaipur, Jaipur, Rajasthan, India*

Abstract: This chapter provides an extensive description of a Junction-less Double Gate-Tunnel Field Effect Transistor (JL-DG-TFET) digital inverter and demonstrates how it outperforms traditional CMOS-based systems. In this inverter, complementary p-type and n-type JL-DG-TFETs are utilized so that better control over leakage current can be achieved, which improves energy efficiency. The JL-DG-TFET is optimized for low power usage, which is achieved due to the steep subthreshold slope, and the absence of junctions further reduces fabrication complexity. SILVACO TCAD and Cadence Virtuoso tools are used for simulations. Simulation results show reduced leakage power, improved noise margin, and minimized average delay. These outcomes illustrate that the JL-DG-TFET-designed inverter is a promising solution for low-power, high-performance circuits and helps address scaling limitations in semiconductor technologies.

Keywords: Boolean-function, Charge plasma, DGTFETs, Junction-less TFETs.

INTRODUCTION

For more than four decades, CMOS technology has been the foundation of digital circuit design, known for its low cost, low power consumption, good performance, reliability, and wide range of applications [1 - 3]. However, as semiconductor technology advances, conventional CMOS devices face serious challenges, including increased leakage current, higher power consumption, and performance limitations [4 - 6]. These challenges primarily arise from the physical limitations of CMOS scaling, which hinder future improvements in energy efficiency and switching speed. For this reason, researchers are actively

* **Corresponding author Arun Kumar Sharma:** Department of Electronics and Communication Engineering, Malaviya National Institute of Technology Jaipur, Jaipur, Rajasthan, India; E-mail: 2020rec9017@mnit.ac.in

Girdhar Gopal, Meena Panchore, Arun Kishor Johar & Tarun Varma (Eds.)

working on alternative transistor technologies to meet the requirements of next-generation digital circuits.

Among available alternative technologies, tunnel field-effect transistors (TFETs) are attracting particular attention because they have the potential to operate below the conventional 60 mV/decade subthreshold limit of MOSFETs.

This characteristic permits the TFET-based devices to improve energy efficiency and reduce overall power consumption [7 - 12]. TFET operates mainly *via* BTBT tunneling, through which charge carriers are transported. Due to BTBT-based transport, we achieve reduced leakage current, improved switching speed, and ultra-low power operation. Due to these properties, TFETs are suitable for miniaturized, energy-efficient electronic devices.

As a significant advancement in TFET technology, a charge-plasma-based junctionless TFET (JL-TFET) has been developed, eliminating the need for a p–n junction [13]. JL-TFET uses a uniform doping profile, which simplifies device fabrication and enables strong electrostatic control. As a result, short-channel effects are reduced, enabling device scaling at the nanometer level [14]. Furthermore, the use of high-k material with an optimized gate work function improves the JL-TFET performance. As a result, precise control over the tunneling barrier is obtai- ned, leading to increased ON-state current and reduced OFF-state current [15].

Material selection plays an important role in optimizing JL-DG-TFET performance. For example, low-band-gap materials such as Ge (germanium) and SiGe, an alloy of Si and Ge, increase the tunnelling probability of the device when added to the source region, thereby increasing device efficiency [16 - 18]. Such an advancement or improvement on JL-DG-TFET makes this device an alternative to CMOS, providing low power dissipation, a simplified device architecture, and better energy efficiency. In this work, an inverter circuit is proposed using complementary n-type and p-type JL-DG-TFETs. The proposed inverter's transistor characteristics are analysed and simulated by SILVACO TCAD, and inverters are analysed by Cadence Virtuoso tools [19]. Our simulation results clearly show that the inverter circuit has a low leakage power, less noise margins, and less delay; therefore, this circuit can be used for future possible solutions.

JL-DG-TFET also has advantages over conventional TFETs, such as the ability to operate at low voltages and to scale devices more effectively while maintaining its characteristics. The steep subthreshold slope of JL-DG-TFETs benefits inverter circuits, which are essential parts of digital electronics. This feature enables the device to operate at lower voltages without sacrificing switching speed, in line with the growing need for high-speed, energy-efficient digital systems.

The rest of the parts are set up in the following order:

- **Part II** delves into the device design and simulation setup, emphasizing the JL-DG-TFET configuration and the tools used.
- **Part III** details the architecture and operation of the inverter, focusing on the complementary use of n-type and p-type JL-DG-TFETs to enhance performance.
- **Section IV** represents and analyzes the simulation outcomes, highlighting improvements in leakage power, noise margins, and delay when compared to conventional designs.

DEVICE DESIGN AND WORKING OF JL-DG-TFET

JL-DG-TFET is made of undoped or mildly doped semiconductor materials, unlike typical transistors, which have strongly doped p-n junctions. This makes the JL-TFET fabrication process easier, improves device uniformity, and promotes device dependability. The absence of abrupt doping changes in the channel region lowers device variability and improves carrier transport.

In JL-TFET, the source and drain regions are also lightly doped or undoped, forming a uniform, continuous doping profile with the channel. This homogeneity minimizes abrupt changes in the electric field, reduces scattering effects, and lowers leakage currents. The gate electrode plays a pivotal role in device functionality by modulating the channel potential and regulating the width of the tunneling barrier. To enhance electrostatic control and reduce dielectric leakage, a high-k dielectric material is placed between the gate electrode and the channel.

Fig. (**1**) illustrates a cross-sectional structure of the Junction-less Dual-Gate TFET (JL-DG-TFET), which may be utilised with both n-type and p-type transistors by simply adjusting the work function of the drain and source. In this configuration, both gates are linked together to increase the ON-state current. To optimise the BTBT process, we utilised a 2 nm SiO_2 gate oxide and a 1 nm HfO_2 high-k dielectric. The simulation parameters for this design are a 50 nm gate length (LG), a 10 nm silicon body thickness (tsi), and a 1.0 V supply voltage (V_{DD}) [13]. In the construction of an n-type DG-TFET, the drain area is created by doping the intrinsic silicon body with electrons at a N^+ concentration, forming the “n” drain. Hafnium, with a work function of 3.9 eV, serves as the drain electrode material. The source region, referred to as the “p” source, is doped with holes at a P+ concentration, and platinum, with a work function of 5.93 eV, is used as the source electrode material. A p-type DG-TFET uses platinum (work function 5.93 eV) as the drain electrode and creates the “p” drain by doping the drain area with holes at P^+ concentration levels. In contrast, hafnium (work function: 3.9 eV) is utilised as the source electrode material, and the source region, also known as the "n" source, is enriched with electrons at N^+ concentration levels. Table **1** shows the remaining simulation parameters for n-type and p-type JL-DG-TFETs.

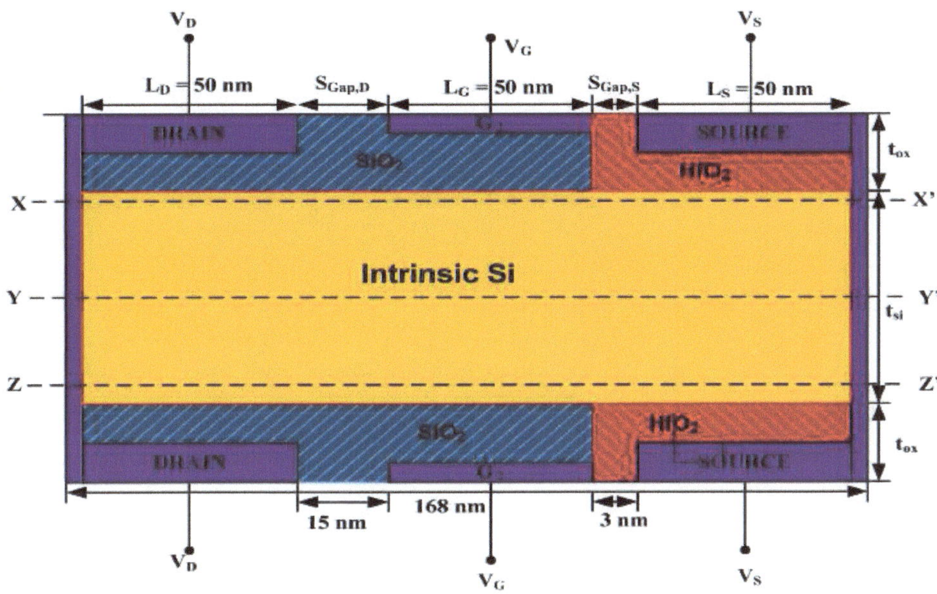

Fig. (1). Cross-section structure of JL-DG-TFET.

Table 1. -JL-DG-TFET simulation parameters.

Parameters	N-type JL-DGTFET	P-type JL-DGTFET
Supply Voltage (V_{DD}) (V)	1.0	1.0
Gate length L_G (nm)	50	50
Si body thickness (t_{si}) (nm)	10	10
Gate dielectric thickness (t_{ox}) (nm)	2	2
Spacer thickness gate to source $L_{gap,\,s}$	3	3
Spacer thickness gate to drain $L_{gap,\,D}$	15	15
Substrate (p-type) (atoms/cm^3)	1 x10^{15}	1 x 10^{15}
Work function of gate V_G, φ (eV)	**4.3**	**5.1**

The fundamental operating principle of the JL-TFET is based on the quantum tunneling mechanism. When the required voltage is applied to the gate, the source–channel barrier is reduced, allowing electrons to tunnel to the drain through a thin barrier. The gate voltage directly controls the barrier height. The tunneling mechanism of the n-type JL-DG-TFET is illustrated in Fig. (**2a**). When a positive gate voltage is applied to the n-type JL-DG-TFET, the tunneling barrier width is reduced, due to which the probability of electron tunneling from the source valence band to the channel conduction band increases, and the electrons finally reach the drain. Similarly, the tunnelling process in a p-type JL-DG-TFET is presented in Fig. (**2b**). In this instance, the tunnelling barrier for holes is lowered by applying a negative gate voltage, which allows them to tunnel from the conduction band of the source to the valence band of the channel and then to the drain. The transfer characteristics of n-type and p-type JL-DG-TFETs can be seen in Fig. (**3a**) and Fig. (**3b**), as well. The findings show that both devices operate efficiently by striking a desired balance between ON-state current (I_{on}) and OFF-state current (I_{off}). Interestingly, the JL-DG-TFETs exhibit a remarkable ON-to-OFF current ratio (I_{ON}/I_{OFF}) of roughly 1012, indicating their suitability for high-performance, low-power applications.

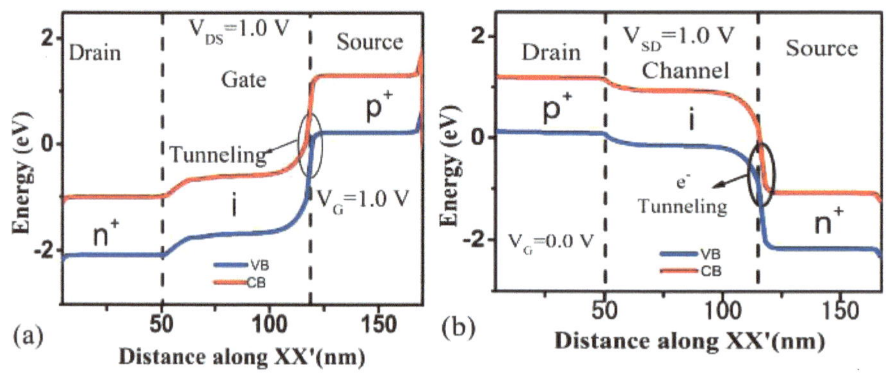

Fig. (2). (**a**) and (**b**) BTBT processes in n-type and p-type JL-DG-TFETs respectively.

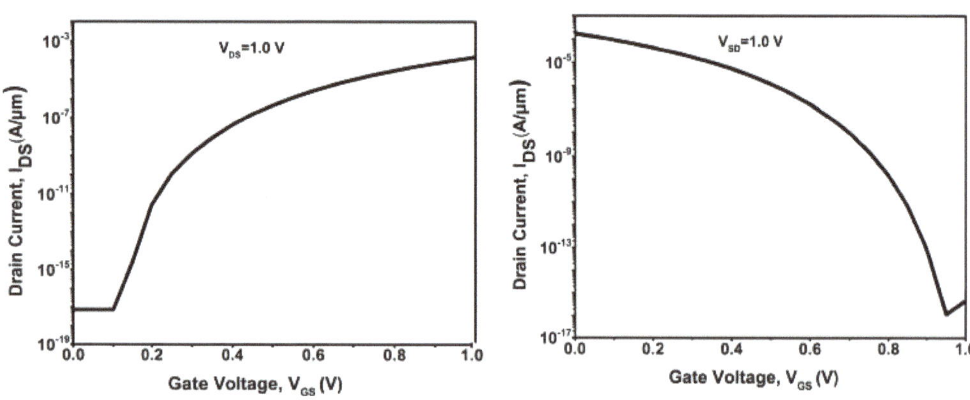

Fig. (3). (**a**) and (**b**) Transfer characteristics of n-type and p-type JL-DG-TFETs, respectively.

Fig. (**4**) shows the circuit schematic of the JL-DG-TFET-based inverter designed using Cadence Virtuoso. A digital logic inverter based on JL-DG-TFET is developed by combining n-type and p-type JL-TFETs in a complementary configuration, similar to the design of a conventional CMOS inverter. The noise margin of an inverter is crucial for reliable digital signal processing. JL-DG-TFETs, with their sharp subthreshold slope, are less prone to noise, resulting in better noise margins. However, achieving the desired noise margin may require optimizing the JL-DG-TFET size, which could impact the circuit density and integration levels. Additionally, JL-DG-TFET-based inverters exhibit significantly lower leakage current compared to their CMOS counterparts.

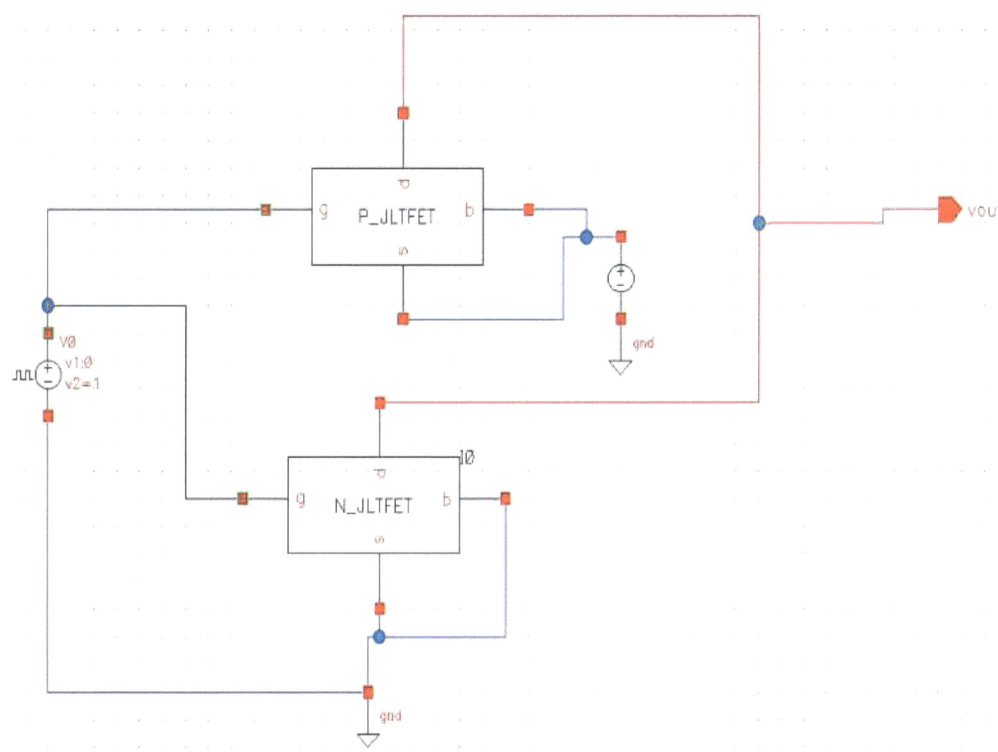

Fig. (4). Schematic of JL-DG-TFET-based inverter circuit in cadence virtuoso.

The inverter design process began with the simulation of n-type and p-type JL-DG-TFETs using the Silvaco TCAD tool. Initially, we analysed the transistor's electrical behaviour, especially focusing on the drain current *versus* gate voltage. This simulation result was very important for further design of the inverter based on JL-DG-TFET. The outcomes of these simulations were used to create a lookup table, a precomputed data sheet that shows the relationship between drain current and gate voltage. In the next phase, the JL-DG-TFET Verilog-A model was developed using Cadence Virtuoso. Verilog-A is a hardware description language suitable for analog systems and helps model transistor behaviour. The lookup table, typically stored in a ".tbl" file, was incorporated into the Verilog-A models so that the transistors reflect the actual results from TCAD simulations. After creating the Verilog-A models, their accuracy was validated by comparing drain current *vs* gate voltage characteristics obtained from TCAD simulations and Verilog-A simulations. The close agreement of these results confirms the reliability of the models. Once validated, these models were used to design the inverter circuit based on the proposed JL-DG-TFET. The inverter is a fundamental logic circuit, whose output is the complement of the input; in other words, when the input is low, the output is high, and when the input is high, the output is low. The performance of the inverter circuit was thoroughly examined to ensure correct operation and reliability.

INVERTER ARCHITECTURE AND SYMBOL

Fig. (**5**) presents a schematic representation of an inverter built using n-type and p-type JL-DG-TFETs, following a configuration similar to that of a conventional CMOS inverter. The pull-up network, labelled P1, consists of a p-type JL-D- -TFET, while the pull-down network, labelled N1, uses an n-type JL-DG-TFET. The inverter functions by producing an output that is the logical inverse of the input signal. The N-type JL-DG-TFET turns on when the input voltage is high and pulls the output to a low voltage. Conversely, the P-type JL-DG-TFET turns on when the input voltage is low and drives the output voltage to a high output [20].

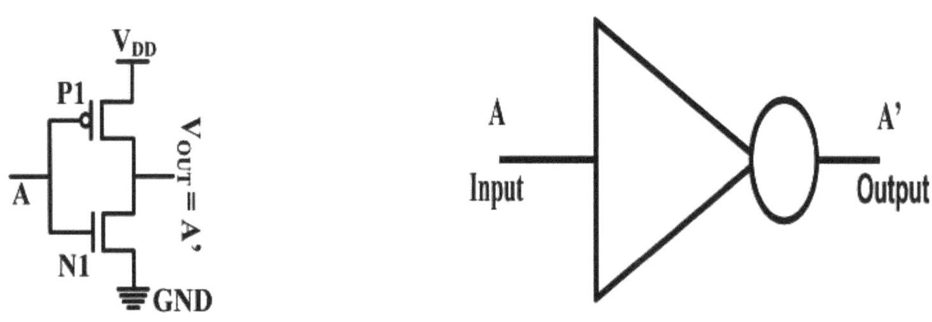

(a) Architecture of an Inverter

(b) Symbol of an Inverter

Fig. (5). (**a**) Architecture and (**b**) symbol of an inverter.

This ensures logic input inversion and complementary switching stability, as in conventional CMOS, where the p-type transistor (P1) is made larger than the n-type (N1) transistor to keep power consumption low, maintain comparable rise and fall times, and improve drive strength. Therefore, in this design, the width of transistor P1 is kept double than (N1) transistor. This sizing technique ensures efficient, symmetrical power switching, resulting in lower power consumption and improved circuit reliability [21]. To evaluate inverter circuit performance, we have evaluated some key metrics like DC and transient analyses, voltage transfer characteristics, noise margin, leakage current, and average delay. These evaluations confirm the device's enhanced power efficiency and dependable logic inversion.

RESULTS AND DISCUSSION

A JL-DG-TFET with a gate length of 50 nm is used to build an inverter circuit in Cadence Virtuoso to evaluate its performance. Simulation results are recorded and visualized in graphical form, which highlight the superior performance of the JL-DG-TFET. The proposed inverter has superior noise margins, reduced leakage current, and faster delay times than traditional CMOS inverters. Voltage transfer characteristics and transient response are presented in Figs. (**6** and **7**), where they can be thoroughly evaluated. These evaluations ensure that this design can be used for future applications.

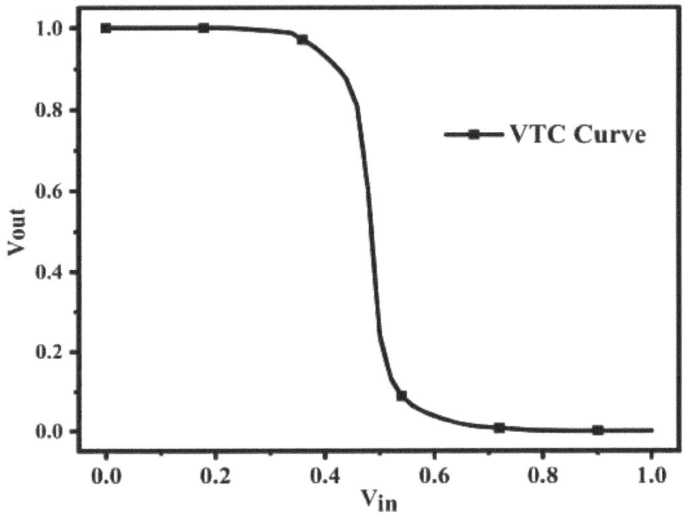

Fig. (6). VTC of JL-DG-TFET-based inverter.

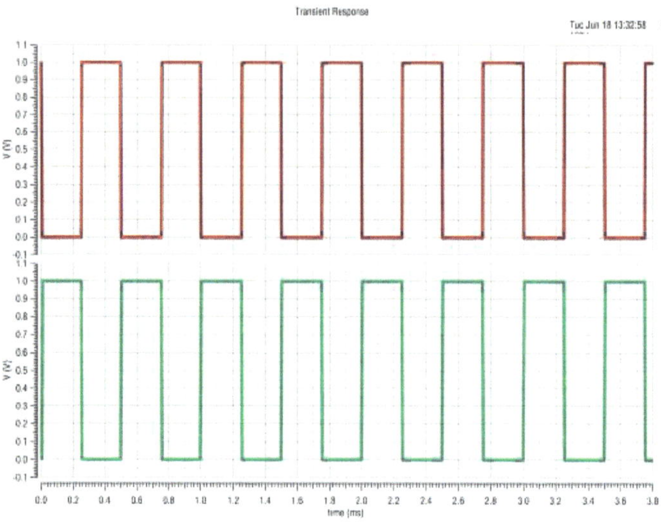

Fig. (7). Transient response of an inverter.

Comparison of the Noise Margin of JL-DG-TFET based inverter with TFET-based Inverter

In this part, we compared the noise margin of a traditional TFET inverter with that of a JL-DG-TFET-based inverter. Our analysis reveals that the JL-DG-TFET offers superior noise margins compared to the conventional TFET. Table **2** and Table **3** below present a detailed comparison of the noise margins for both inverters.

Table 2. Comparison table of noise margin of TFET and JL-DG-TFET-based inverter.

VDD	TFET [22]		V_{DD} (V)	JL-DGTFET (This Work)	
	N_{MH} (mV)	N_{ML}(mV)		N_{MH} (mV)	N_{ML} (mV)
1.5 V	630	620	1.5 V	547.829	493.485
1.8 V	749	749	1.8 V	667.477	627.42
2.0 V	870	800	2.0 V	795.67	712.88

Table 3. Comparison of JL-DG-TFET-based inverter with 22 nm FinFET.

Parameters	22 nm Fin-FET [23]	50 nm JL-DGTFET (This Work)
Delay	45 ps	2.56 μs
Noise Margin (high)	600 mV	375.477 mV
Noise Margin (high)	800 mV	331.528 mV

Leakage Current: The small amount of current that passes through a transistor when it is off and is referred to as OFF- current. We have simulated the same circuit to evaluate the leakage current in both n-type and p-type JL-DG-Tunnel FETs. Figs. (**8** and **9**) illustrate the leakage for both p-TFET and n-TFET, respectively. When compared to conventional CMOS technology, the JL-DGTFET exhibits significantly lower leakage current, making the JL-DGTFET-based inverter a superior choice in terms of minimizing leakage.

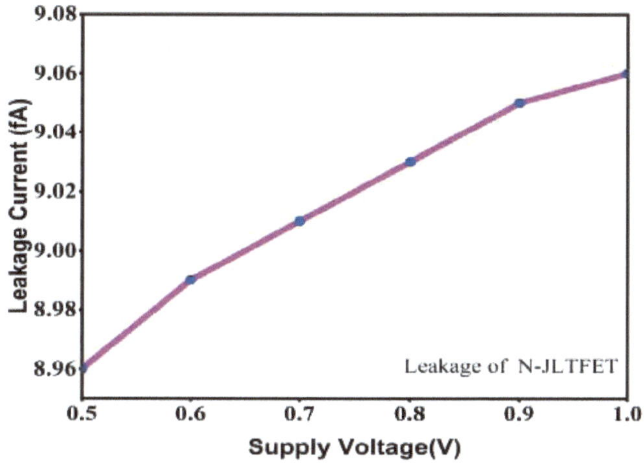

Fig. (8). Leakage current of N-type JL-DG-TFET for different V_{DD}.

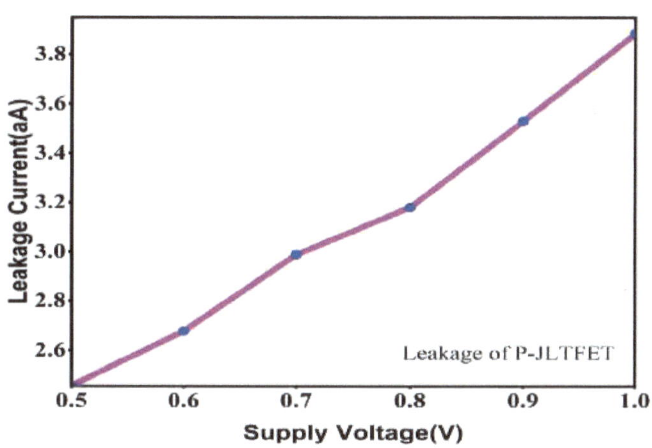

Fig. (9). Leakage current of P-type JL-DG-TFET for different V_{DD}.

Average Delay

Definition: The average delay is the time required of inverter to change state, specifically, from high to low or low to high.

Components: This metric is further broken down into rise time (low to high) and fall time (high to low).

Average Delay Time = 2.425 microsecond = 2.425×10^{-6} second

Importance: Lower delay times are essential for fast and efficient signal processing in modern electronic systems.

CONCLUSION

This study demonstrates the design and realization of a JL-DG-TFET-based digital logic inverter, which shows its potential as a high-performance and low-power alternative to conventional CMOS inverters. By using an n-type and p-type JL-DG-TFETs in a complementary configuration, the inverter obtained efficient logic inversion, and the rise and fall times have been optimized through size adjustments of the p-type transistor. Key performance metrics, such as sharp voltage transfer characteristics, high noise margins (NMH = 375.477 mV, NML = 331.528 mV), low leakage power, and an average delay of 2.425 microseconds, are included, which emphasize its reliability and energy efficiency. As a result, the JL-DG-TFET-based inverter presents a promising solution for next-generation digital circuits, addressing the challenges of power consumption and performance in highly scaled semiconductor devices.

AI DISCLOSURE

During the preparation of the work, the author(s) used AI tools and the Grammarly tool to enhance the writing style and readability of the content. The authors clarify that AI was not involved in generating the substantive content and that they take full responsibility for the publication.

REFERENCES

[1] N.H.E. Weste, and D. Harris, *CMOS VLSI Design: A Circuits and Systems Perspective* 4th ed. Addison-Wesley, 2010.

[2] T. Sakurai, "Low-power CMOS design through VTH control and low-swing circuits", *IEEE Solid-State Circuits Mag.,* vol. 10, no. 4, pp. 68-77, 2001.

[3] S.M. Kang, and Y. Leblebici, *CMOS Digital Integrated Circuits: Analysis and Design* 3rd ed. McGraw-Hill, 2003.

[4] International Technology Roadmap for Semiconductors (ITRS), *Emerging Research Devices,* 2013.

[5] K. Roy, S. Mukhopadhyay, and H. Mahmoodi-Meimand, "Leakage current mechanisms and leakage reduction techniques in deep-submicrometer CMOS circuits", *Proc. IEEE,* vol. 91, no. 2, pp. 305-327, 2003.
[http://dx.doi.org/10.1109/JPROC.2002.808156]

[6] K.P. Cheung, "On the 60 mV/dec @300 K limit for MOSFET subthreshold swing", *Proc. 2010 Int. Symp. VLSI Technology, Systems and Applications*. Hsinchu, Taiwan, pp. 72–73, Jun. 2010.

[http://dx.doi.org/10.1109/VTSA.2010.5488941]

[7] J. Appenzeller, Y.M. Lin, J. Knoch, and P. Avouris, "Band-to-band tunneling in carbon nanotube field-effect transistors", *Phys. Rev. Lett.,* vol. 93, no. 19, p. 196805, 2004. [http://dx.doi.org/10.1103/PhysRevLett.93.196805] [PMID: 15600865]

[8] Byung-Gook Park, B-G. Park, J.D. Lee, and T-J.K. Liu, "Tunneling field-effect transistors (TFETs) with subthreshold swing (SS) less than 60 mV/dec", *IEEE Electron Device Lett.,* vol. 28, no. 8, pp. 743-745, 2007. [http://dx.doi.org/10.1109/LED.2007.901273]

[9] U.E. Avci, D.H. Morris, and I.A. Young, "Tunnel field-effect transistors: Prospects and challenges", *IEEE J. Electron Devices Soc.,* vol. 3, no. 3, pp. 88-95, 2015. [http://dx.doi.org/10.1109/JEDS.2015.2390591]

[10] Y. Khatami, and K. Banerjee, "Steep subthreshold slope n- and p-type tunnel-FET devices for low-power and energy-efficient digital circuits", *IEEE Trans. Electron Dev.,* vol. 56, no. 11, pp. 2752-2761, 2009. [http://dx.doi.org/10.1109/TED.2009.2030831]

[11] K. Bernstein, R.K. Cavin III, W. Porod, A. Seabaugh, and J. Welser, "Device and architecture outlook for beyond CMOS switches", *Proc. IEEE,* vol. 98, no. 12, pp. 2169-2184, 2010. [http://dx.doi.org/10.1109/JPROC.2010.2066530]

[12] A.C. Seabaugh, and Q. Zhang, "Low-voltage tunnel transistors for beyond CMOS logic", *Proc. IEEE,* vol. 98, no. 12, pp. 2095-2110, 2010. [http://dx.doi.org/10.1109/JPROC.2010.2070470]

[13] M.J. Kumar, and S. Janardhanan, "Doping-less tunnel field-effect transistor: Design and investigation", *IEEE Trans. Electron Dev.,* vol. 60, no. 10, pp. 3285-3290, 2013. [http://dx.doi.org/10.1109/TED.2013.2276888]

[14] B. Shah, P. Singh, A. Raman, and N. P. Singh, "Design and investigation of junction-less TFET (JL-TFET) for the realization of logic gates", *Nano,* vol. 20, no. 07, p. 2450160, 2025.

[15] C. R. Allemang, E. M. Anderson, X. Gao, C. Arose, J. P. Mendez, T. A. Weingartner, D. M. Campbell, A. J. Muhowski, T. M. Lu, and S. Misra, "Next-generation tunnel FETs: exploring material perspectives and areal tunneling configurations", *Mater. Quantum Technol.,* vol. 5, no. 4, p. 042002, 2025.

[16] R. Rooyackers, A. Vandooren, A. S. Verhulst, A. M. Walke, E. Simoen, K. Devriendt, and S. Lo-Corotondo, "Ge-source vertical tunnel FETs using a novel replacement-source integration scheme", *IEEE Trans. Electron Devices,* vol. 61, no. 12, pp. 4032-4039, 2014.

[17] A. Pal, K. Agashiwala, J. Jiang, D. Zhang, T. Chavan, A. Kumar, C. H. Yeh, W. Cao, and K. Banerjee, "Two-dimensional materials enabled next-generation low-energy compute and connectivity", *MRS Bull.,* vol. 46, no. 12, pp. 1211-1228, 2021.

[18] B. Ghosh, and M. W. Akram, "Junctionless tunnel field effect transistor", *IEEE Electron Device Lett.,* vol. 34, no. 5, pp. 584-586, 2013.

[19] Silvaco International, Atlas Device Simulation Software [Online]. Santa Clara, CA, USA, 2014. Available: https://www.eng.buffalo.edu/~wie/silvaco/atlas_user_manual.pdf

[20] S. Bhattacharya, and S.L. Tripathi, "Implementation of low power inverter using Si1−XGeX pocket n-and p-channel junction-less double gate TFET", *Silicon,* vol. 14, no. 14, pp. 9129-9142, 2022. [http://dx.doi.org/10.1007/s12633-021-01628-w]

[21] Y. Khatami, and K. Banerjee, "Steep subthreshold slope n- and p-type tunnel-FET devices for low-power and energy-efficient digital circuits", *IEEE Trans. Electron Dev.,* vol. 56, no. 11, pp. 2752-2761, 2009. [http://dx.doi.org/10.1109/TED.2009.2030831]

[22] M. Yadav, "Performance analysis of tunnel FET-based ring oscillator using Sentaurus TCAD", *Silicon,* vol. 14, no. 13, pp. 7769-7779, 2022. [http://dx.doi.org/10.1007/s12633-021-01509-2]

[23] R. Hajare, C. Lakshminarayana, S.C. Sumanth, and A. Anish, "Design and evaluation of FinFET-based digital circuits for high-speed ICs", *Proc. 2015 Int. Conf. Emerging Research in Electronics, Computer Science and Technology (ICERECT),* IEEE, pp. 162-167, 2015. [http://dx.doi.org/10.1109/ERECT.2015.7499006]

SUBJECT INDEX

A

B

C

D

E

Girdhar Gopal, Meena Panchore, Arun Kishor Johar & Tarun Varma (Eds.)

N

O

P

Q

R

S

T

V

W

www.ingramcontent.com/pod-product-compliance
Lightning Source LLC
LaVergne TN
LVHW070946210926
848482LV00002B/117

* 9 7 9 8 8 9 8 8 1 5 8 1 3 *